Discovering the Okapi

ANIMALS, HISTORY, CULTURE
Harriet Ritvo, *Series Editor*

Discovering the Okapi

*Western Science, Indigenous Knowledge, and
the Search for a Rainforest Enigma*

SIMON POOLEY

Johns Hopkins University Press
Baltimore

2 4 6 8 9 7 5 3 1

Johns Hopkins University Press
2715 North Charles Street
Baltimore, Maryland 21218
www.press.jhu.edu

Library of Congress Cataloging-in-Publication Data
Names: Pooley, Simon, author.
Title: Discovering the okapi : Western science, Indigenous knowledge,
and the search for a rainforest enigma / Simon Pooley.
Description: Baltimore : Johns Hopkins University Press, 2025. |
Series: Animals, history, culture | Includes bibliographical references and index.
Identifiers: LCCN 2024061686 (print) | LCCN 2024061687 (ebook) |
ISBN 9781421452487 (paperback) | ISBN 9781421452494 (ebook)
Subjects: LCSH: Okapi. | Animals—Africa—Identification. |
Captive wild animals—Western countries.
Classification: LCC QL737.U56 P66 2025 (print) | LCC QL737.U56 (ebook) |
DDC 599.638—dc23/eng/20250221
LC record available at https://lccn.loc.gov/2024061686
LC ebook record available at https://lccn.loc.gov/2024061687

A catalog record for this book is available from the British Library.

Special discounts are available for bulk purchases of this book.
For more information, please contact Special Sales at specialsales@jh.edu.

EU GPSR Authorized Representative
LOGOS EUROPE, 9 rue Nicolas Poussin, 17000, La Rochelle, France
E-mail: Contact@logoseurope.eu

For Suz,
and in memory of Buta and Toto

CONTENTS

Discovering the Okapi is supported by an online supplement. It provides some further notes on sources and topics touched on in the book. Also included are five appendices referred to in the text that more fully address other claims to priority in discovering the okapi, thinking on ruminant evolution at the time okapi taxonomy was being settled, the British Natural History Museum's okapi atlas, other African names for okapi, and the diversity of pygmy peoples and their relations with African farmers in the Congo basin region. The additional materials for this book can be found by visiting the specific book page on press. jhu.edu.

Soon I shall be in the land of these strange animals, the okapi of
the Ituri forest on the equator.

—*John Bond, Moçâmedes, May 1957*

This book originated at two widely spaced moments in my life. The first was in boyhood, when a Swiss herpetologist friend, René Honegger, visited my father Tony at the Ndumo Game Reserve in northern Zululand, South Africa, bringing an illustrated encyclopedia of animals as a present. In it was a photograph of the remarkable okapi, with its pale giraffid head, deep red-brown pelt, and zebra-striped legs, all of which made a big impression on me.

The second moment followed a period of being unable to travel following an intense bout of dengue fever picked up researching human-crocodile interactions in Gujarat, India. The aftermath left me unable to think, remember, function. For six months I drifted in and out of self-awareness, my internal landscapes as trackless as what early maps of the Congo labeled "inaccessible country covered with dense forest."[1]

When I emerged a few months prior to the COVID-19 lockdowns of 2020–21, I sought solace in visits to the London Zoo in my home city. When the zoo reopened for a while in summer 2020, I haunted the Africa section, including the okapi enclosure next to the giraffe house. I also visited the library and archives of the Zoological Society of London because I had begun thinking about historical legacies of conservation science and practice following a period of heightened public debate in the United Kingdom over legacies of slavery and other colonial interventions.[2]

At this juncture, Zoological Society of London librarian Ann Sylph mentioned a book to me about Obaysch, the first hippo to arrive at the zoo. She also observed that the zoo had a long history with okapi. Reminded of my childhood enthusiasm, I looked up the literature, finding to my surprise that the most recent book in English attempting an overview was a short, illustrated volume published in 1999.[3]

It also struck me that the species name unusually combines African and European names. This resonated because I had recently observed in a paper on human-wildlife coexistence that conservationists were torn between rushing to implement Western scientific approaches to save crashing biodiversity, and acknowledging a need to engage with Indigenous peoples' conservation efforts, knowledge systems, and ways of valuing nature.

These objectives are often presented as contradictory: as demanding that researchers maintain their disciplinary approaches and authority while acknowledging their own values and limitations and engaging with incongruent ways of knowing the world. They must be scientifically rigorous while remaining open to alternative framings, methodologies, and terminologies. This, it is supposed, will be conceptually difficult and require extra time to achieve, time which is in short supply as we face a planetary environmental emergency.[4,5]

Having grown up in game reserves in South Africa, however, I had seen scientists and conservationists, including my parents, work closely with local African experts in natural history, including Thonga field naturalist Sigia Gumede.[6] I doubted that reconciling Western and non-Western knowledge systems was as excruciatingly difficult and complicated as claimed. I suspected rather that while non-Western contributions were often important to the discovery of new facts and ideas, they were subsequently filtered out in the process of scientific knowledge making.

The search for the okapi, I realized, provided an emblematic story of how much the Western[7] world's discovery of this enigmatic rainforest giraffid—as well as Western collectors' and scientists' descriptions of it in scientific terms—depended on local ecological knowledge. By writing a history of the discovery, collection, and display of okapi (dead and alive) alongside the history of scientific knowledge making about it, I could address these themes in a coherent way. Further, I could try to research and write the book I had always wanted to read about okapi.

In writing this book, I soon realized why so few have attempted a more in-depth overview. I had to decide on what I could and should not attempt to do.

First and most importantly, I decided that while the histories of the pygmy and other local African peoples are essential to this story, I would not pretend to speak for them in any way. Rather, I reflect on Western representations of these African peoples and the consequences of such representations for Western knowledge making about okapi and for okapi conservation. This is one reason I have retained the use of the word "pygmy" despite agonizing over its associations and connotations. Earlier drafts of this book mentioned only "forest people" instead. I considered the collective term "Batwa" for peoples in the region where most okapi were captured, but it is unfamiliar and may be confused with a term for a particular clan of the peoples in question.

The term "pygmy" is apparently disliked by some of those peoples so named, but used proactively by others. I have considered the arguments carefully, and have decided to stay with the historical and familiar term, for reasons to be elaborated upon further in chapter 13. These include the fact that it is used in almost all of the literature I am analyzing, that my discussion engages critically with Western conceptions of these peoples, and because not all forest peoples are pygmies, and not all pygmies live in forests. I offer my apologies to anyone so described who may read or hear of this book, and feel offended by my use of this term.[8]

I am acutely aware that I have not spoken with locals from the region. I found no literature written by local African researchers and scholars on my core themes, though I read novels with related social and ecological themes by Congolese novelists, and my understanding of African environmental thinking is informed by African environmental philosophers, as will be apparent in chapter 15 in particular.

That said, this book recovers, through careful reading of expedition diaries, newspapers, and other published and unpublished materials, how much was learned by Westerners from local Africans. It demonstrates how much Westerners depended on local Africans for their collecting and knowledge making. It acknowledges the ideas, information, and expertise of people written out of scientific literature and popular media accounts about okapi.

Wherever possible, I have identified the contributions of individual Africans rather than make vague references to Indigenous or local people collectively. These characters and voices should be welcomed into discussions of knowledge making and conservation of okapi. Future biographical research on these individuals, including intriguing figures like José Lopes and Juan de Medina, would be welcome. It may best be performed by researchers from the region.

Second, while I have written a lot about different ways of knowing okapi, I do not attempt to say what it might be like to be an okapi. This might seem obvious, even absurd to mention, but it shouldn't be forgotten that much about the nature of the singular being at the core of all the observations and speculations about okapi that I narrate in this book remains elusive. This is despite the sustained examinations of Western science and even despite the expertise of those Africans who have lived alongside the okapi in the rainforests of the Congo basin.

Third, I quickly realized that writings about the okapi are spread across numerous countries and collections in several languages I do not read well. This may explain the paucity of comprehensive books in English. I have done the best I can through the kindness of colleagues I thanked in my Acknowledgments, who know the literatures in French, Flemish, and German (for instance) far better than I do. I also relied extensively on translation software.

I have neither visited the home habitats of the okapi nor met the peoples who live alongside it. A more thorough history of okapi conservation in the Democratic Republic of Congo remains to be told. More could be written based on archives in Belgium, the Democratic Republic of Congo, France, Germany, Italy, the United States, and elsewhere, as well as interviews with those working on the frontline of okapi conservation in the Democratic Republic of Congo. My book aims to reveal the breadth of topics and sources, and encourage scholars working in those countries to pursue this further.

Fourth, in ranging across so many topics relating to okapi, from art history to paleontology to zoology, I have strayed into disciplines less familiar to me. No doubt experts on aspects of this history will be annoyed by omissions and inaccuracies (which I have worked hard to avoid): my apologies in advance for any you find, and I encourage you to publish on these subjects! Certainly, any errors and crucial omissions are my responsibility entirely, not those of people who have been generous with their time and expertise in answering my queries.

I have been surprised at the paucity of material in English on Belgian history and Belgian colonial activities in the Congo. While mindful of the complexity of the history of this vast country, I have had to make judgments about how much information the reader needs. Here, as on other topics, like the history of wildlife conservation in the region, captive care of okapis, or zoo history, I have of necessity risked offering breadth at the expense of depth.

Finally, for the reasons I have just outlined, this book cannot be the definitive history of all aspects of the okapi. My aim has been to offer a thorough outline and an intriguing invitation to further investigations. It is a personal account of a long and fascinating history informed by themes that have interested me, some since childhood.

My identity as a white male who grew up in a family of conservationists and natural scientists living in remote protected areas in Africa and who now resides in London, and wrote this book during a historical moment in which new ideas about biodiversity conservation and human relationships with nature are being broached, has shaped my writing. This has helped me in many respects, but has also limited me in certain ways and exposed my biases. There are many other voices to be heard regarding the story of the okapi. I look forward to hearing what they have to say.

This is also a very personal book woven into what happened to me—and in some respects to many of us—at the time I began thinking about it. In the first months following the intense fevers characteristic of dengue fever, I lapsed into a complete brain fog (*postviral fatigue syndrome* is the suitably vague term). Emerging toward the end of 2019, I became aware that my short-term memory and capacity to concentrate were virtually nonexistent. I feared never being able to return to work, but over the course of another three months using diaries and a Fitbit and making many missteps, I was able to return to my job. Other things have taken longer.

The idea of writing a book gave me hope and offered a challenge. Could I manage it? Could I get my head around a huge and disparate new area of research, organize it, and retrieve a narrative and a message from it? Bringing the okapi into focus from the obscurity (for me) of the rainforests of the past and the present was my road back out of my own darkness and brain fog. My quest was for a clear head and a heart of brightness.

In the final stages of writing this book, serendipity seemed to favor me. My mother, Elsa, visited us from South Africa. She had been looking through papers left to her by her mother, including an envelope of letters from her father, John Bond. A journalist and historian, he had traveled widely in southern and central Africa, reporting on environmental and agricultural subjects.

On May 19, 1957, in the year in which he unexpectedly died, John sent my ten-year-old mother a postcard from Angola. On the front, a black-and-white photograph showed a touching parental scene of a mother okapi and her calf about to touch noses. The image was of very good quality, and the postcard had been produced by a company based in Leopoldville in the Belgian Congo (now Kinshasa, in the Democratic Republic of Congo).

On the reverse of the postcard, my grandfather began his message with "My dearest Elsa, soon I shall be in the land of these strange animals, the okapi of the Ituri forest on the equator." I do not know if John ever saw an okapi, but I know he traveled to the Belgian Congo and flew over its vast rainforests.

It seemed remarkable that my grandfather had chosen this postcard in 1957 and sent it to my mother in Southern Rhodesia (Zimbabwe) from Mossamedes (Moçâmedes) in Angola, where it lay forgotten in an envelope until that moment in my own history with okapi.

This book is my own postcard to the future. It shares what I have learned about the okapi and the history of knowledge about them so far. It emerged from difficult times. Above all, I hope it will raise awareness about okapi and provide ideas and questions that will stimulate further efforts to understand and enable these singular, enigmatic animals to flourish, both in captivity and in the rainforests of the Congo.

I hope that my ancestors and yours will know that okapi flourish unseen and unheeded deep in the rainforests of the Congo basin.

Discovering the Okapi

Introduction

While researching and writing this book, I have been repeatedly surprised by how little-known the okapi is outside of the Democratic Republic of Congo. This once internationally famous animal, eagerly sought by Western natural history museums and zoos, has retreated back into the obscurity from which it emerged 125 years ago. Even those who have heard of it are unsure of its appearance or what sort of beast it is. Few I spoke to knew it was the only living relative of the giraffe. The use of alternative common names like "forest giraffe" or "zebra giraffe" has further confused matters. Shown a photograph, most guessed it was an equid that was probably related to the zebra. Almost no one could tell me where it lives in the wild. Very few people indeed have seen an okapi, alive or dead—and trust me, if you encountered this wondrously strange and striking animal in life, you would not forget it.

I saw my first live okapi in London Zoo. No okapi fundi myself, I read the notes on display at the okapi enclosures on diet, habitat, threats, and conservation status (evolutionarily distinct and globally threatened). A signboard describes the okapi as "a shy cousin of the giraffe . . . unknown to Western science until 1901, when a Zoological Society of London fellow discovered it," and notes that "the okapi can clean its own ears using its 35cm long tongue."

A silent video featuring subtitles inside an enclosure providing views of the internal quarters of the okapi at the zoo notes they are cousins of the giraffe, which dwell in the Ituri rainforest in the Democratic Republic of Congo. According to the video, their tongues are 15.5 inches (40 centimeters) long, and they have huge mobile ears that warn them of danger and long eyelashes that protect them from grit. Their "striped backsides" serve as camouflage in the forest, and calves are able to follow their mothers' white-striped derrieres through the gloom (their front upper legs are also striped). Despite these adaptations and their remote home ranges in the northeastern Democratic Republic of Congo in Africa's Congo basin, okapi urgently require protection. The Zoological Society of London is helping to do that in situ in Virunga National Park, the video says (although that was no longer the case at the time of writing due to the security situation), and also ex situ by breeding okapi in captivity (footage shows a mother and calf at the zoo).[1]

The next obvious place for me to scout for okapi was London's famous Natural History Museum, where I was nonplussed to find no physical trace of it in the mammals section. There were giraffes, including taxidermized and fossil remains, and a cast of the skulls of two ancestors of modern giraffids, *Samotherium boissieri* and *Canthumeryx sirtensis.* There was in fact more information about extinct giraffids than the okapi. At first, I could find only one tiny outline sketch (no markings) of an okapi with a note to the effect that the okapi is one of the two surviving giraffids that inhabit the dense forests of West Africa (in fact it inhabits Central Africa).[2]

Surely there must be some physical trace of an okapi, I thought, in this, the very museum where its taxonomy was decided, its status as a new species was confirmed, and the first taxidermized okapi in Britain was displayed? Eventually, I found a display board titled "Rare" featuring four captioned photos of animals, one of which was an okapi standing on a road next to a grass verge. The caption reads "This strange relative of the giraffe was first discovered in Africa in 1901. You can see an okapi in the African mammals exhibition, on the first floor of the Museum."[3]

The thing is, you can't. When I visited in 2024, those okapi were long gone. Decades ago, the British Museum's aging taxidermized okapi vanished like Lewis Carroll's Cheshire Cat, leaving just this reference to tantalize the questing few who seek it out.

Scope and Intentions

I have explained my personal aims and my hopes for raising the public profile of the okapi in the preface to this book. The book offers a history of the search for the okapi—literal, intellectual, and metaphorical—while also embodying the challenges of that search.

In a few places, I reflect on the process of my research and on progress in understanding okapi. As it evolved, my research expanded into many more regions, both disciplinary and geographical, than I had anticipated. Myriad characters and events concealed by the published accounts of a sensational discovery and the formal processes of scientific knowledge creation emerged, unbraiding the typical linear scientific discovery story of the singular heroic explorer whose findings are confirmed and certified by metropolitan experts.

I show that the man credited with being the European discoverer of the okapi, Harry Johnston, although this attribution was itself contested, had been accompanied by significant others, including Henry Morton Stanley of Congo fame (or infamy), kidnapped Mbuti pygmy people and their would-be German impresario, Belgian colonial officials in the Congo, Philip Sclater of the Zoological Society of London, and Ray Lankester of the British Natural History Museum.

Telling the subsequent story of how the okapi was debated and described by science required me to engage with the histories of zoology and taxonomy and with the conventions of naming species. Following its discovery, scientists asked whether it was an equid or a giraffid, wondered if it could be a "living fossil," or a "missing link," and whether its discovery would simplify or complicate schemas for explaining the evolution of ruminants. These debates unfolded in the context of national rivalries between scientific societies, territoriality over physical evidence, and political battles within the great institutions of British natural history science.

Reading scientific and popular reports and seeing the first speculative images of this visually striking but mysterious and evasive new African mammal took me on an unexpected journey into scientific illustration, taxidermy, photography, and wildlife films. The first artists' images shaped the scientific as much as the public understanding of the as-yet unseen okapi for decades. I was led to consider European influences on African art and vice versa, and I trace a trajectory from alleged ancient Egyptian okapi rock carvings to artwork by Andy Warhol.

This book is about the natural history explorers, hunters, and animal collectors who sought okapi and reveals the expedition partners, go-betweens, local hunters, trappers, traditional leaders, and others they depended on. Many of the key Western figures are little known and yet have fascinating histories.

The contributions of the better-known male figures familiar to aficionados of stirring tales of jungle exploration in the Congo like anthropologist Patrick Putnam, hunter Percy Powell-Cotton and Italian American explorer Attilio Gatti, are complemented by those of Putnam's wife, Anne Eisner, Percy's and Attilio's wives, Hannah and Ellen, and the intrepid Delia Akeley.

A key aim of the book is the recovery of the hidden histories and contributions of Africans central to the story of the discovery and capture of the okapi by (or for) Westerners. This includes the indispensable contributions of traditional leaders like Kotu-Kotu and Akegne, African hunters including Abawe, Aposho, and Etumba Mingi, and pygmy hunters like Kalumé and Makulu-kulu.

Understanding the history of searches for the okapi, the first captures, and how these developed into an official capture and export operation requires understanding the colonization of the vast country that is now the Democratic Republic of Congo and the interplay of colonial powers in the region. Famous journeys of geographical exploration by Europeans play a part, alongside Stanley's disastrous attempt to rescue Emin Pasha from the Mahdi. So did several natural history and anthropological expeditions led by Americans, Britons, Germans, Italians, and other Westerners.

The zoological, economic, and cultural institutions of colonial powers influenced the course of okapi collecting and distribution. Faced with rising criticism of the inhumane strategies employed in rubber collection in his Congo colony, King Leopold II of the Belgians used okapi as diplomatic currency to win favor with European royalty, governments, and US institutions like the American Museum of Natural History.

Western zoos, aided by Belgian colonial governments eager to promote their colony, colluded to export okapi to the West. This continued for seventy years, despite horrific mortality rates, until air travel cut journey times, vets began to understand the key health challenges, and zookeepers developed ways to keep okapi (and later breed them) safely in zoos.

Zoo scientists and zookeepers like Heini Hediger, Agatha Gijzen, and the charismatic Bernhard Grzimek enter the story of the okapi in the 1950s. During this period, Western zoos reoriented their missions and began focusing on wildlife conservation. In the Ituri forest in the Belgian Congo, local Congolese men like Jean de Medina managed the capture of okapi for export. Independence from colonial rule in 1960 disrupted these operations and linkages, and episodic political instability and violence in the Democratic Republic of Congo has shaped research and conservation efforts ever since. The difficulties and dangers of conserving okapi in situ in the Democratic Republic of Congo are considerable, and I salute those brave souls doing this important work.

By the time I had straightened out the okapi discovery story and put together an overview of the development and current status of okapi science, it had become apparent that mainstream narratives about both the region's wildlife and its peoples have been seriously distorted by Western paradigms about scientific authority, Indigenous or local knowledge, and conceptions of African peoples, especially the pygmy peoples of the Congo basin.

These distorting lenses extended beyond the region's wildlife and peoples, to the rainforest, and so chapter 14 renders these notions explicit and offers a critique of erroneous ideas about them, based on the latest available knowledge. Neither the rainforests nor the African inhabitants of the Congo basin are "primeval," and all have deep, interconnected histories in which Western interference has played a significant part in recent centuries. This critique, along with African perspectives on the environment and locals' views on conservation, informs my assessment of the current state of okapi conservation in chapter 15. Some differences between African and Western conceptions of human-wildlife coexistence and natural resource use are explored.

In short, the journey I undertook in writing this history of the okapi required me to engage with histories of science, anthropology, colonialism, the environment, ethology, scientific exploration and collection, natural history institutions, museums, and taxidermy and with zoo science and zoo keeping, wildlife conservation, and art and media representations of okapi. My understanding has been enriched by exchanges with academics, archivists, conservationists, museum curators and researchers, zookeepers, and others. I have encountered once famous, now mothballed taxidermized okapis and met and stroked Edi, a live okapi.

From this work of bricolage I have assembled as rounded a picture of Westerners' attempts to find, shoot, capture, trade, describe, categorize, represent, exhibit, keep, understand, and conserve the okapi as possible.

In the conclusion, I briefly suggest how Western scientists and knowledge platforms can include all of those involved in producing knowledge of okapi and engage better with Indigenous and local experts. Natural history museums and zoos can also acknowledge these histories, which are usually filtered out in their displays. This book supports such efforts through providing names, narratives, and images of key African individuals who have played a significant role in the history of the okapi, all of which can be shared with communities in the home territories of okapi and thus digitally repatriated. The stories of the individual okapis related in this book, wild and captive, could also be shared more widely.

Throughout this book, I have aimed to keep in mind the limits of our knowledge about this remarkable and still enigmatic being, the okapi. These limits pose challenges for keeping okapi in captivity and should encourage us to continue to study both zoo and wild okapi. Ultimately, okapi are emblematic of the limits of what we can know about other beings while also presenting us with reasons to improve our ways of learning about our world. More pluralistic approaches to knowledge making can help us humans to more deeply value and respect the beings we share our planet with, and relate better to them.

The Structure of the Book

Chapters 11 and 12 of this book share what has been learned about okapi since they were discovered by Westerners. They are placed deep within the book, as the okapi is its heart—the object of all the strivings of the artists, hunters, trappers, collectors, researchers, conservationists, and others discussed in these pages.

These two chapters are preceded by histories of the institutions and individuals involved in the discovery and subsequent searches for the okapi, its description, categorization, capture, and exhibition (dead and alive) by Westerners (chapters 1–4); non-Western and Western visual representations of okapi (chapters 5 and 6); and the first expeditions to capture live okapi, official capture and export operations, and zoo conservation and display (chapters 7–10). The chapters following those focused on okapi biology and behavior explore what was learned from Congolese locals about okapi and how they themselves (and their knowledge) have been represented (or obscured) by Westerners (chapters 13–14). Chapter 14 also explores the history of Central Africa's rainforests. This informs discussion of some consequences of these ideas for okapi conservation (chapter 15).

Scientific Authority and Metropolitan Knowledge Institutions

One day in the late 1890s, Mbuti pygmies killed an okapi in the Ituri rainforest, west of Lake Albert in Central Africa. They expertly butchered it to preserve the richly patterned hide and returned with this and the meat to their camp. Sections of hide were bartered to African soldiers working for King Leopold II of the Belgians' private Congo colony. One of the soldiers fashioned a prized striped section of the okapi hide into a handsome pair of bandoliers.

These bandoliers were bought by an Englishman, Harry Johnston, in July 1900. He sent them over eight hundred miles east across rainforest and open savannas to the Indian Ocean coast. From Mombasa, they were shipped northward around the horn of Africa, across the Red Sea and through the Suez Canal, west across the Mediterranean and into the choppy gray Atlantic, finally arriving at the sea-reach of the Thames in November 1900. They traveled up the old river, gathered in from one of the uttermost ends of the earth, following the same course as Sir Francis Drake, John Franklin, and so many maritime adventurers before them, to arrive in London, capital of Victorian Britain at the zenith of its power.

And so the first evidence of the existence of the elusive okapi that arrived in England was a pair of bandoliers cut and fashioned by African hands. They were first shown at a meeting of the Zoological Society of London (or ZSL) in November 1900. To this day, the type specimen—that is, the foundational evidence for the Western scientific description and naming of the okapi—is a pair of tasseled Congolese bandoliers stored in the Natural History Museum in London.[1]

This chapter introduces the two most important institutions and the major players involved in the discovery of the okapi by Western science: the Zoological Society of London and its zoo in Regents Park and the Natural History Museum in South Kensington. These were Great Britain's preeminent institutions for the study of natural history at the close of the nineteenth century.

Natural history studies and the biological sciences were not carried out in a rarefied, objective realm of calm contemplation of the regions of the world being opened up to Western eyes by European imperialism. The biological sciences as we know them today evolved through a period of tumultuous expansion and contestation in the 1800s.

An emerging body of scientists studying the natural world struggled to free itself from an establishment dominated by aristocrats and clergy who regarded natural history as a middle-class pastime inferior to the higher pursuits of literature, antiquities, theology, and the arts. Natural scientists fought to establish new societies, institutions, and positions of influence that could further their interest in the natural world. They ran into the same resistance as Queen Victoria's reforming husband and consort, Prince Albert did, but they lacked his power and influence.

The term "biology" came into use in this period, and an effort was made to establish the study of living things on an equivalent footing to that of the prestigious physical sciences. Starting in the 1870s in Britain, merely collecting specimens and classifying species was no longer sufficient; biologists now also aimed to investigate and explain the internal structures of life, how they evolved over evolutionary time, and how they developed over the lifespans of individuals. Morphological and later physiological research (how the parts of the body function) was mainly carried out in laboratories and dissecting rooms. Field studies became marginalized and, unfortunately, so did the study of animal behavior.

This was an opportune time for the professionalization of the life sciences, spurred on by competition between the European imperial powers. The British empire was at its zenith and specimens poured in from the field into its zoos and museums that required description and classification, which was now based on the study of a species' internal structure. Claims to scientific discovery and the authority and prestige associated with the naming and classification of important new finds were fiercely contested by the gatekeepers of the emerging institutions.

Science is not just an objective field of authoritative knowledge produced exclusively by rigorous processes of empirical study, comparison, methodological discipline, and peer review. Its products and processes are also shaped by the aspirations and personalities of individuals and the historical contexts in which scientists pursue their careers. The history of the discovery, naming, and classification of the okapi is emblematic of such processes of natural history knowledge making in the West.

The Zoological Society of London and Its Zoo

In 1820s England, zoologists began motivating for the establishment of a society more focused on their interests. Thus, when Sir Stamford Raffles proposed forming a zoological society with a collection of live animals for study, there was significant support in London.

Unlike the Royal Society and the Linnean Society, which had partially catered to the zoological sciences, the Zoological Society of London established a collection of live animals. Raffles, founder of Singapore and a keen animal collector, had been much impressed by a visit to the menagerie in the Jardin des Plantes in Paris. He reached out to wealthy landowners who kept notable collections on their estates such as the thirteenth Earl of Derby. In addition to seeking to provide live animals for zoological study, the society also aimed to introduce useful animals to Britain that could advance the economy.

An important relationship developed between zoologists at the Zoological Society of London and at the Natural History Museum, especially after the museum moved to South Kensington. Both became centers of taxonomic expertise, and both encouraged the collection and sending of specimens back to London from across the globe in a period of intensifying globalization. The zoo housed living specimens (see fig. 1.1) and provided opportunities to dissect animals that died, and the museum provided a reliable and organized storage facility for specimens, including type specimens. Its scientific staff had almost unparalleled specimen collections at its disposal and so became experts on the morphological description and taxonomical categorization of species.[2]

Philip Lutley Sclater, Secretary of the Zoological Society of London

In 1859, Philip Lutley Sclater was elected secretary of the Zoological Society of London at the age of thirty, a position he would hold until 1902. This was one of the most sought-after posts in British zoology because the society was at the

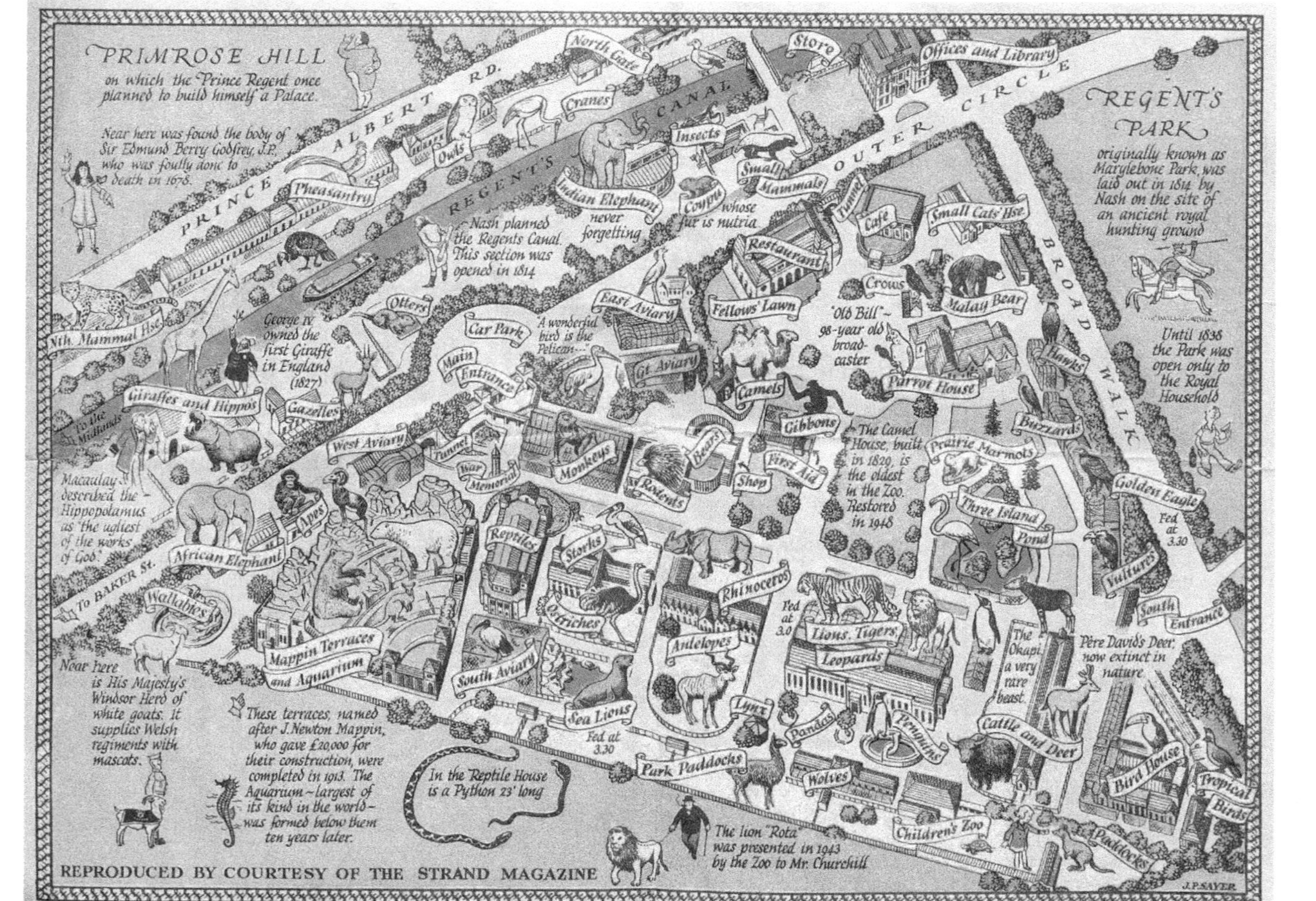

PRIMROSE HILL
on which the Prince Regent once planned to build himself a Palace.
Near here was found the body of Sir Edmund Berry Godfrey, J.P. who was foully done to death in 1678.
REGENT'S PARK
originally known as Marylebone Park, was laid out in 1814 by Nash on the site of an ancient royal hunting ground
Until 1838 the Park was open only to the Royal Household
ALBERT RD.
PRINCE REGENT'S CANAL
OUTER CIRCLE
BROAD WALK
North Gate
Store
Offices and Library
Cranes
Pheasantry
Owls
Indian Elephant never forgetting
Insects
Coypu whose fur is nutria
Small Mammals
Tunnel
Cafe
Small Cats' Hse.
Nth. Mammal Hse.
Otters
Restaurant
Crows
Malay Bear
Nash planned the Regents Canal. This section was opened in 1814
George IV owned the first Giraffe in England (1827)
East Aviary
Fellows' Lawn
"Old Bill" 98-year old broad-caster
Hawks
Car Park
'A wonderful bird is the Pelican...'
Parrot House
To the Midlands
Giraffes and Hippos
Gazelles
Main Entrance
Gt. Aviary
Camels
Buzzards
Macaulay described the Hippopotamus as 'the ugliest of the works of God'.
West Aviary
Tunnel
War Memorial
Monkeys
Bears
Gibbons
Prairie Marmots
The Camel House, built in 1829, is the oldest in the Zoo. Restored in 1948
First Aid
Shop
Rodents
Golden Eagle
Fed at 3.30
To BAKER St.
African Elephant
Apes
Reptiles
Storks
Rhinoceros
Three Island Pond
Vultures
Wallabies
Ostriches
Fed at 3.0
Lions, Tigers, Leopards
South Entrance
Near here is His Majesty's Windsor Herd of white goats. It supplies Welsh regiments with mascots.
Mappin Terraces and Aquarium
South Aviary
Antelopes
Lynx
The Okapi, a very rare beast.
Père David's Deer, now extinct in nature
These terraces, named after J. Newton Mappin, who gave £20,000 for their construction, were completed in 1913. The Aquarium — largest of its kind in the world — was formed below them ten years later.
Sea Lions
Fed at 3.30
In the Reptile House is a Python 23' long
Park Paddocks
Pandas
Penguins
Cattle and Deer
Wolves
Children's Zoo
Bird House
Tropical Birds
Paddocks
The lion "Rota" was presented in 1943 by the Zoo to Mr. Churchill
REPRODUCED BY COURTESY OF THE STRAND MAGAZINE
J.P. SAYER.

heart of zoological research and was connected to an international network of scientists and collectors. Sclater put the society's finances in order, and oversaw the development of its scientific publications and library.

Sclater, a friend of Thomas Huxley, set up the office of prosector (the word "prosector" is Huxley's invention) to ensure that animals that died were preserved and studied. In the prosector's laboratory, skeletons were articulated and body parts preserved in spirits, while skins, horns and feathers, and similar features were sold to taxidermists. Preserved remains were supplied to museums across England. Eminent comparative anatomists and medical specialists in London were supplied with material for dissection, and the results were published to advance zoological science, including work by Richard Owen on giraffe remains. The knowledge gained, the specimens made accessible, and the new pool of experts would be important for the discovery and description of the okapi. Johnston had studied giraffe skulls in London, and as a result knew what he was looking at when he first saw an okapi skull: a giraffid.[3]

Sclater encouraged a worldwide network of explorers, collectors, and contributors, including Johnston. During his career, he met and interacted with most of the British explorers and prominent zoologists of his time and was responsible for the scientific naming of hundreds of remarkable species of birds and mammals discovered during his period of tenure.

Sclater worked in a domain divided between the explorers and collectors observing animals in the wild, scientists mostly studying their remains after death in the metropole, and the keepers whose job it was to look after live animals in captivity. To this day, the divide between interacting with and studying animals in situ in the wild, or ex situ in captivity remains, with different sets of people acquiring largely unrelated kinds of training and experience with the same species, their knowledge seldom traveling across these divides. The Zoological Society of London as an organization spans and links up these domains and therefore has the potential to be greater than the sum of its parts.

This book recounts the history of okapis across these domains, considering their journeys from encounters with humans in the wild through their being hunted, captured, and transported to their exhibition alive in zoos or dead in museums. Their lives in captivity and their afterlives as museum specimens—

Figure 1.1. (*opposite*) 1936 map of London Zoo by J. P. Sayer. Okapi, "a very rare beast," were kept with cattle and deer (*bottom right*). They were later moved to be next to their relatives, the giraffe (*center left*). The zoo was between okapi in 1936, Congo having died in 1935 and Buta still to arrive in July 1937.

their "unnatural histories" as Nigel Rothfels calls them—are critical to how they have been received, represented, and, ultimately, mobilized for conservation.[4]

Sclater was succeeded in 1902 by Chalmers Mitchell, a distinguished zoologist of the old style who focused on structural zoology and phylogeny. The Zoological Society of London began to fall behind the Royal Society, which was more alert to advances in experimental biology. Until the 1900s, the study of wild animals was focused on taxonomy and comparative anatomy and carried out using dead animals.

Physiology, or the study of living animals' body parts and their functions, was a later development. Scientists pursuing behavioral studies or ethology (studying live animals) were still struggling to have this field recognized as a discipline in the mid-twentieth century (T. H. Huxley's grandson Julian was a key British figure in this effort). That ethology was unable to gain a disciplinary footing, along with the undeniable difficulties of accessing wild okapis and high mortality rates of captive ones, may account for the scanty knowledge of okapi behavior and physiology that persisted into the second half of the twentieth century.[5]

The Empire and Its Limits

London Zoo and the Zoological Society of London can be usefully studied in their imperial context as a node and a beneficiary of imperial networks, particularly in the later Victorian era, which was more explicitly jingoistic—although even then, British influence was often diplomatic and indirect. Those the zoo engaged to procure or look after exotic animals did not necessarily do so to serve the British empire. Various non-European peoples had realized there was a market in selling local wildlife to Europeans and developed businesses to take advantage of that market.

When Johnston first reached the mouth of the Congo, for example, he encountered local Africans who caught and tamed colorful birds such as waxbills and parrots, mammals such as mandrills and bush babies, and sundry reptiles and then sold them to merchants at Banana or to passengers aboard Dutch and English steamers.[6]

London Zoo depended heavily on private businesses for its stock. Europeans like the famed Hamburg-based animal dealer Carl Hagenbeck (who also exhibited non-Europeans in "ethnographic displays") and Charles Jamrach (also from Hamburg), who took over his father Jacob's business in London, set themselves up to supply zoos and circuses across Europe and in the United States with exotic animals. They sent out collecting expeditions and bought from local agents, taking advantage of faster travel times to Europe that steam ships made possible along with the opening of the Suez Canal in 1869.[7]

Good examples of the complicated commodity chains servicing the exotic animal trade include London Zoo's attempts to secure giraffe (the okapi's only living relative), the first of which arrived in 1836, and Obaysch, the zoo's first hippopotamus. Obaysch arrived in 1850 aboard a specially modified steamer fitted with a four-hundred-gallon iron tub for the five-hundred-pound youngster. All of these exotic beasts, so important for the survival of the zoo (the paying public flocked to see them), were captured by Africans and transported to the zoo in the care of African keepers, who initially lived with them at the zoo.[8]

Similarly, while the scientific networks and the process of procuring wildlife established by Sclater and others at the Zoological Society of London benefited from British imperial connections, this doesn't encompass all the networks and alliances involved in the story of the okapi. Harry Johnston's career in the Foreign Office put him in a position to follow up rumors of the existence of the okapi, but the impetus for his journey into the Congo forests came from Belgian officials and the dubious activities of a German impresario. Nationalistic competitiveness certainly featured in the discovery story but so did international collaborations. The story featured individuals and organizations from Britain, Belgium, France, Germany, Italy, and the United States, and nationals of countries including Cabo Verde, Portugal, Sweden, and Switzerland.

Okapi were found in the territory of the Independent State of the Congo, also known as the Congo Free State (1885–1908, thereafter Belgian Congo until independence from colonial rule in 1960) with the cooperation of local colonial officials and go-betweens. Hunting or collecting okapi required permission from Belgian authorities beginning in December 1901 formalized from September 1902. In the first years, okapi could only be found and caught by local Africans, who were not under British rule, many of whom had never heard of Great Britain. As anthropologist Colin Turnbull notes in *The Mbuti Pygmies: Change and Adaptation*, many of these local hunters, notably pygmy peoples, "had little or no concept of what a nation was, or of where the capital of their particular nation lay, or of who made up the government."[9]

While locals in the Congo were buffeted by massive and violent change at the hands of Europeans and in particular King Leopold II of the Belgians's brutal extractive colonial regime, their history and agency—then and now—are not best understood by absorbing them into frameworks centered on British or Western colonial power. This is because they responded, as Turnbull argues, in resilient ways that preserved important elements of their own distinctive ways of living with other humans and in the natural world.[10]

A Natural History Museum for Britain

By the 1830s, the public galleries housing natural history collections in the first British Museum building in Bloomsbury (an area of London south of Euston Station in the Borough of Camden) had become cluttered, dreary, and disordered. This is unsurprising given that leadership of the Department of Natural History was decided through patronage by a board of forty-eight trustees almost entirely ignorant of science. Members were nearly all aristocrats and senior clergy, and museum appointments were overseen by three trustees, the archbishop of Canterbury, the lord chancellor, and the speaker of the House of Commons. Despite repeated parliamentary enquiries urging reform of this system, no changes were made during the time the natural history collection was lodged in Bloomsbury.[11]

Finally, after Robert Smirke was commissioned to build a new "temple of the arts" (no mention of the sciences) in Bloomsbury, a much larger, Greek Revival style building was completed in 1846. Once installed, natural history collections occupied a third of the ground floor space, and the Elgin marbles were located adjacent to an "insect room."[12]

In this period, artifacts and natural history specimens in the form of appropriations, gifts, plunder, and surveys from Britain's expanding overseas territories and spheres of influence worldwide and from archeological digs and collecting expeditions in Egypt and the Middle East poured in. The keepers of the museum's departments were soon competing again for space, the excellent new keeper of zoology John Edward Gray having considerably enlarged his collection.[13]

The appointment of Antonio Panizzi as principal librarian (head of the museum) in 1856 brought the growing tensions between the library and antiquities departments (the priority for most trustees) and natural history into the open. Panizzi was dismayed by the popularity of natural history with the public who were visiting the museum in ever greater numbers to view the natural wonders rather than the antiquities or the library. Panizzi was said to have a thoroughgoing contempt for "men of science."[14]

Fortunately, a new position of superintendent of the natural history collection was created in the same year, and the man who would successfully advocate for the magnificent, separate museum in South Kensington was appointed. Richard Owen opened his period of tenure with three public interventions calling for more space for his poorly lit and difficult to access collection (*The Times* described it as "dull, close, crowded, and very fatiguing"). He cannily won over William Gladstone, who went on to become chancellor of the exchequer, to his cause.[15]

Owen calculated that he needed a ten-acre building (40,468 square meters) or five acres over two stories, to accommodate and properly show a national collection. This would be cheaper to acquire in the suburbs than in central London. "I love Bloomsbury much," he quipped, "but I love five acres more." While existing provincial or university collections, such as those of Cambridge and Oxford, could show small objects in some detail, he maintained that only a grand national museum could show, for example, the main genera and species of whale and the megafauna of the empire's growing tropical territories, adding that indeed it was the job of a national museum to show off such prizes. The natural sciences required their own temple to house them all.[16]

The museum took nearly twenty years to arrange and complete, opening to the public on a new site in South Kensington (then a suburb west of London) on April 18, 1881. The final zoology collections to go on display opened to the public in May 1886, by which time 394 cartloads of specimens and materials had been trundled down to Kensington from Bloomsbury.

Designed by an emerging star of British architecture, Alfred Waterhouse, the museum was an instant landmark. On first opening, the building measured 675 feet (205 meters) across the front (south facade facing Cromwell Road), with 192-foot- (59-meter-) tall towers. All external facades and interior wall surfaces were fashioned entirely from terracotta. The richness of the collections is matched by a terracotta menagerie clinging to arches, columns, parapets, and ventilation covers, with mostly extinct animals in the east wing (look out for bats though) and living animals in the west wing.[17]

The mission of the museum at this time can be understood through the arrangement of the material in the departments or groups into three series. There was an elementary or introductory series for each group that showed typical structural features and so forth. The exhibited systematic series showed the most important types of the various groups using systematically arranged, well-preserved specimens. The reserve or study systematic series allowed scientists to study the full diversity of the collection and explore the finer points of anatomical distinctions, variations, and geographic ranges.[18]

The first two series aimed to disseminate scientific knowledge and the third to advance it. As an early guidebook noted, "It is to this part of the collection that zoologists and botanists resort to compare and name the animals and plants collected in expeditions sent to explore unknown lands." A lot of individual remains are required to judge whether a species is new or whether it is a subspecies, whether remains are those of a juvenile animal or an adult, or whether sexual dimorphism has to be taken into account (e.g., differences between males and females in terms of size, color, or presence of horns). In-depth comparisons are needed to judge whether the variations are individual or species level and to determine which taxonomic groups a new species should be placed in.[19]

The Natural History Museum would play an important part in the discovery, description, classification, naming, and exhibition of the okapi. By the early 1900s, the museum was literally overflowing with African mammals that had been sent back from British territories and colonies and from other far-flung parts of the continent by explorers or hunters. These were exhibited in new display areas spilling out into the corridors facing the main hall on the first floor. It was here that the first okapi remains to reach Britain would be shown—in the east corridor, near the giraffes.[20]

Ray Lankester, Director, 1898–1907

The key personalities at the Natural History Museum for the purposes of this history are Edwin Ray Lankester (known as Ray) and Charles Immanuel Forsyth Major (see chapter 2). Lankester became the third head of the museum in 1898, succeeding Richard Owen and Sir William Flower. A protégé of Thomas Huxley, he had been Linacre Professor of Human and Comparative Anatomy in Oxford when he accepted the post.[21]

Lankester took on the role in a turbulent time, and the appointment of a distinguished and senior academic to the post was greeted with general relief following a vacancy of nearly a year. Supporters of the Natural History Museum feared that the trustees and director of the British Museum in Bloomsbury (still in overall administrative and financial control) intended to "curtail the authority and to lessen the dignity of its head." This was in fact the case: unbeknown to Lankester, in 1898 a standing committee of the trustees had gone back on an earlier decision of 1885 conferring comparative independence on the position of director of the Natural History Museum. Instead, the principal librarian would have overall authority. Lankester also had big shoes to fill (Flower, a strong but also more tactful man, was a very successful director) and a difficult challenge on his hands in ensuring his department both educated the public and advanced science.[22]

Lankester did not lack confidence: he was a large, vigorous, and forceful man (see fig. 1.2). He came from a middle-class rather than an aristocratic background and was not religious, which meant that his relations with his institutional su-

Figure 1.2. Punch cartoon of Ray Lankester astride an okapi (November 12, 1902).

periors would be tempestuous. His nemesis would be the director and principal librarian of the British Museum and secretary to the trustees at the time, Sir Edward Maunde Thompson, who wanted control of the entire institution including the Natural History Museum. He and nearly half of the trustees had opposed Lankester's appointment.[23]

While the trustees were concerned about the appointment, and even Lankester's mentor Huxley (no shrinking violet himself) worried that his protégé could be a little hot headed, the public seems to have taken a liking to Lankester, and he was respected by zoologists in Britain and across the rest of Europe. The *Candid Friend* described him as "the bulky and excellent biologist" with "a head like a benevolent biscuit tin," whose tendency to not suffer fools lightly and to speak his mind meant those who didn't know him better saw him as "pugnacious and irascible." However, he was trusted by his students, and was acute and generous in recognizing talent and industry in others. He also told very good stories.[24]

Scientific Authority, Knowledge Institutions, and the Naming of Species

Natural scientists were trying to establish themselves in a period characterized by overseas expansion and technological development on an epic scale and yet still dominated in its institutions by an establishment comprising mostly aristocrats and senior clergy. There was a rising cadre of professionals, largely men, of middle-class origins taking up senior positions in the new scientific and professional institutions. What this meant was that questions of institutional and scientific authority were keenly contested.

The problem for the emerging metropolitan centers of natural science in the period of interest here (1860s–circa 1914), was the extent of the dependence of the institutionalized, respected authorities on evidence and narratives from the (far-flung) field. The armchair zoologists presiding over museum collections were reliant on but also distrustful of the adventurers and explorers sending back (or worse, bringing with them, along with their untutored opinions) specimens and stories from regions as yet little described by Western science. Metropolitan European experts' authority was vested in their training and academic credentials, and hard-won positions in the great imperial institutions that catalogued and explained the world beyond Europe's shores.

In the 1860s, Paul Du Chaillu's stories about his travels and discoveries in West and Central Africa created a public sensation, most famously his self-proclaimed discovery of live gorillas and, later, his confirmation of the existence of pygmies. Metropolitan scientists argued that information reported in such accounts by untrained observers "may be truth, but . . . is not evidence." In other words, these accounts could describe actual events or animals but were not to be taken as true interpretations of what had been observed. The cause of authors like Du Chaillu was not helped by their publishers' tendency to encourage sensationalist accounts of their travels, such as Du Chaillu's tales of violent encounters with savage gorillas.[25]

But criticisms of explorers' narratives and factual claims were not simply the cool, methodical critiques of objective scientists either. They were colored by professional jealousies, class snobbery like that encountered by the Welsh explorer Henry Morton Stanley and by Harry Johnston, and even racism. Du Chaillu was rumored to have a "mixed-race" mother, which allegedly reflected poorly on his "character" and hence the trustworthiness of his scientific findings.[26]

The relationship between European explorers and collectors (whether they were travelers or resident in distant countries) and the experts of the metropole could be very productive. So long as the former confined their ambitions to observations, collection, and narratives of their adventures, metropolitan experts back in what Bruno Latour calls "centers of calculation" could be supportive, providing the necessary expertise in identifying and classifying their findings. Notable discoveries could earn such adventurers medals or corresponding memberships in learned societies. They were invited to share accounts of their exploits at prestigious speaking events. Many new species were named after their European discoverers in the field. However, claims to scientific discoveries and theoretical speculations by those outside the elite scientific community attracted criticism and ostracism from the experts, especially those made in popular books that evaded the strictures of peer review.[27]

Naming a species after their collector was perhaps the highest honor that could be bestowed, although the use of eponyms (e.g., *Okapia johnstoni*) in naming new discoveries to commemorate not just collectors but also royalty, donors, officials, or the person naming the species or genus, was not uncontroversial. By the nineteenth century, the Linnaean binomial system for naming new species had been adopted in Britain, though this had yet to replace the chaotic profusion of common names that had gone before. Arguments persisted over correct naming practices, laced with class snobbery over correct usages of Greek and Latin and references to the classics.[28]

It is a truism that discoveries of most new species outside of Europe were only discoveries for the West. The names and details of the local informants and experts directing Europeans to such interesting species or collecting them on behalf of Europeans are absent from scientific presentations and publications. So are those of skilled animal handlers who cared for live specimens in captivity and who in some cases accompanied them back to Europe. It is thus not surprising that few species were named using Indigenous names (the okapi is a rare exception), and vanishingly few were named after non-Europeans or even after the places where they were collected.

Famous cases of non-European animal handlers in Europe include Atir and the giraffe sent by the Ottoman viceroy of Egypt to Paris in 1826. In 1836, three Nubian keepers—whom we only know as Cabas, Omar, and Abdalah—accompanied the Zoological Society of London's four giraffes (who they had named Guib-Allah, Mabrouk, Selim, and Zaida) to London and remained with them at the zoo over the summer, along with the French trader Thibaut who procured the animals. Hamet Safi Cannana accompanied Obaysch the hippo to the zoo in 1850. Scant details remain of their biographies or contributions, and these handful of cases are unusual in being recorded at all.[29]

Thus not just the identity but also the knowledge and expertise of local peoples was obscured, omitted, and (like that of the untutored European collectors), sidelined from science. Today their input still remains largely unacknowledged; this failure to recognize their contributions is part of a wide-ranging silence (or amnesia) about the intellectual legacy of Western colonialism. However, it is worth noting that collectors and explorers like Harry Johnston and Paul Du Chaillu knew that their field observations would be minutely pored over by armchair experts back in the metropole. Given the scientific racism (following the publication of Darwin's theories of evolution) of the later 1800s, attributing field knowledge and discoveries to Indigenous peoples was unlikely to strengthen the case for taking such findings seriously. In certain cases, then, arrogance or a desire to claim discoveries as all their own might not be the only reason some Western explorers failed to name their local informants when reporting their discoveries.

Indeed, what is surprising is the degree to which some explorers and collectors *did* acknowledge Indigenous peoples and sometimes specific individuals in their narratives of journeys of exploration and collection. There is evidence available, for those willing to look, to aid the recovery of the historical contributions of these individuals.

The importance of "the British overseas" for the development of zoological knowledge in Britain was belatedly acknowledged in the publication of a collection of papers from a 1976 conference organized by the Zoological Society of London on the history of the society. The book focused on the careers of ten such overseas members, beginning with the Society's founder Sir Thomas Stamford Raffles and concluding with Harry Johnston. However, the role of Indigenous and non-European persons was completely overlooked.

It is only in past thirty years that historians have moved beyond acknowledging non-Western knowledge in a general, nonspecific way and begun exploring individual contributions. Examples include recognition of botanical identifications by the Malayali physician Itty Achuthan, the work of the Guinean botanist and healer Kwasímukámba in Suriname, and the contributions of Alfred Russel Wallace's Malay servant, guide, and expert collector, Ali. In 2023, James Poskett published *Horizons: The Global Origins of Modern Science*, a popular history of non-Western thinkers' contributions to the advancement of science.[30]

In this book, I recover the names of some of those Africans who guided Westerners to the okapi and shared their knowledge of its habits. It is not solely a list of Westerners and non-Westerners, however; there is a more complicated web of relationships and identities to uncover. The focus of this book is on the okapi and the relationships between the explorers and collectors, locals, indigenous experts and go-betweens, metropolitan knowledge institutions, and state and colonial authorities and conservationists involved in finding, describing, studying, exhibiting, trading, and trying to save it. It explores the politics of knowledge-making shaping this emblematic story of natural history discovery.

The okapi was in the process removed from the ecological, cultural, and social contexts of its discovery, and relocated to Western taxonomic systems, natural history museums, and zoos. Removed from their richly entangled life in the Congo, okapi arrived as usually singular, lonely strangers in the zoos and museums of the West. Congo, the first okapi to arrive in London, was delivered thirty-five years after the first evidence for the existence of okapi was sent back from the Congo to Sclater. He soon died.

Congo's journey presents a mirror image of Joseph Conrad's description of Marlow's voyage from London to the Congo in *Heart of Darkness*. A rainforest giraffid, he was shipped down the river Congo and across the seas to sail up the Thames and arrive in a place of darkness for him: a treeless, fenced paddock in a cold climate; and like many okapis to come, he was soon relocated to a mausoleum for dead fauna.

CHAPTER TWO

Discovery of the Okapi

The Belgian officers . . . knew the okapi perfectly well, having frequently seen its dead body brought in by natives for eating. . . . [C]alling forward several of their native militia, they made the men show all the . . . parts of their equipment made out of the striped skin of the okapi.

—*Harry Johnston, 1902*

The discovery of the okapi was a collective experience rather than a heroic individual one and happened as a result of a series of events commencing in the 1880s rather than as a singular event in July 1900 when Harry Johnston secured physical evidence of the okapi's existence for Europeans.[1]

Johnston was not the first to explore the natural history of the region, and the experiences, speculations, and written accounts of Europeans who had traveled to the region before him were influential. Most notable was Welsh explorer Henry Morton Stanley's calamitous expedition to rescue Emin Pasha during which he received the first clue to the existence of the okapi. Little has been said about the circumstances in which Stanley heard about the animal that turned out to be the okapi. An investigation of these circumstances also illuminates the colonial context of Johnston's quest.

Despite Stanley's brief note in an appendix and notwithstanding later claims made for other travelers in the region, it was certainly Harry Johnston who set out to find proof of the existence of the okapi and who succeeded. Given his role, I offer an outline of his biography up to the point of discovery, followed by an account of the first steps that validated his discovery as a novel one as well as the initial speculations about the taxonomical categorization of the animal. Once the okapi went on display and its taxonomic status was confirmed in scientific journals, controversies blew up about who had really discovered it and delivered the confirmatory scientific evidence. These controversies are emblematic of the kinds of arguments around priority, authority, and reputation characteristic of the period.

Many versions of the discovery of the okapi exist, most lacking in detail or mixing up or omitting significant dates, persons, and events. Several falsehoods were circulated about the contributions of particular individuals. It was part of my own journey out of the fog of dengue fever to straighten out this story insofar as possible.

Explaining why the discovery of the okapi was so sensational to zoologists (and the general public) at the time requires delving into more technical details. At first, I had thought the ferment was down to it being a remarkable new large mammal, but as I read through the early reports in zoological journals, I realized that it was greeted with such excitement more specifically because it was a new (living) giraffid. Reading the taxonomical literature revealed that the discovery was big news because of the okapi's taxonomic standing, which might provide evidence for (or upset) then cutting-edge ideas about the evolutionary relationships of the ungulate mammals. Was the okapi a living fossil, a missing link, or a forest giraffe?

Discovery: Rumors of an African Unicorn

The clue to the existence of the okapi that led to its discovery for Western science was the observation of Mbuti people ("pygmies") to Stanley that his riding ass (or donkey) reminded them very much of an animal resident in the rainforests of the northeastern Congo. Stanley's steed was apparently the first such animal known to ("discovered by") the Mbuti and thus a subject of great fascination.

The single mention Stanley makes of an animal that could be an okapi doesn't reference the actual exchange in which he learned of it. It appears in an appendix to volume 2 of *In Darkest Africa* (1890), his account of his expedition to rescue Emin Pasha (1887–89), which Stanley chose to carry out by traveling all the way around the continent from Zanzibar and up the Congo River from the west coast, even though the pasha was in Equatoria (today South Sudan). This expedition merits brief explanation, as it was the reason Stanley was in the Congo when he was given this clue about the existence of the okapi.[2]

Emin Pasha (whose real name was Eduard Schnitzer) was a Prussian who on failing to get a license to practice medicine in Germany had set up a practice in

Albania, taken up with the regional pasha's wife, returned with her and her children and slaves to Germany after her husband died, then abandoned them and fled to Africa where he set up as an Arab doctor in Khartoum (he was a brilliant linguist) in 1875. Here he was recruited as a doctor by General Charles "China" Gordon and in 1878 made British governor of the southern Sudanese province of Equatoria.[3]

Gordon subsequently left Africa but returned in 1884 to arrange an orderly retreat from the Sudan, where the British occupying forces were under attack from the growing army of "the Mahdi" (or "expected one," Muhammad Ahmad, from Dongola). The general was killed by the Mahdi's forces in Khartoum in January 1885, following a long siege. This left Emin as the last of Gordon's men holding out against the Mahdists (or dervishes). After letters were published in England about Emin's plight, the public howled for a relief column to be sent in time to assist him (the British government had been too slow to save Gordon). When the government failed to respond, it was Stanley who set out to relieve Emin in January 1887, funded by public subscription.[4]

In a curious twist, Emin, a keen naturalist who (we learn from a posthumous collection of his letters and journals) meticulously prepared birds and other specimens to send to the Natural History Museum in London and corresponded with Philip Sclater at the Zoological Society of London, had (according to Harry Johnston) apparently heard of a type of zebra when he was in the northeastern Congo basin forests between 1883 and 1888. Johnston briefly noted this in a review of an edited volume of Emin's zoological records, published in German twenty years after Johnston's discovery of evidence of the existence of the okapi.[5]

There is, however, no mention of this Congo zebra in the selection of Emin's travel notes published in 1888. Much later, Alfred Pease claimed a Sudanese in southern Kordofan told him in 1906 that Emin's army had hunted and eaten okapi during their retreat from the Mahdi's forces through Equatoria. It is odd, though, that Emin didn't mention this in his notes or to Stanley (who recorded his discussions on natural history with Emin).[6]

Johnston's later claim (reported by Ray Lankester in November 1901) that "Stanley and his companions occasionally caught sight of the Okapi when they traversed the region of the Congo forest on the western side of the Semliki River" is very likely apocryphal. Johnston claimed he learned this from a conversation with Stanley in 1901, *after* the okapi had become a cause célèbre. Stanley never refers to seeing an okapi-like animal in his books nor does he describe the animal. In fact, Stanley remarks that "moving as we did [in a large caravan] everything of a mild or timid nature fled before us." While small parties sent out to hunt meat could have had more success (though Stanley still thought these would be too noisy), Stanley forbade them because he feared they would get lost in the rainforest, or worse, be killed (and possibly eaten) by hostile local Africans.[7]

Stanley does not mention okapi in his published account or in his exploration diaries of his first journey down the Congo to the Atlantic. He does note that in Manyema, his "riding-asses were the first ever seen . . . and obtained more admiration than even we Europeans" and that "hundreds of natives ran up to us at each village in the greatest excitement to behold the strange long-eared animals." Later, Wané-Mpungu people, Maruna warriors, and the king of Ntamo and his wives were similarly impressed by the asses. However, none of these were pygmies, and there is no mention of comparisons with wild animals.[8]

Stanley distinguishes between asses and donkeys, referring to "riding-asses" on his first Congo expedition. Wild assess are native to Africa, and donkeys were domesticated from them in around 7,000 BP and spreading into Eurasia from around 4,500 BP. The histories of asses and donkeys has been little investigated until recently, as unlike the horse, they are the beasts of burden of the poor. It is poignant to me that the critically endangered yet little lamented African wild ass (*Equus africanus*)—of which today about two hundred adults survive in the wild in the Horn of Africa—may have been the animal that reminded rainforest-dwelling Africans of the okapi.[9]

Stanley devotes a chapter of his book *In Darkest Africa* to "The Great Central African Forest," providing a vivid sixteen-page description of its vegetation, and devotes five pages to its mammals, birds, reptiles, and insects as well as including an account of its peoples. Nothing like an okapi is mentioned.[10]

Stanley's descriptions of pygmy peoples and his interactions with them are mostly linked with his time in the rainforest (October–December 1888). Disappointingly, there is no mention of the incident in which pygmies noted the resemblance of the okapi to his donkeys (throughout *In Darkest Africa*, Stanley refers to "Zanzibari donkeys," though he once also refers to "a dainty Zanzibar ass").[11]

The chronology is also a puzzle, as by mid-September 1887, the expedition had only three Zanzibari donkeys left, and on October 15, 1887, Stanley shot his remaining donkey, which was starving, and distributed the meat to his hungry men. The only Mbuti (whom Stanley refers to as "Wambutti") person he seems to have met prior to this was a young woman originally from "north of the Ituri."[12] She was exhibited to him on September 17 as the captive of a chief named Ugarrowwa. He mentions no conversation with her but only describes her attractive appearance. She remains the likeliest source, however, as Stanley's expedition saw their first (deserted) pygmy villages on October 30, 1887. The next Mbuti he met was the wife of the chief of Indekaru, on February 18, 1888, and he also captured five pygmies in April 1888, but he had no donkeys by this stage. So, it remains a puzzle how it was that an Mbuti ("Wambutti") person made the comparison between a donkey and an okapi to Stanley sometime between September 1887 and December 1888. Hopefully, further archival work will shed more light on this.[13]

Other claims to prior knowledge of the okapi would inevitably emerge after its discovery by Johnston had been publicized. Claims were made for the German explorers Wilhelm Junker and Franz Stuhlmann, and French ornithologist Auguste Ménégaux claimed that the French explorer and soldier Colonel Jean-Baptiste Marchand had seen one in the Bahr el-Ghazal (South Sudan) in 1898. It is extremely unlikely that Marchand saw one, as South Sudan is well outside of natural okapi range (and habitat).[14]

Whatever these Europeans may have seen or noted prior to 1900, nobody had pursued these stories any further. Johnston's interest, however, was piqued by Stanley's brief note in appendix B of *In Darkest Africa,* which states that "the Wambutti knew a donkey and called it 'atti.' They say that they sometimes catch them in pits. What they can find to eat is a wonder. They eat leaves."[15] Johnston found this curious because to his knowledge no European had seen an equine animal in the forests west of the Nile. Horses are grazers and plains animals, not browsers (they do not eat leaves).

Johnston followed this up partly because as a boy he had been inspired by Philip Gosse's curious book *The Romance of Natural History* (1860). The book describes the known mammals of the world but also includes a chapter speculating on rumors of fantastical as yet undiscovered beasts including several accounts of the existence of the unicorn in Africa. It is a precursor of "cryptozoology," today considered a pseudoscience by mainstream zoologists.[16]

Gosse was writing in a period of plentiful new zoological discoveries for Europeans that are described, for example, in Charles Darwin's journals of his voyage aboard the *HMS Beagle* (1839), in Alfred Russel Wallace's papers describing his discoveries in the Malay Archipelago in the 1850s (prior to publication of his best-selling book), and in Paul Du Chaillu's sensational accounts of the gorilla and other unusual Central African animals.[17]

Gosse singles out Central Africa as a likely hotspot for new discoveries. So when Johnston read Stanley's note on the donkey-like *atti*, he was intrigued by the idea of an animal that unlike all known equines was not "partial to treeless, grassy plains" lived in "the depths of the mightiest forest of the world," and he wondered if it could be the mythical unicorn Gosse suggested might inhabit the region.[18]

Harry Johnston: Renaissance Man or Jack of Many Trades?

Harry Hamilton Johnston is best known to historians as a colonial administrator and author of comprehensive books on British colonial territories during the height of the European "scramble for Africa" and to African linguists for his work on Bantu languages. He is often used as a primary source in these fields, but he has been little explored in his own right. His contributions to natural history have been neglected, even though he was awarded silver and gold medals by the Zoological Society of London. Once the intimate of famous explorers and political grandees including Lord Salisbury, Theodore Roosevelt, and Cecil Rhodes, and feted by learned societies, he was utterly forgotten when he died in 1927. Historians regard him as a footnote—partly due to a spat between his first two (and only) biographers.[19]

Johnston wrote pioneering histories of several African regions, some of which remain influential. His taxonomies of African races were shaped by his experiences as a young man exposed to the Darwinian revolution in life sciences in London that embraced natural history and "race science" within an evolutionary framework. This revolution in scientific thinking meshed with an imperial project to discover, order, and control overseas territories.

Johnston's deep learning gained from field experience and from conversing and reading in many languages set him apart and made him useful. His accounts of his travels included a wealth of historical, zoological, botanical, and ethnographic detail. However, while popular with colleagues and subordinates, a tendency to show off his knowledge, along with an irreverence for authority, made him an object of envy and ridicule among his professional peers and superiors. Johnston's diminutive stature and social background provided easy targets. Thomas Pakenham (Lord Longford in title and stature) described Harry as "unmistakably middle class" and an "*enfant terrible* of five foot three inches."[20]

As a child Johnston showed a keen interest in natural history and in drawing and painting. In 1874, at sixteen, he enrolled at King's College London to study European languages; at the same time, he also studied painting at the South Lambeth Art School. By then he had already been drawing animals in the Zoological Gardens in Regents Park (London Zoo) for two years. He was able to get into the gardens regularly with a student ticket; Sclater, then secretary of the Zoological Society of London—who became a lifelong friend and important ally of

Johnston's—granted student tickets to enable young people to draw the collection during weekdays.[21]

Johnston's drawing attracted the attention of Alfred Henry Garrod, lecturer on comparative anatomy at King's College London and prosector (dissector of dead bodies for anatomical examination and demonstration) for the Zoological Society of London from 1871 to 1879. Garrod granted the boy access to his prosectorium, where along with other boys and young men (and a handful of women who visited on separate days), Johnston learned not just to draw but "to dissect" and also acquired knowledge of "the structure of birds, beasts and reptiles." Through art, he learned his science.[22]

Johnston earned pocket money illustrating Garrod's scientific papers, and Garrod introduced him to Sir William Flower. Flower—a surgeon, professor of comparative anatomy, and curator of the Hunterian Museum of the Royal College of Surgeons (who later succeeded Owen as director of the Natural History Museum)—allowed Johnston to study the Hunterian collection. Thus, although he lacked formal scientific training, Johnston possessed detailed firsthand knowledge of animal (including human) anatomy. He had learned to observe carefully and precisely, which served him well when he turned to natural history explorations and descriptions. The ideas Johnston imbibed about the evolutionary relationships of human "races" were rather less helpful.[23]

Through a friend at London Zoo, Johnston was invited to join Dermot Bourke, seventh Earl of Mayo, on an expedition into the interior of Angola that got underway in 1882.[24] Johnston, now in his early twenties, traveled inland as far as Humbe on the Kunene before returning to the coast to take up an offer to meet Stanley and visit "the Congo, then a land of scarcely solved mystery."[25] Stanley was by then opening up the Congo (1879–84) in the service of King Leopold II of the Belgians. Along with many prominent Europeans, Stanley had been taken in by the king and led to believe he was on a mission to bring civilization to the region and end the slave trade.[26]

Belgium, a new country created in 1831, was a liberal parliamentary democracy divided between a Dutch-speaking north and a French-speaking south that was united to a degree as a nation by a constitutional monarchy, with Leopold of Saxe-Coburg as first monarch. Leopold I and II felt constrained by the constitution and modest size and resources of their kingdom and sought to expand the country's influence domestically and internationally by creating an empire. In this quest, Leopold II succeeded in the Congo.

In reality, though, the Independent State of the Congo was an elaborate subterfuge. Instead of being an international philanthropic initiative designed to bring stability and prosperity to the Congo by introducing Christianity, commerce, and civilization to it and by ending the slave trade and establishing free trade along the river, the operation was run by Leopold's personal commercial company (disguised as an international association) and intended to establish his ownership over African lands and extract as much ivory (and later rubber) and other natural resources as possible and by whatever means.

By the time Johnston visited the territory in 1883, Leopold was only a year away from gaining international recognition for his duplicitously named "Independent State." It would be established with its own flag, following a series of bilateral treaties signed at the Berlin Conference, at which European colonial powers carved up African territories between them (1884–85). This was the culmination of a long series of maneuvers by Leopold dating back to his establishment of the International African Association in 1876.[27]

Johnston spent eight months exploring the Congo; he met Stanley twice in April 1883 and with his assistance traveled upriver as far as Bolobó. Johnston's descriptions of the region were published in his first book, *The River Congo* (1884), which established his reputation as an explorer and Africanist. He became a close friend of Stanley's after the explorer settled in England in 1890 and was a bearer of Stanley's catafalque at his funeral. It is possible this thin-skinned man of middle-class origins identified with Stanley's struggles with the British establishment that owed to his own (much more) humble origins.[28]

Following publication of his Congo volumes, Johnston completed an exploration of the Mount Kilimanjaro region in the name of science and treaty making, the former often an excuse for the latter, for which he learned surveying. This initiated an eighteen-year career with the Foreign Office, with postings to the Niger Delta (1885–88), British Central Africa (Nyasaland, 1889–96) for which he was knighted in 1896, Tunis (1897–99), and finally the Uganda Protectorate (1899–1901).[29]

Johnston hesitated over this last assignment, framed as a two-year special commissionership during which he would be expected to steady the ship in the protectorate. The territory had been unstable owing to a rebellion of Sudanese troops in 1897, difficulties with the Bunyoro Kingdom in the west, and the attempts of the French to undermine British interests. Further, he and his wife, Winifred, were enjoying their life in Tunis, and she would not be able to accompany him to Uganda. The job would require much travel and would be dangerous at times, and he feared a recurrence of blackwater fever (a complication of malaria, named for the resulting blood in the urine). He had first contracted blackwater fever in British Cameroons and again in British Central Africa

(Johnston's cure was champagne, lemonade, and quinine). Those of us who have endured the rigors of tropical fevers know that recurrences can be worse than the initial infection, and even fatal.[30]

Ultimately, Johnston could not resist the opportunity to complete his research on Bantu languages (and publish a definitive book) and to "find many things that were new in the African fauna and flora." In sum, Johnston was aware of "the risks to life and the possibility of failure," but it was not in his nature to turn down this offer. After a holiday in Venice with Winifred, he set sail for Zanzibar from Marseilles in late 1899.[31]

Kidnapped Pygmies and Evidence of the Okapi

In early 1900, in post as special commissioner in the Uganda Protectorate, Johnston was contacted by Belgian officials in the Independent State of the Congo, requesting his assistance in apprehending a German who had kidnapped a group of "Congo Pygmies" in the Ituri forest. This man was fleeing with them to the east coast, intending to exhibit them at the 1900 Paris Exhibition. He had requested and been refused permission to acquire the pygmies.[32]

Johnston agreed to apprehend the fugitive, who was found, tried, and sentenced with a heavy fine. Several of the "Bambute Pygmies"—Mbuti people—had already escaped; the remaining seven were sent to Johnston at Entebbe in March. At Entebbe they lived in the parklike area behind Johnston's house where he kept wild animals. Johnston later claimed they had confirmed the existence of the okapi to him, pointing to a live mule and a zebra skin.[33]

Johnston next traveled west to survey the Rwenzori Mountains, accompanied by the Mbuti men and the photographer and taxidermist W. G. Doggett (sent out by Sclater). By this time, Johnston was an established collector and author on the zoology, botany, ethnography, languages, and geography of several African regions. The Zoological Society of London had awarded him its silver medal in 1894 for facilitating collections in British Central Africa (now Malawi) through his own efforts and those of Alexander Whyte, a Zoological Society–approved collector. They sent rich materials back to England (including a hippo head shot by Sclater's son William), which were the basis for numerous publications by scientists at both the Society and the Natural History Museum.[34]

In July 1900, Johnston's expedition crossed the Semliki into the Ituri forest in the Independent State of the Congo. Here they visited the Belgian commanding officer Lieutenant Meura and his second in command, a Swede named Karl Eriksson, at Mbeni Station (Fort Mbeni). According to a later visitor, Cuthbert Christy, this was an important station sited on high ground overlooking the river valley. It had the full length of the snowcapped Rwenzori Mountains for a backdrop, and Christy thought it "one of the most beautiful spots" in Central Africa.[35]

Johnston's pretext for this visit was to return the kidnapped Mbuti to the Ituri Forest. In volume one of his book *The Uganda Protectorate*, he recalls that the Belgians confirmed what the Mbuti had told him about the mysterious forest equid, describing it as a "creature of the horse tribe, but with large ass-like ears, a slender muzzle, and . . . more than one hoof," like a hipparion. If such an animal existed, it would be a sensational find, as the hipparion was believed to be the (three-toed) extinct ancestor of modern horses.

Although often credited as the sole discoverer of the okapi, Johnston always acknowledged the Mbuti's prior knowledge of the animal, and likewise recognized that "the Belgian officers . . . knew the okapi perfectly well, having frequently seen its dead body brought in by natives for eating." However, as, according to the Belgians, okapi were caught in pitfalls and arrived at European settlements either as decorative strips of skin or as meat for consumption but never in the form of an entire animal, in fact most Belgians didn't know what an okapi looked like.[36]

Bad Luck and Bandoliers

It was agreed that Johnston and his party would accompany the Mbuti home to "see them definitely repatriated," a journey that would take them through the region where specimens of these strange creatures were occasionally obtained by local African soldiers working for the Belgians. Johnston "entered the forest with the keenest anticipation of discovery," but when the Mbuti showed him okapi footprints, he dismissed them because they were "two-toed," and he was looking for a track resembling that of a donkey or horse (an equid, with one "toe" or "wall" on the hoof). In fact, okapi are even-toed ungulates (artiodactyls) like giraffes, antelope, and deer.[37]

In several days of searching in the "almost unbreathable" atmosphere of the forest—"with its Turkish bath heat . . . and its powerful smell of decaying, rotting vegetation" reminiscent of "Miocene times" and "scarcely suitable for the modern type of real humanity"—the closest Johnston came to an okapi was in the form of Congolese soldiers wearing okapi hide bandoliers, encountered in a village. He bought two bandoliers to send home as evidence that he "was really on the tracks of a new creature, even if it only turned out to be a forest-dwelling zebra."[38]

After only a few days, Johnston's entire caravan except himself was suffering from fevers, and he decided to return across the Semliki into the Uganda

Protectorate. To console Johnston for his great disappointment, Lieutenant Meura promised to acquire a complete skin and send it to him. In the interim, when he arrived at Fort Portal in western Uganda in August 1900, Johnston sent the okapi bandoliers off to Sclater in London.[39]

Sclater received the okapi bandoliers in November 1900 and read out a letter from Johnston on his remarkable "new horse" at a Zoological Society of London meeting. In his letter, Johnston gives both his version of the Mbuti form "o'api" (the apostrophe standing for a gasping sound, or Arabic *k*) and the local Bambuba form "okapi." Thus, November 20, 1900, is the date on which the word "okapi" was introduced into English. Sclater exhibited the bandoliers at the December 18, 1900, meeting. This was the first evidence of the existence of the okapi to be shown in Britain (see fig. 2.1).[40]

That just two bandoliers fashioned from the skin of an unseen okapi served as the first evidence of the existence of a new mammal is remarkable.

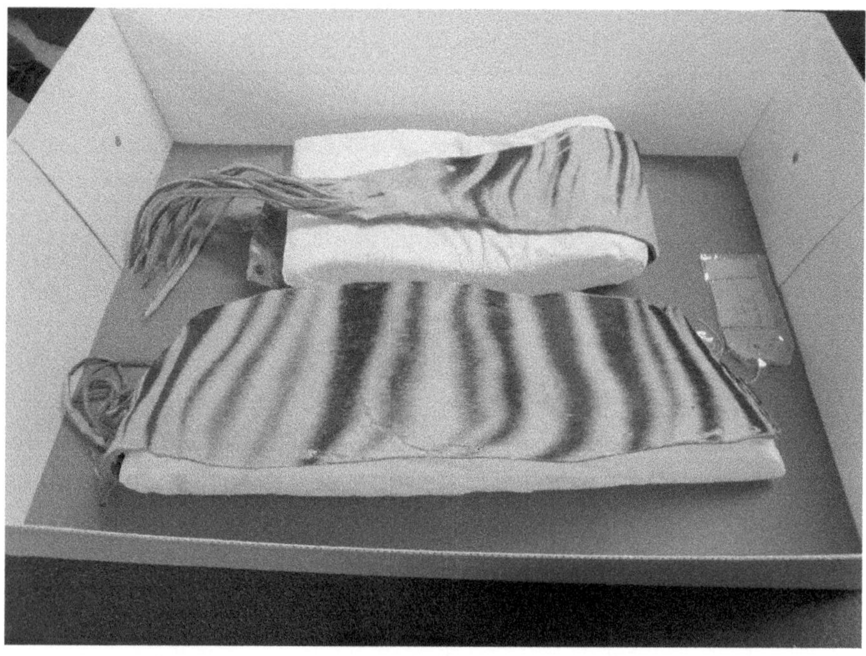

Figure 2.1. The type specimen okapi bandoliers (syntypes of *Okapia johnstoni* Sclater, 1901) stored at the British Museum of Natural History. A small label attached to the larger bandolier reads "*Equus* (?) *johnstoni*, Sclater = *Okapia johnstoni*" (© Simon Pooley, by permission of the trustees of the Natural History Museum).

Until the 1890s, zoological mammal collections comprised specimens pickled in spirits or stuffed with their skulls included inside the stuffed and mounted skins (taxidermized). However, Michael Rogers Oldfield Thomas, who was in charge of mammals at the British Natural History Museum, discovered from reading a series of books on the North American Fauna published by the United States Department of Agriculture that it was possible to work with only prepared, dried skins, each accompanied by its cleaned skull. This approach had been pioneered by C. Hart Merriam for the Biological Survey launched in the US in 1889. Oldfield Thomas adopted this approach at the Natural History Museum, which proved enormously influential as he commenced a survey of the world's mammals outside of North America. This, then, was the evidence required to confirm whether the okapi was a new species: complete skins and a skull.[41]

Descriptions and Speculations: Scientific and Press Reception in England

The Times reported on Johnston's travels in Uganda and the Congo in December 1900, including on his discovery of "a most remarkable species of horse or zebra" in the Congo forests. The article noted that these animals were called the "o'api" by pygmy peoples and the "okapi" by Bantu peoples. This information had first been sent to the Royal Geographical Society, who published it in the January 1901 issue of its journal. It was this journal publication that first alerted officials of the Independent State of the Congo to this sensational discovery on their territory. They immediately gave orders to their agents in the Haut-Ituri zone in the Orientale province to capture okapi and send them to the Congo Museum (from 1908, the Museum of the Belgian Congo) in Tervuren. (It seems King Leopold II, a keen reader of *The Times,* had missed the story.[42])

Based on the scanty evidence of Johnston's informants in the Congo and these strips of skin, Sclater described the new mammal as possibly a kind of zebra at a Zoological Society of London meeting in February 1901. In an article published that same month Sclater proposes naming it "*Equus* (?) *Johnstoni, sp. nov.*" after its discoverer ("until better specimens are obtained"). He follows this proposed name with a brief flourish of Latin describing its appearance and habitat, thereby layering a veneer of confident expertise onto this meager knowledge ("in sylvis fluvio Semliki adjacentibus" indeed!). Sclater quotes Johnston on his difficulties in trying to locate the animal and its African names and includes Johnston's incorrect speculation that the okapi is "dun-coloured or dark grey" on the upper parts of its body (he'd only seen strips from the legs). Assuming it belonged in the genus *Equus,* Sclater doesn't include an African name in his initial proposal for a scientific name.[43]

While at Eldama Ravine Station on the eastern border of the Uganda Protectorate in early March 1901, Johnston received an almost entire okapi skin and two skulls obtained for him by "the truly kind Lieutenant Meura." However, as Meura had since died of blackwater fever, they had been forwarded by Karl Eriksson (a circumstance that had consequences for controversies around priority and naming).[44]

In sending Johnston these okapi remains, Eriksson contravened the Independent State of the Congo's instructions to send all significant natural history remains to the museum in Tervuren. Perhaps he felt more compelled to fulfill the promise of his dead colleague, and he had of course met Johnston himself, rather than send the remains to a museum with which he had no formal connection. Johnston also learned from Eriksson that (to the Swede's surprise) the okapi had cloven hoofs (frustratingly, the hoofs had disappeared by the time the skin reached Entebbe, from where the remains were forwarded to Johnston at Eldama Ravine).[45]

Johnston realized—examining the skulls and based on knowledge he'd picked up from Garrod years before—that the bilobed lower canines (see fig. 2.2) meant "the beast was a near relation of the giraffe." Johnston knew of recent discoveries of fossil remains of giraffe-like animals and speculated that the okapi was a surviving relative of the *Helladotherium*, an early (long extinct) giraffe-like animal of what he referred to as the Tertiary Epoch (he meant the Miocene Epoch of the Tertiary Period) found in fossilized form in Asia Minor, Greece, and India. This was the conclusion he communicated to Sclater and Lankester, along with the proposal that it be placed in the genus *Helladotherium*.[46]

On March 31, 1901, Johnston wrote to Sclater that he had studied the skulls and then drawn what he imagined "this wonderful new creature" looked like,

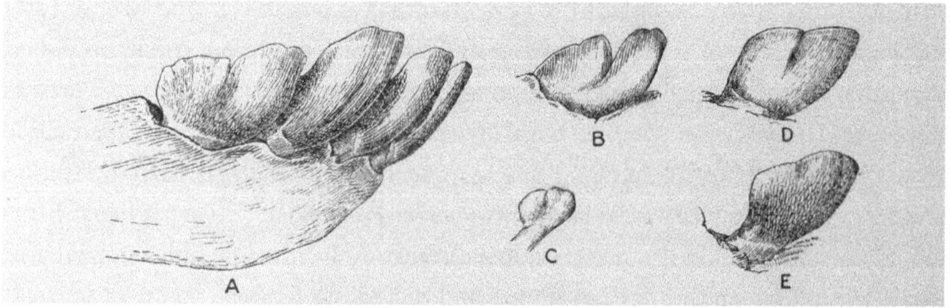

Figure 2.2. The bilobed lower canines characteristic of the giraffids (E. Ray Lankester, "On *Okapia*, a New Genus of *Giraffidae*, from Central Africa," *Transactions of the Zoological Society of London* 16, no. 6 [1902]: 288).

having observed the skin "while it still retained some indications of the shape of the animal" and having "questioned various Congo natives . . . as to the shape and appearance of the animal." He also painted a scene with two okapi. Sclater related this information at the May 7, 1901, meeting of the Zoological Society of London, where he showed Johnston's original watercolor painting (see front cover). Although the skin and skulls had not yet arrived in England, Sclater was convinced that Johnston "had made a most important discovery."[47]

Sclater was now of the opinion that "the animal portrayed . . . was, of course, not a Zebra, nor even a member of the family Equidae." While it would be necessary for experts at the British Natural History Museum to confirm what it was, Johnston might be correct that it was "allied to" the *Helladotherium* or another extinct form "allied to the Giraffe." In addition to the teeth, it was possibly significant that the animal stood slightly higher at the withers (shoulders) than at the hindquarters and that the skulls showed the "degenerated" remains of horn-cores in a similar position to those of a giraffe.[48]

Sclater quotes at great length from an article in *The Times* based on the letter Johnston (clearly aware of the scientific and public interest of his find) had sent to the newspaper. Entitled "A New Mammal," this article was published on the same day Sclater presented Johnston's painting at the Zoological Society of London and provides more details of the discovery and appearance of the okapi. The article does not use the name "okapi" regularly, though it notes "okapi" was the African name (there were many African names—see appendix 4). This *Times* article asserts that Johnston had discovered "a mammal . . . entirely new to science." It was "distantly related" to the giraffe but more excitingly, it was possibly a "living representative of the *Helladotherium*."[49]

A second article in *The Times* entitled "*Ex Africa semper aliquid novi*" ("out of Africa, always something new") reports on a Royal Society conversazione held two days prior to publication of the "New Mammal" article, at which Johnston's bandoliers were exhibited together with his painting. The article recalls Stanley's mentioning having "heard . . . from the dwarfs of the Semliki Forest" of such an animal, and claims that Stanley had urged Johnston to investigate.

This story remarks that zoologists had at first thought that "the animal" (still not "the okapi") might be a kind of zebra but now proposed it was allied to the giraffe and related to the *Helladotherium*. The sharp-eyed reporter notes that "originally called by Dr Sclater *Equus Johnstoni*, it now appears to be called *Helladotherium Johnstoni*, according to the label attached to the fragments exhibited at the Royal Society."[50]

Living Fossil or Missing Link?

The sensation caused by the discovery of the okapi was considerably augmented by the initial idea based on limited evidence that it was a primitive kind of horse, the multi-toed hipparion. On receipt of the skulls and skin, Johnston postulated equally sensationally that it might be a "living representative of the *Helladotherium*"—or a primitive missing link between the extinct *Helladotherium* and the modern giraffe.[51]

The idea of a missing link between extinct and living creatures, or less and more evolved ones, had been something of an obsession for nineteenth-century zoologists (and anthropologists: pygmies were once alleged to be a missing link between primates and *Homo sapiens*). Thus, Johnston's suggestion that the new animal could be a *Helladotherium* was a spectacular "coup de theatre," according to Auguste Lameere, professor of systematic zoology at the University of Brussels. Lameere wrote the first detailed consideration of the bearings of all this on the okapi's taxonomic status and evolutionary lineage in an article published in French in November 1902.[52]

As Lameere remarks, "We were all . . . entitled to believe in a new incarnation of the famous Sea Serpent, since the animal was described by the hand of a master [Johnston] and since zoologists had been waiting for some months already for the confirmation of a discovery that they did not, however, suspect of being quite so sensational."[53]

A series of exciting paleontological discoveries made from the mid-1800s had transformed the classification of ungulate mammals (hoofed, four-legged herbivores). This was almost wholly based on morphology, that is, on the shapes and sizes of the fossilized hard body parts that are usually all that remains of extinct animals. The *Helladotherium* was regarded as an extinct genus (related group of species) of the Sivatherine giraffids—giant, short-necked giraffes with ox-like rumps, sturdy legs, and a variety of differently shaped fur-covered forehead protrusions. They were discovered in Miocene deposits in mainland Greece by Jean Albert Gaudry, a French geologist and paleontologist.[54]

While conducting a government scientific survey in Greece in 1853, Gaudry was directed to a remarkable collection of fossil bones at Pikermi, a village between Athens and Marathon. He spent the next twelve years working on these bones and in 1860 published the first attempt to systematically arrange extinct animals in linear fashion concluding with their present-day ancestors. This classic volume proved very influential in attempts to trace evolutionary lineages from earlier periods, through the late Miocene, to the present. His blueprint for detailed comparisons, focused on feet and teeth, was influential for later, similar studies.[55]

Another important paleontologist working on fossil mammals in this period played a central role in the description and classification of the okapi. Charles Immanuel Forsyth Major was born in Scotland but moved to Constantinople (Istanbul) as an infant. He went on to study in Switzerland, Germany, and Italy, graduating as doctor of medicine in Basel in 1868, where he had been encouraged by Karl Ludwig Rütimeyer to pursue studies of fossil mammals. His studies of fossil remains in Italy showed him that it was possible to distinguish later Pliocene Epoch (circa 3.6–2.58 million years ago) mammals from those of the early Pleistocene Epoch (2.58–1.8 million years ago).[56]

Forsyth Major gave up his medical practice in 1886, when he was in his forties, to dedicate himself full-time to research, beginning with an exploration of the Pliocene mammalian bones on the Greek island of Samos. Part of his collection was bought by the British Museum. He assembled another important collection of similar remains in the Carrara mountains in Italy that was also bought by the British Museum. In 1893, he followed these collections to the British Natural History Museum in London, where he was temporarily employed (he doesn't appear on published staff lists) in cataloguing these fossil mammals until 1908.[57]

Forsyth Major's discovery of an extinct giraffid on the Greek island of Samos, placed in the genus named *Samotherium*, was important because he was at the British Museum when the okapi was discovered. He was therefore well placed to consider its evolutionary relationship to this extinct giraffid. As a result of this paleontological expertise on extinct giraffids, in 1902 the government of the Independent State of the Congo approached Forsyth Major about writing the definitive monograph of the okapi.[58]

Johnston had suggested that the skull of the okapi resembled that of the *Helladotherium*, and he even suggested naming it *Helladotherium tigrinum* (presumably for the stripes, but, still, it was an odd choice for an herbivore). By the beginning of the twentieth century, *Helladotherium* was regarded as an extinct "brother" to *Samotherium*, with no known modern ancestors. Members of the genus *Samotherium* were regarded as the ancestors of modern giraffes, though were later redesignated as a separate subfamily.[59]

The first skull of an okapi to arrive in Europe was therefore examined with great interest and described in minute detail by Lankester at the Natural History Museum. The skull resembled the elongated, simple skull of a deer, with the teeth of a giraffid as Johnston had described. It appeared to have bumps where the horns of a giraffe appear, though the specimen had no horns. Lankester agreed it was a giraffid, but important differences between *Helladotherium* skulls and giraffid skulls—as well as similarities to *Samotherium* skulls—led him to assign the okapi a new genus.[60]

Once more skeletons became available, it would be ascertained that okapi adults of both sexes can have horns (or, rather, ossicones); one pair on the frontal bumps of the skull (a third "bump" on the base of the nasal bones has no ossicone). These horns are covered with hair, like those of the giraffe. However, male and female giraffes have horns of the same dimensions, whereas okapi female horns are either very little developed and barely visible, or absent. Some living giraffe subspecies, like their giraffid ancestors, have more than two horns. *Samotherium* differed from okapi in that only males had horns, but like okapi, they had just two.[61]

What to conclude from all these similarities and differences? Lameere concurs with Forsyth Major that "the okapi is a step towards the giraffe" and "a little less primitive than the *Samotheriums*." However, he argues that this didn't mean the okapi was the direct ancestor of the giraffe: in the greater development of its ears, for example, he claims it had evolved further than the giraffe. He concludes his survey of the evolutionary relationships of the okapi with the suggestion that "if it were possible to compare the Giraffe to a Parisienne at the end of the century, the Okapi would be a provincial cousin." [62]

Confirmation and Classification

In a letter published in *The Times* on June 18, 1901, Lankester announces that "I have this afternoon received and unpacked the case shipped at Mombasa on April 19, containing the skin and two skulls of the remarkable new giraffe-like animal obtained from the Semliki forest by Sir Harry Johnston, and sent by him to me" (publicizing Lankester's link to "Sir Harry Johnston's New Beast"). He notes breathlessly that "I write without loss of time to say that the specimens have arrived in perfect safety, and they fully . . . bear out Sir Harry Johnston's statements and inferences." While agreeing that Johnston was justified in assimilating the animal to the extinct *Helladotherium*, Lankester asserts that "the 'Okapi' (the native name) . . . must be placed in a new genus." The signs were clear it was not horse-like but had cloven hooves.[63]

The contents of the May 7, 1901, letter published in *The Times* and the knowledge that a complete skin and skulls had arrived in London in June, was communicated to American zoologists in the *News Bulletin of the Zoological Society of New York* in July 1901. The bulletin lauds "the discovery of a new animal in

Central Africa which must rank as one of the zoological wonders of recent years." It credits Eriksson for collecting and forwarding it to Johnston (implying Eriksson had primacy) "for the British Museum."[64]

At a June 18, 1901, Zoological Society of London meeting, Lankester established the genus *Okapia* "for the reception of the Okapi." At first, he had called it *Ocapia*, but by June he had decided on the hard *k* to avoid confusion over the correct pronunciation. This he explained in a letter to Sclater, noting the convenient "liberty" of "the use of K for the hard sound of C" with a quip about "Ciceronian" not "Kikeronian." Lankester also published a brief note on the skin and skulls.[65]

In a long paper "received and read" at the Zoological Society in November 1901 and published much later, in 1902, Lankester confirms that the okapi "stands close to" the *Helladotherium* of the Miocene as a hornless representative of the once extensive family of Giraffidae but also "differs in important characters of generic value from it" (see fig. 2.3). He now provides a very detailed description of the skull, teeth, and skin of the okapi, speculating on its evolutionary place and development (describing many features as "primitive") and notes developments in the skull not present in *Helladotherium* but present in *Samotherium*.[66]

Persistent identifications of the okapi as potentially a "missing link" or "living fossil" with "primitive" characteristics, hanging on in the primeval rainforests of Central Africa, have not been helpful to how it has been perceived in the West. As Jack Ashby has argued for the platypus, being pegged at a lower rung on the notional evolutionary ladder of being has negative connotations. It implies that animals so ranked are of a lower intelligence.[67]

The giraffids evolved in the Lower Miocene (23.03–15.97 million years ago), and some of the oldest distinctly classified as giraffid were members of the subfamily Paleotraginae, which included species of *Samotherium* and *Palaeotragus*, among other genera. About the size of red deer and possessing many deer characteristics (with whom they shared an ancestor), they had normal-length necks and legs, and paired forehead structures covered with skin. In the first decades after its discovery, arguments ensued over the nature of the relationship between *Okapia* and modern giraffes (genus *Giraffa*), and whether *Okapia* was a standalone subfamily of the family Giraffidae or a genus of the subfamily Palaeotraginae.[68]

Lankester's paper was the first in-depth scientific paper on the okapi in English in a scientific journal, and it was to remain (with its 1902 updates) the most comprehensive for many years. In it, Lankester acknowledges that important information he includes derived from his having been able to examine the first okapi

Text-fig. 2.

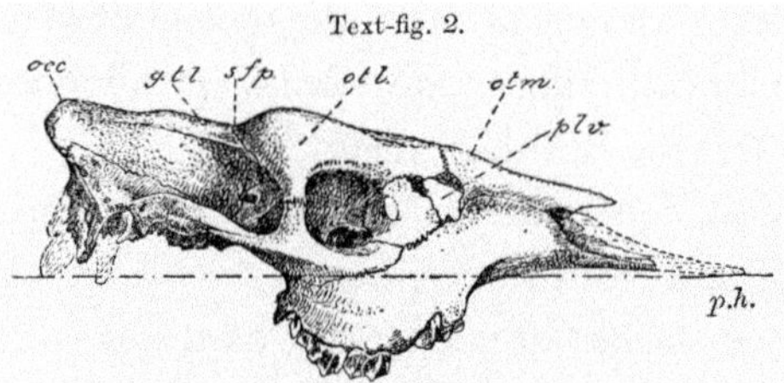

Lateral view of the skull of the Okapi, the individual being about two-thirds grown, to compare with the similar figure of a Giraffe's skull (text-fig. 3).

occ., occipital crest; *g.t.l.*, position of the lateral parietal tumescence and epiphysis of the Giraffe, absent here; *sfp.*, fronto-parietal suture; *otl.*, the lateral frontal tumescence of the Okapi, absent in Giraffe; *otm.*, the median basinasal tumescence of the Okapi; *plv.*, prælacrymal vacuity; *p.h.*, palatine horizontal.

Text-fig. 3.

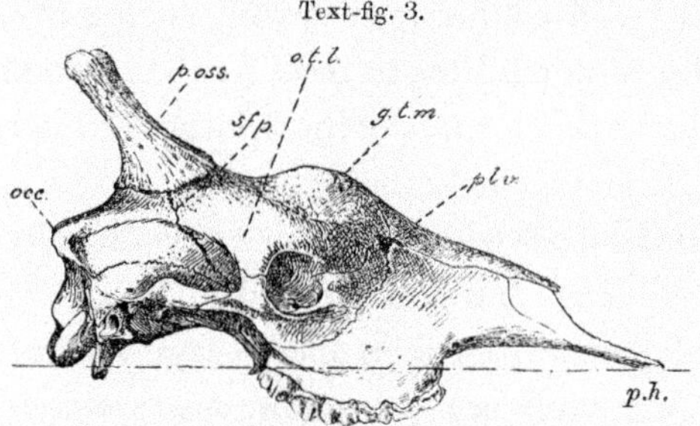

Lateral view of the skull of a Giraffe, about two-thirds grown, to compare with the similar view of the Okapi's skull, which also appears to have belonged to an individual which had completed only two-thirds of its growth.

Text-fig. 4.

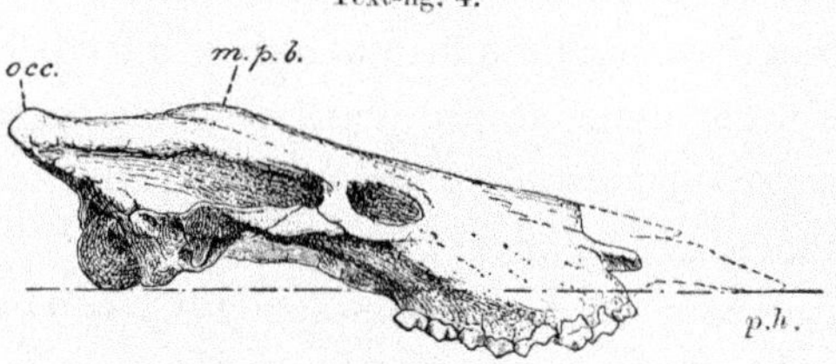

Lateral view of the skull of *Helladotherium*, from the Miocene of Pikermi: after Gaudry.

Figure 2.3. The skulls of (*top to bottom*) an okapi, a giraffe, and *Helladotherium* (E. Ray Lankester, "On *Okapia*, a New Genus of *Giraffidae*, from Central Africa," *Transactions of the Zoological Society of London* 16, no. 6 [1902]: 284, 285).

skull sent to Brussels, courtesy of Forsyth Major, who had subsequently brought it to the Natural History Museum in London. Further, while he claims he did not want to "anticipate the interesting results which will be given in Dr Forsyth Major's full account of the new specimens," Lankester conveys "a few of the more striking facts." The sketch of the skull he includes in the paper had been "prepared under [Forsyth Major's] direction."[69]

Lankester cautions that given the limited physical evidence and lack of descriptions by Europeans—although by October 1901 a specimen mounted by the renowned taxidermist Rowland Ward could be viewed at the Natural History Museum—"the proportions assigned to the various parts of the animal" and the claims regarding the "carriage of its head and neck are . . . conjectural and determined by the proportions of the skin only." Basically, Lankester suggests it was unclear whether it had the posture of an antelope or a giraffe. At this early stage, he proposes that the genus *Okapia* contains only a single species, *Okapia johnstoni*.[70]

Priority and Possession

Several differing claims were made about which European first came to know of the okapi's existence (discovery), who first confirmed (mainly via body parts) its existence, and who integrated it into scientific knowledge (by naming it and establishing its taxonomic relationships). Sclater, Lankester, Johnston himself, and some newspapers claimed Johnston had first discovered and confirmed its existence, though they acknowledged earlier rumors of its existence. Johnston admitted the assistance of Belgian officers in finding the first evidence, the bandoliers, but said they thought it was a type of horse, which he did as well until he examined the first skull procured for him by Meura, but forwarded by Eriksson.

In December 1901, the periodical the *Field* noted three recently discovered large mammals, Prevalsky's (Przewalski's) horse, Grant's zebra, and a "newly discovered animal allied to the giraffe, and named after the native name *Okapi*." The skin and skulls were "obtained by Lieutenant Erickson [*sic*], whose services have been unduly ignored in the discovery, which has been attributed to others." The *Field*, like the New York Zoological Society, apparently knew nothing of Meura, and questioned Johnston's priority. In 1909, the German explorer Duke Adolf Friedrich of Mecklenburg-Schwerin would bluntly assert that "Lieutenant Erikson . . . discovered the existence of . . . 'Okapi.'"[71]

At a June 3, 1902, Zoological Society of London meeting, the Belgian-born but naturalized British zoologist George Albert Boulenger (1858–1937), who was employed at the Natural History Museum in London, exhibited a strap of okapi skin first received at the Benedictine Abbey of Maredsous near Dinant in Belgium in December 1899. As the editorial in the *Proceedings of the Zoological Society of London* remarks, this was "thus some time previous to the arrival in this country of the piece of skin on which '*Equus johnstoni*' . . . had been founded. This had been obtained by M. E. Vincart, a lieutenant in the service of the Congo State, in the Mangbettu country."[72]

Indeed, in July 1901, the New York Zoological Society speculated that Belgian naturalists would now want to know "just why the Belgian officers of the Independent State of the Congo who told Sir Harry Johnston about the animal never took the trouble to send a skin of it to the Brussels Museum." This was the same thought that immediately occurred to officials of the Independent State of the Congo.[73]

What (or Who) Is in a Name?

Johnston was one of the very few people to be awarded the Zoological Society of London's gold medal. Oldfield Thomas put his name forward in 1902 for his continuing zoological investigations and contributions to the Society's living collection certainly, but specifically for his discovery of the okapi (thus cementing it as his discovery), which was recognized in the species name.

Johnston was well aware of the jealousies and politics surrounding scientific naming. In his memoir, he remarks that he knew that he wasn't a professional zoologist, "though my studies and knowledge of anatomy might almost entitle me to be called so." Indeed, in the opinion of scientists like Auguste Lameere, Johnston was a bona fide zoologist. The author of the first monograph on the okapi, Julien Fraipont, would compliment him on his knowledge and acuity in immediately realizing on examining an okapi skull that it was a kind of giraffid.[74]

However, Johnston lacked formal training, was not working for a major natural history institution, and while he had published well-received books containing detailed natural history information and had collected new species that were named after him, he was evidently not regarded as a proper scientist in Britain. Chapter 1 shows that zoology was going through challenging times becoming established within the scientific and museum establishments in England, so this mattered.

Johnston resented "the behaviour of certain zoologists, both at Regent's Park," where the Zoological Society of London is located, "and at the Natural History Museum," which he attributed to jealousy of the discoveries that he had made and presented to the museum. In his autobiography, he claims that he had always made it clear that several such discoveries were more the result of luck than discernment, and he also had not exaggerated their importance. He also recalls a particular (unnamed) professor writing "an angry article" arguing that the "revelation of the Okapi's existence was a nothing in importance compared to the discovery of a new death-dealing microbe, a new nutritive fish, a new fossil bird." While he quite agreed, he did not "deduce from this that I ought to have been actually silent on the subject, and leave the okapi to be found by somebody else."[75]

Johnston's cause was not helped by his discovery of the five-horned giraffe shortly after he had tracked down proof of the existence of the okapi in early 1901: "So angry were the Museum officials at my course of luck, that they ultimately named the sub-species after Lord Rothschild" (Walter, Second Baron Rothschild, procured many specimens on his own collecting expeditions and through those he funded for others).

The problem was the Zoological Society of London's "tendency to name everything new" that Johnston sent back when he was governing British Central Africa (now Malawi) "after me merely because I had sent the specimens home." Johnston protested that he had clarified who had shot or procured each specimen sent.[76]

When it came to the okapi, Johnston recalled in the 1920s that "it was difficult to ignore the original impulse and action of Sir E. Ray Lankester and Dr Sclater, in naming it '*Ocapia johnstoni*,' on the receipt of the first specimens." However, the fact that examination of the very few skulls and skins available in 1902 suggested there might be more than one species of okapi gave the museum a way to put Johnston "on a side track."[77]

Okapis Proliferate

The first of the okapi remains to arrive in Brussels on orders of the Independent State of the Congo, the skin of a "mid-adult" and the skeleton of another individual, were delivered in May 1902. They were examined by Forsyth Major, summoned to Brussels on the strength of his expertise in fossil giraffids. He published short notes on these remains in *Belgique Coloniale* in May and June and in the *Proceedings of the Zoological Society of London* on June 3.

The big news was that adult okapis had horns: the females' were small and covered by skin, while the males' were larger and exposed at the tips. In Forsyth Major's opinion, the fact that the okapi had horns, together with their positioning, placed the okapi at an intermediate evolutionary point between the extinct *Samotherium* and the modern giraffe. He denied it was a "degenerate giraffe."[78]

The remains and skeleton of a large hornless okapi arrived in Brussels in October 1902, sent by Second Lieutenant Léoni. Forsyth Major published a fourth note titled "New Information on the Okapi" in the November 1902 issue of *Belgique Coloniale* in which he concludes that the specimens in the museum in Tervuren were of a different species to those at the Natural History Museum in London.

Forsyth Major dedicated this "new" species to "Commander Liebrecht." *Okapia liebrechtsi* was presumably named after Charles Liebrechts (1858–1938), a Belgian soldier, explorer, and administrator. Based in Équateur province (1885–89) during his time in the Congo, Liebrechts was later to wield great influence as secretary general of the department for internal affairs of the Independent State of the Congo, based in Belgium. It is likely he was commemorated because of his role in ensuring the delivery of scientific materials to Belgium from the Congo until it was taken over by the Belgian state in 1908. It also allowed Forsyth Major to recognize the Belgian authorities administering the king's colony, who had cause to be aggrieved by the manner of the acquisition of okapi remains, and by the naming of this sensational new species by British scientists in British publications. Forsyth Major described this new hornless specimen in the November *Proceedings of the Zoological Society of London*, comparing okapi to the giraffe and the fossil giraffids.[79]

It was in this period that the Independent State of the Congo commissioned Forsyth Major to write a definitive monograph on the okapi, to be published in the transactions of the Museum at Tervuren, and put all the materials sent to them from the Congo at his disposal. As Forsyth Major was working at the British Natural History Museum cataloguing the fossil mammal collection, the Independent State of the Congo's materials followed Johnston's okapi remains to South Kensington in London, the domain of Lankester, its director. This raises a question over the precise relationship between Forsyth Major's investigations and expertise and Lankester's role in describing, exhibiting, and establishing a genus for the okapi. Lankester was a formidable and respected scientist, albeit a very busy one, and while his work was very wide ranging, his specialism was not mammals. It is not hard to imagine that following delivery of the sensational remains from Johnston in 1901, he seized the opportunity to make the most of the in-house (but not formally attached) expertise of Forsyth Major.

Just as important as access to okapi specimens delivered to the Natural History Museum directly by Johnston, especially at this early stage, was the additional treasure trove of okapi remains delivered by the Independent State of the Congo for Forsyth Major's work on its commissioned monograph. Strictly speaking, Lankester shouldn't have had access to these, but he acknowledged being "allowed" to look at these remains by Forsyth Major at the Natural History Museum in 1902 (was Forsyth Major in any position to refuse?).[80]

In an appendix to his 1901 paper, added in late May 1902 before it was finally published, Lankester concludes that it seemed "probable" that there were two distinct species of okapi. This was based on Forsyth Major's assessment of the new specimen received in Brussels in May from Mawambi in northeastern Congo, which he compared with the remains sent by Johnston that were mounted in the Natural History Museum in South Kensington. Lankester knew his claim about there being two species of okapi was a hasty conclusion, noting that "certainty as to the distinctness of these two species . . . if they should prove to be, as now supposed, distinct, can only be based on the examination of a large series of specimens." However, he presumably did not want to risk losing out on naming the species.[81]

Matters risked getting muddled, as Lankester followed Forsyth Major and asserted that the first material (bandoliers) sent by Johnston in 1900 and named *Okapia johnstoni* by Sclater represented the same species as that examined by Forsyth Major in Brussels in May 1902 that he named *Okapia liebrechtsi* after he examined a second specimen there in October 1902. According to Forsyth Major, these remains differed from those sent by Johnston in 1901 (an almost entire skin and two skulls)—which Lankester now assigned a new species name, *Okapia erikssoni* "after the Belgian officer Mr Eriksson."[82] Lankester thus joined Sclater and Forsyth Major in having his name attached to the okapi through having named a species, of which there were now three: *Okapia johnstoni* (Sclater), *Okapia liebrechtsi* (Forsyth Major), and *Okapia erikssoni* (Lankester).

What nobody yet realized, was that every okapi has a unique pattern of stripes, and as Fraipont observes in his 1907 monograph, these are not identical on pairs of the legs of the same okapi either. At the time, the fact that the Tervuren specimen had a darker color and many more stripes than the Natural History Museum one seemed significant. Lameere thought there may be a parallel with zebra, as during this period zoologists believed that species with more numerous stripes were more primitive. Further, the relative sizes of the few available okapi specimens (they weren't sure yet how old they all were) was confusing. Was the observed variation merely due to age or sex or was it a significant difference indicating more than one species?[83]

Initially, according to Lameere, based on the remains of two specimens that arrived in Tervuren in May 1902, zoologists concluded that there was a smaller species with horns, possibly in both males and females, like giraffe. Of the Tervuren specimens, one was described by zoologists as a smallish, horned male, and the other (skin only) as probably a horned female. The London specimen—also small—was possibly a female or a male yet to develop horns due to youth, but it was also lighter in hue and had less stripes.[84]

The larger specimen that arrived in Tervuren in October 1902 was possibly a separate species. As it was larger than both the other specimens but had no horns, it was possibly a hornless male. According to the evolutionary scheme of the time, the larger "pollarded" okapi species had lost its horns as it no longer needed them for defense, while the smaller more "ornate" one still needed horns.

However, as it turns out, Johnston's specimen was not fully grown, as Lankester had suspected from the development of its teeth. It was also a female, not a male, as they had first guessed. The scientists had made the reasonable assumption that male okapis would be larger than females, as is the case with giraffes (and most African ruminants). It was only later discovered that adult female okapis are larger than males.

In 1903, Lankester published a study comparing the specimen that Baron Walter de Rothschild had obtained for his museum in Tring (provenance unknown, its legality questionable give the restrictions on capturing okapi that had been imposed by then) with the mounted specimen based on the skin and skulls sent by Johnston to him at the Natural History Museum. He now thought that Johnston's okapi was probably a subadult female and that the Tring specimen was "a *smaller* specimen of an apparently adult female" (his italics). He focused on the arrangement of the vortices and the direction of the hairs on the fronto-parietel region (the forehead, more or less), finding significant differences with the specimen mounted in the Natural History Museum (now known to be a subadult female).

While admitting it was not known to what degree such differences in hair growth might indicate "distinctness of race or species" (race equates to subspecies) Lankester clearly implies that such differences along with the supposed size difference supported his naming the South Kensington specimen *Okapia erikssoni*. Lankester advocated keeping *Okapia johnstoni* for the specimen in Tring.[85]

What was urgently needed, to arrive at a more secure taxonomical categorization for the okapi, was further evidence. Once the Belgian colonial authorities became aware of the sensational new mammal living only in their territory, this began to pour in. The task, described in chapter 3, was to make sense of all these remains and provide a definitive description and taxonomical categorization of the okapi in a detailed scientific monograph.

CHAPTER THREE

Settling Okapi Taxonomy, and the First Monograph

The most interesting thing about the giraffe is the okapi.

—*E. Ray Lankester, 1909*

By 1903, there were three recognized species of okapi. This had more to do with the level of excitement generated by its discovery, combined with the paucity of evidence, than with sound scientific judgement. There was great popular and scientific interest in the okapi and demand for publications in newspapers, magazines, and scientific journals was palpable. But aside from a few short popular articles and scientific notes, only one full journal paper by E. Ray Lankester had been published in English and only one by Auguste Lameere in French, both in late 1902. Charles Forsyth Major's officially commissioned monograph was therefore eagerly anticipated.[1]

By 1910, however, this monograph had not appeared, which puzzled me greatly as I did research for this book given the advantages of expertise and access to specimens that Forsyth Major and Lankester had at the British Natural History Museum and their prominent role in the early description, naming, and categorization of the species. I was excited to find the *Monograph of the Okapi* published by the British Natural History Museum in 1910 in its library collection therefore, only to discover that it is not actually a monograph. It is an illustrated and captioned atlas of okapi remains with no substantial text. This likewise puzzled me because the discovery of the okapi had been a sensation in the early 1900s, and the museum had been the first in the world to exhibit it.

The answers came in two parts. The first was the untold history of the failure of Forsyth Major to deliver a manuscript and of Belgian paleontologist

Julien Fraipont's taking over of the project. Untold in English, that is, and only, so far as I know, explained in the foreword to Fraipont's difficult-to-access French monograph. I managed to track down a copy in the Wellcome Collection library in London, and later I bought an unbound copy from a secondhand book dealer in Amsterdam.

More fragments of this story emerged after I had pored through obituaries, annual reports, staff lists, and other materials in the Natural History Museum's library and archive. Here I sat at long tables surrounded by the sloped foam cushions, paperweight chains, and other paraphernalia of archival work, eagerly working through the trolley loads of boxed-up documents that archivist Kathryn Rooke dug out for me. I worked against a background of floor-to-ceiling wooden bookshelves in the presence of a slightly daunting white life-size statue of Charles Darwin reading (no pressure).

I learned more about Forsyth Major, his precarious position at the museum, and his difficulties in writing up his considerable volumes of research in his final years there. One photo of him survives, a brooding image from 1903 showing an introspective old scholar, seated sideways on to the camera, looking down at his hands that clasp one knee. His Victorian suit and distinguished white beard are offset by a Turkish fez and the cigar in his left hand.

Reading the yellowing newspaper and periodical clippings pasted into huge red leather-bound volumes from the museum's archive, I found extensive materials documenting the stormy final years of Ray Lankester. It gradually became clear why Fraipont published the monograph and why neither Forsyth Major nor (the second part of the story) Lankester did.

Fraipont's monograph was the first on the okapi and still stands as the benchmark historical document for Western science on the species. In addition to providing detailed descriptions and summary of the knowledge of the time, it includes an incredible assemblage of anatomical drawings of okapi skeletons and external features. It deserves further exploration, certainly by scholars publishing in English.

The first decade of the twentieth century was a crucial one for the settling of okapi taxonomy. The proliferation of named species and then the retraction to recognizing just one is central to this story. The lack of substantial books on the okapi after Fraipont's monograph, aside from Agatha Gijzen's 1959 German monograph and a slim 1999 book by American zookeepers, remains surprising to me. This book is intended to begin to fill this gap.

Fraipont Succeeds Forsyth Major

By 1904, Forsyth Major had produced only a few published notes (his last in November 1902) and had commissioned fifteen plates and two watercolors for his monograph. This must have been galling for the officials of the Independent State of the Congo, as by 1904, numerous articles on the okapi had been published in scientific and popular journals in Belgium, France, Germany, and the United Kingdom. Every recent encyclopedia and dictionary included an article on the okapi.[2]

In frustration, the government of the Independent State rescinded its contract and recovered the materials it had sent to Forsyth Major at the Natural History Museum in London. The secretary for the interior then asked Julien Fraipont, professor of paleontology at the University of Liège and a member of the commission of the Museum of Tervuren, to write the monograph.

Fraipont was a slim man with a patrician nose, handlebar moustache, and neatly clipped beard whose hair was swept to the side off a high forehead. Having worked on Neanderthal fossils, extinct worms and mollusks, and the systematics of protozoa, he was not perhaps an obvious candidate, but he was associated with the museum and was a respected paleontologist.[3]

Fraipont first attempted to help Forsyth Major salvage his reputation (as he saw it) and offered to collaborate to finish the book free of charge, but while Forsyth Major initially responded positively to the idea, he stopped answering Fraipont's letters after January 1905. In late 1905, Fraipont gave up and went ahead with the book himself.

Fraipont was a collegiate researcher, and his acknowledgments are generous, noting prior work by Forsyth Major and Lankester and crediting the discovery of the okapi to Harry Johnston, who he quotes at length. He offers his "affectionate thanks" to a network of scholars across Belgium and France, particularly for their having given him access to giraffe skeletons for comparison with okapis. He is also at pains to acknowledge the contributions to zoological science of "the dedicated officers, functionaries, and agents" of the Independent State of the Congo, "who have been our pioneers and collaborators in central Africa."[4]

Fraipont was given full access to all the materials sent to the museum in Tervuren and to further materials and information from Congo state agents as these came in. The remains of twenty-eight okapi were known to have reached Europe by the time Fraipont published his book in 1907, plus remains of an unknown number taken to England by Boyd Alexander (including one he had mounted by Rowland Ward, which he presented to the Natural History Museum). Fraipont himself had access to the remains of twelve okapi, including skins of adult males and females, among which were old individuals, and also young okapi, two of the latter known to be female. He had access to eleven skulls and seven skeletons of adult and young okapi. He also consulted photographs of the remains of four adults (two male, two female) that had been gifted to foreign governments by the independent state.[5]

Fraipont surveyed the literature on okapi, which by 1906–7 already included a healthy number of articles. Richard Lydekker, reviewing new publications on mammals for the *Zoological Record* in 1901, notes that "in zoological circles the first year of the twentieth century will probably be known as the 'okapi year'" but huffily adds that "of even greater importance than the okapi is the discovery in Egypt of Eocene mammals." In the *Zoological Record* for 1902, Lydekker recognizes Lankester and Forsyth Major's publications on okapi as "noteworthy events of the year." Over the subsequent decade Lydekker would report on publications on okapi by Belgian, British, French, German, Spanish, and Swiss academics.[6]

As Fraipont makes clear in his account of the materials consulted for his monograph, he had access to no soft parts. Anatomical studies were limited to examination of skins and bones. Fraipont combined a close study of these with a review of the descriptions Johnston and Lankester provided for the London specimens, Forsyth Major's assessment of the first specimens to arrive at the museum at Tervuren, Auguste Ménégaux's account of the specimen in Paris's Muséum National d'Histoire Naturelle, and Antonio Carruccio's report on the specimen sent to the museum of the Rome University. Fraipont also surveyed the handwritten documents of agents of the Independent State of the Congo. All of this led him to the conclusion that there was, in fact, only one species of okapi, *Okapia johnstoni*.[7]

In Fraipont's opinion the differences in color and pattern of the available okapi skins were the result of individual differences, differences due to age or gender, or differences in the mode and state of preservation of the remains. He similarly attributes differences in shape and size of skulls and teeth to individual variations or characteristics of gender or age. A quick appraisal of the photographs of mounted okapis included in Fraipont's book reveals notable differences in head and body shape, and posture, which are partly the result of the attempts of individual taxidermists to create lifelike versions of an animal they had never seen in the flesh or even in photographs.[8]

Fraipont confirms that only male okapi have two small conical horns of variable size (three to five inches, or eight to thirteen centimeters). These conical or columnar bone structures made up of ossified cartilage covered in skin and hair as in the giraffe are known as ossicones. On male okapi, the free end of an ossicone is bare up to about 0.4 inches (one centimeter), like an antler. Females may have ossicones beneath the skin, but no projecting horns; as Lankester observed, the hairs form two characteristic vortices in the place occupied by horns in males.[9]

Fraipont offers an incredibly (or excruciatingly, depending on your inclination) detailed description of every scrap of the head, the "very beautiful" hide, tail, and hooves of the okapi. His account does include endearing details, such as that "in both sexes we also notice a small whirlpool of hair between the ears (occipital whirlpool)" and "another in the middle of the forehead (frontal vortex)." He was very impressed by the "brilliance and vivacity of the colors" of the coat of the okapi, "its polychromy and the presence of sharp creamy-white stripes on the top of the legs with the reddish-brown background."[10]

The bulk of the book (pages 33–80) covers the skeleton, piece by piece, which Fraipont documents with illustrations. He compares the skeleton with giraffe bones and also compares the measurements of okapi foreleg and hindleg bones made by Gaudry, Forsyth Major, and Carruccio with his own measurements, concluding that they all agreed that in the okapi and *Samotherium* species, the front leg is only very slightly shorter than the hind leg, whereas in the giraffe the foreleg is notably longer. In most ruminants (including antelopes, cattle, deer, and sheep), the hind legs are longer than the forelegs. In terms of leg length, the okapi-like members of *Samotherium* apparently occupied an intermediate position between true ruminants and the giraffes.

Fraipont's long section on okapi bones is followed by a short section on the habits of the okapi, its preferred habitat, and distribution in the rainforests of the Congo (see fig. 11.1). For this section Fraipont drew on "information from district chiefs, post chiefs, explorers and natives."[11] Fraipont also summarized the scanty and sometimes contradictory information about okapi behaviour reported by employees of the Independent State of the Congo. This information, together with Fraipont's synthesis of the knowledge of okapi distribution in 1907, is discussed in detail in chapter 11.

Fraipont ends his book by stating that the genus *Okapia* describes a giraffid on account of its horns and frontal bumps and other details of the skull and teeth, although he notes that certain features were less developed in the okapi than in the giraffe. He agrees with Forsyth Major that the okapi was no "degenerative giraffe" but was rather a member of the Giraffidae family that possessed characteristics typical of ordinary ruminants. It was less primitive than *Samotherium*, occupying an intermediate position between it and *Giraffa*, the true (modern, long-necked) giraffes. Fraipont concludes that the okapi was a giraffe-like animal that had yet to gain a long neck and longer forelegs and that there was only one species: *Okapia johnstoni* (Sclater).[12]

The final section of Fraipont's monograph comprises 111 full-page plates, mostly showing okapi bones, but including four color plates of okapi in naturalistic settings. These are discussed in chapter 4.

A British Okapi Monograph?

While Fraipont had in a sense preempted the British scientists with the publication of his book as a result of Forsyth Major's failure to complete his commission, Fraipont's book is entirely in French. The Natural History Museum Library appears not to have received a copy, or, perhaps it didn't order one. Strangely, too, the book doesn't appear in the detailed annual survey of zoological publications in the main European languages, the *Zoological Record*, though many other publications on okapi do. Were the editors waiting for Natural History Museum scientists to complete a definitive monograph in English?

I have speculated that Lankester was in a way poaching on Forsyth Major's turf in writing about the exciting new species. Johnston had sent his first substantial evidence to the Natural History Museum in London but not, as far as I can discover, specifically to Lankester. I have discovered the circumstances around the withdrawal of Forsyth Major's contract but I have not definitively determined why he did not complete it. However, a paragraph in an (appreciative) obituary for him published in the journal *Nature* is suggestive: "In his later years . . . Dr Major found increasing difficulty and diffidence in preparing his results for publication, although his researches were pursued with accustomed diligence. Much of his valuable work on rodents and on the relationship between the fossil *Samotherium*, which he discovered in Samos, and the existing okapi of the Congo Forest, is thus unfortunately lost to science."[13]

Lankester would have known of Forsyth Major's difficulties; he even notes in a 1907 paper that "Dr Major does not himself propose to publish anything further at present on the okapi." It is possible Lankester had already represented a certain amount of both of their work in his long, detailed, and well-illustrated 1902 paper in which he acknowledges his debts to Forsyth Major.

To be fair to Forsyth Major, in the early 1900s he was busy publishing his work on fossil mammals from Samos and his more recent work on fossils on Madagascar. He also had a job to complete in classifying collections at the Natural History Museum. He would have been sixty in 1904, when he finally lost his contract to produce the okapi monograph, and four years later he retired on a small pension, spending most of his remaining years collecting and studying mammal remains on Corsica. This left Lankester to pull together a monograph on the okapi for the museum and English-speaking world. However, he had troubles of his own.[14]

Ray Lankester: Is a Man of Genius Too Old at Fifty-Nine?

On August 1, 1906, on the day he gave the presidential address to the British Association for the Advancement of Science, it was leaked in the press that Lankester had been informed that he would be retired on turning sixty, on May 15, 1907. There followed two weeks of outraged criticism of this decision of the trustees of the British Museum (it remains murky who exactly made the decision) in national newspapers across England.

Lankester was offered no explanation and given no opportunity to plead his case for staying on (Richard Owen remained in his post until he was eighty and William Flower until he was sixty-seven). The decision appeared vindictive, as he was also informed he would be awarded a paltry annual pension of three hundred pounds sterling. Newspaper headlines included "Is a Man of Genius Too Old at 59?" and "Turned Out: Nation's Reward for a Great Scientist."[15]

Despite prolonged public calls for an explanation for their decision, the trustees refused to comment. In response to widespread speculation in the media, in a letter to *The Times* of August 8, 1906, Lankester published as full an account as he could of the facts. The letter revealed the archaic system of control over the museum and renewed public criticism of the idea of putting a librarian and a standing committee of trustees—mostly without relevant experience—in charge of a natural history museum.

While it seemed that Lankester might have won the day when he continued in his post beyond May 1907, on December 30 of that year it was announced that he had retired. (The principal librarian, Edward Maunde Thompson, incidentally, retired on grounds of ill health in 1909 at the age of sixty-nine.) The only positive outcome for Lankester was that he had been awarded a more generous pension.

Lankester's final scientific publications on the okapi included a paper on okapi ossicones published in the *Proceedings of the Zoological Society of London* in 1907, a paper on okapi and giraffe neck vertebra in 1908, and a two-page descriptive letter on "supposed horn-sheaths of an okapi" in *Nature* in 1915. The paper on rudimentary antlers in the okapi is important because it acknowledges limitations in the available evidence, including uncertainties about the gender of okapi remains brought in by local Africans and the possibility that skins and skulls from different animals got mixed up, for answering the question of how many species there were. Perhaps most significant was his apparent decision (by omission) that there was only clear evidence for two species, a decision he based on observation of consistent differences between skulls and not on variations in skin patterns, and that these two species were *Okapia johnstoni* (Sclater) and *Okapia liebrechtsi* (F. Major). Lankester did admit that the differences between the skulls of these two species might relate to gender: *Okapia johnstoni* may actually represent the female of the species and *Okapia liebrechtsi* the male.[16]

As Fraipont notes in an appendix written after he received Lankester's paper late in the publication process of his monograph, despite distinguishing two consistent types of skulls attributed to *Okapia johnstoni* types and *Okapia liebrechtsi* types ("types" is a more noncommittal word than "species"), Lankester does not offer "an explicit opinion about the existence of several species of okapi." Indeed, in a September 1907 article in the *Illustrated London News* about the first photograph of a living okapi, Lankester attributes this difference in skulls to gender: "My own conclusion is that there is only one [species], though individuals differ a great deal in the striping . . . and the males differ from the females in the size and shape of the skull as well as in the size of the horns."[17]

Lankester's paper on okapi ossicones published in the *Proceedings of the Zoological Society of London* in 1907 describes similarities and differences (notably in position) between giraffe horns (based on the remains of a well-developed embryo first examined by Richard Owen) and okapi horns (based on Lankester's examination of Johnston's and the explorer Boyd Alexander's skulls and one in Paris). This was published alongside a paper discussing hitherto unremarked-on striping in the giraffe embryo that was subsequently also identified in some adult giraffes and that thus suggested that giraffids have a propensity to develop stripes (he does not draw attention to the obvious difference between patterning between the two living giraffids, the spotted or blotched giraffe and the striped okapi). Lankester's 1908 okapi paper points out that their neck vertebrae have more in common with the ox (*Bos taurus*) than the giraffe.[18]

When Lankester retired from the Natural History Museum, he was in theory still going to complete his monograph on the okapi. He had signed off on the budget for this himself in the previous year, and in his *Illustrated London News* article he notes that he was "intending very soon to publish a copiously illustrated account" of the more than twenty specimens of okapi in different museums. As Sidney Harmer's preface to the British Museum's volume explains, however, this had "taken longer than was expected," and so the museum decided to print the illustrations, which had already been prepared by 1908, as an atlas of captioned plates without the full text they were originally intended to accompany. The "monograph" (published in 1910) was still titled *Monograph of the Okapi* but also carried the subtitle *Atlas (of 48 Plates)* below Lankester's name (see appendix 3 on the atlas).[19]

In his review of *Monograph of the Okapi* for *Nature* in December 1910, Johnston complains it is "an incomplete treatment," supposing that the British Museum had not wanted to spend the money on publishing the volume with text "which should have accompanied the mere illustrations." He was not persuaded by Harmer's claim in the preface that the publication of Fraipont's monograph made publishing a full text superfluous and still hoped the text would be published. He notes that several years had elapsed since Fraipont's volume had appeared and that surely new information was available. He also points out that Fraipont's book was "not nearly so accessible to ordinary students of zoology as the British Museum publications" (especially if they didn't read French).[20]

Lankester responded in the January issue of *Nature*, explaining that he had approved the expenditure on the volume including text while still director of the Natural History Museum. The text was missing, he admits, only because he had "not provided such further text." He allows that it would have been better to have avoided the word "monograph," though he maintains that a monograph needn't summarize existing knowledge on a subject, adding that he had originally intended to provide a complete account of the specimens of okapi in the national collection. He nevertheless then offers a rather evasive reason for not writing a more complete book: "The problems which arose in the course of my work could not, in many cases, be satisfactorily solved by the examination of the existing material." He argued that "new observations made upon fresh or living specimens" were necessary in order for anyone to properly work out the distinguishing characteristics of male and female okapi, their geographical variations, and whether there were "distinct races or subspecies."[21]

In a two-page introductory note to his atlas, Lankester concludes that "not only in the striping of the skin, but in regard to many osteological features such as breadth of skull . . . there is a very large range of variation on Okapis, no two specimens being closely alike." He observes that the late development of ossicones (not present at birth, as in giraffes) was a confounding factor in determining the sex of okapi skins and skulls.[22]

Johnston thought that Lankester's preface to his atlas implied that further text might be forthcoming, revealing that in private correspondence with him, Lankester had suggested three years previously that such text already existed. He expresses his disappointment that while the book contained valuable information, it lacked "the deductions to be drawn from these illustrations as to the affinities and systematic position of Okapia: in short, a statement of Sir E. Ray Lankester's personal opinions." He allows, however, that Lankester was "probably quite right" to wait until more was known of okapi musculature and intestines to try to address the various outstanding questions, notably, how many species of okapi to recognize.[23]

In 1906, the now retired Sclater—who had by then examined most of the available specimens received in Europe—declared himself "strongly of opinion that there are no sufficient grounds for considering that there is more than one species of Okapi." Lydekker, in his Rowland Ward volume *The Game Animals of Africa* (1908) and his *Catalogue of the Ungulate Mammals in the British Museum (Natural History)* (1914) mentions only *Okapia johnstoni* (Sclater). In a short article on an okapi exhibited in Rowland Ward's studio in 1910, Lydekker reviews the available evidence on the stature of okapis based on specimens in Madrid, Tervuren, and London and reports on those which did and didn't have horns. He concludes that hornless okapis are females and that they are larger than males and that this accounts for the size differences in specimens and does not indicate the existence of a larger and a smaller species.[24]

Henri Schouteden of the Museum of Belgian Congo agrees with Fraipont that there is only one species of *Okapia* in his 1912 "Notes sur l'okapi." In a 1910 article, Maurice de Rothschild and Henri Neuville likewise suggested that the variable coloring and numbers and patterns of stripes along with variability in the skulls had initially been mistaken for specific characteristics. However, they maintain that considerations of habitat and internal and external anatomical differences would probably lead to a splitting of the species once enough data was available. In the interim, they propose sticking with two types, as specified by the "distinguished scientist" Ray Lankester, namely, *Okapia johnstoni* (Sclater), which had a shorter, wider skull and sometimes no horns, and *Okapia liebrechtsi* (F. Major), which had a narrow, more elongated skull and usually had horns). They also claimed that the hides of the two types differ, *Okapia johnstoni* being predominantly a brighter brown with a darker stripe along the backbone and yellowish-white cheeks with a predominance of black over white on the forelimbs—while *Okapi liebrechtsi* was generally dark brown with a paler dorsal stripe, grayish cheeks, and white predominating over black on the hind limbs.[25]

Ten years after its discovery, then, some notable scientists still felt that okapi taxonomy remained unresolved (see fig. 3.1). Fraipont had gone ahead and writ-

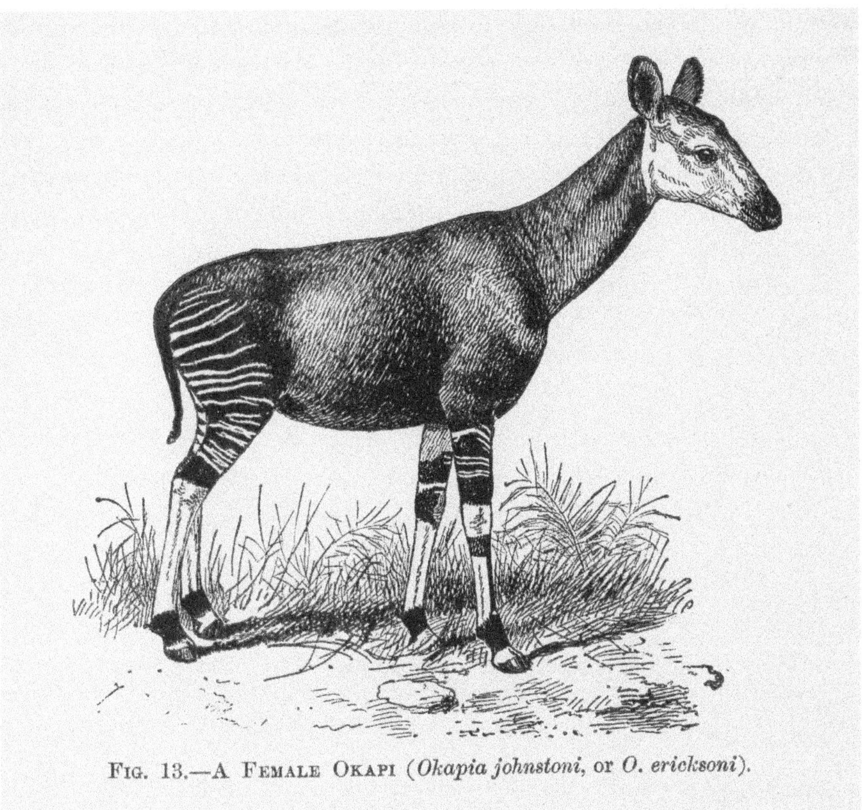

Figure 3.1. Okapi illustrated in the 1906 edition of the British Natural History Museum's general guidebook that still included "*Okapia ericksoni*" (*Okapia erikssoni*, Lankester) as a possible species option. The 1909 edition refers only to *Okapia johnstoni* (© Trustees of the Natural History Museum).

ten a monograph based almost entirely on the durable remains of the animal: hides, horns, hooves, and skeletons; as he notes in his book "so far no soft parts have been sent to the authorities of the Independent State." Rothschild and Neuville had contributed a detailed study too, but Lankester had hesitated. The difficulty of acquiring living okapi meant the hard remains was all scientists could access to work on. Tantalizingly, Albert Sillye, district chief of the Orientale province in the Independent State of the Congo, had acquired a live young okapi in 1903, but it escaped.[26]

As Lankester concluded in 1910, okapi appear to be both strikingly distinctive in appearance, and unusually variable and resistant to neat classification by taxonomists. This book will show that more than a century after its discovery by Western science, questions remain to be definitively answered about its taxonomic relationships and the genetic status of species and possible subspecies. Physically describing a species and assigning it a taxonomic status does not mean that it has been fully understood: the okapi remains in many respects as mysterious as the unicorn Johnston hoped to find in the rainforests of the Congo basin.

CHAPTER FOUR

Possession, Exhibition, and Dissemination

Visitors to the North Hall of the Natural History branch of the
British Museum will soon be able to feast their eyes on the skin of
the new mammal, discovered by Sir Harry Johnston in the Semliki
Forest. . . . For the last fifty years no new discovery of strange
beasts has excited the popular imagination more than the advent of
the striped black-and-white bandoliers.

—Westminster Gazette, *August 1901*

Following the sensational news of the discovery of the okapi, the focus shifted
to the procuring and exhibition of okapi remains. The scanty evidence for the
existence of "the strange new beast" was highly prized and carefully guarded.
Philip Sclater and Ray Lankester and later the government of the Independent
State of the Congo, were well aware of the importance of this rare physical evidence, and a race for okapi parts ensued. First to arrive in Britain were the bandoliers sent by Johnston, then followed hides, skulls, drawings, and paintings,
and then taxidermized okapi—in guessed-at postures—and then finally photographs of the taxidermized okapi.[1]

In this chapter, I explore how the evidence for the existence of okapis was procured, valued, curated, and shared by looking at who possessed, exhibited, and
illustrated the animal. Taking possession occurred in a colonial context requiring political accommodations. I show how the first evidence was prepared, shared
with the scientific community, and exhibited to the media and the public in the
context of Britain. While collecting and categorizing specimens has long been a

focus of scholarly research, the many other dimensions of the material afterlives of animals removed from their natural habitats, reconstituted, and exhibited in unnatural environments have been less often investigated.

Possession: "The Race for Specimens" (1900–10)

The history of acquiring okapis and okapi remains is bound up with the colonial history of the Congo and the politics of securing access to its habitats. For the Anglophone world, this history is linked to the British Natural History Museum's and the American Natural History Museum's dealings with King Leopold II of the Belgians and his colonial state in the Congo.[2]

Currency for a King

When officials of the Independent State of the Congo learned of the existence of the okapi through the report on Harry Johnston's travels in the *Geographical Journal* in January 1901, their Department of the Interior immediately ordered officials in the Haut-Ituri zone in far northeast of the country, west of Lake Albert (see fig. 11.1), to collect okapi and send them to the museum in Tervuren. This museum played an important role in the history of the okapi and merits a brief introduction here.[3]

In the era following the success of the 1851 London Crystal Palace Exhibition, Belgium became one of the most frequent hosts of such world fairs in Europe, favored by its position and excellent railway network. The first to feature a colonial exhibition was held in Antwerp in 1885, and another in Antwerp in 1894 showcased colonial products from the Congo including rubber and ivory that were accompanied by propaganda about the antislavery heroics of colonial officials and missionaries.[4]

The 1897 Brussels International Exhibition included a colonial exhibition housed in a specially built palace on King Leopold II's estate in nearby Tervuren. It was linked to Brussels by a tramway and attracted over a million visitors. It featured fake "native" villages and a village of "civilized" Congolese with a sign reading "Do not feed the blacks. They are already being fed." Belgian exhibitions promoting the Congo as a colony continued until 1958. After a period of discomfort about King Leopold II's legacy of exploitation and human rights abuses between 1905 and 1913, the tenor of the exhibitions shifted back toward celebrating the king and reimagining Belgium's role as a benevolent colonial power.[5]

The 1897 exhibition of zoological and ethnographic objects and displays from the Congo was moved into a new building and opened as a museum in

1898, becoming the first permanent museum of the Congo. It was originally essentially a propaganda tool by which Leopold sought to win support from Europeans and Belgians for his colonial aims. Functioning as both a scientific institute and a museum, it was renamed the Museum of Belgian Congo when the Belgian state took it over in 1908. The collection was moved into a larger building already being built by Leopold II that was inaugurated by his successor King Albert I in 1910. Covering 30,480 square feet (2,831 square meters) with a dome that was 92 feet (28 meters) tall, its French architect Charles Girault modeled it on his Petit Palais in Paris. In current value, it cost 35 million euros ($37.6 million US dollars). The museum was renamed the Royal Museum of the Belgian Congo in 1952 and then Royal Museum of Central Africa following independence of the Democratic Republic of Congo in 1960; today it is known as AfricaMuseum.

One of the museum's key aims after 1910 as part of the Ministry of Colonies was to project an impression of a single coherent Congo colony that fully documented and categorized its peoples, natural resources, and wildlife, all of which were represented as being firmly under Belgian control. Sculptures of the emblematic species of the rainforest, the gorilla, and the okapi featured at the 1935 exhibition.

While it grew its collections, the institution remained essentially unchanged throughout the colonial period, providing a stable image of the colony for the nation. It remained the only official museum of empire in Belgium; the others were privately funded.

Back in 1901, the officials of the Independent State of the Congo had also reminded their agents in the Congo that all interesting animals and other specimens were to be sent to this museum. On May 4, 1902, the first okapi remains arrived in Brussels: the skin of a mid-adult okapi and the skeleton of another individual, killed in the forest near the village of Mundala in M'Beni (upper Ituri). These had been sent by Second Lieutenant Léoni, the chief of the post in Mawambi, who had dispatched them on their long journey to the coast and then on to Brussels on December 5, 1901.[6]

Three days after these first remains arrived in Belgium, the central government instructed the governor-general in Boma to forbid anyone besides agents of the state charged with finding okapi from hunting this prized new discovery. Existing legislation, namely article 5 of a decree of April 29, 1901, made it a legal requirement to hand over any okapi corpse collected on the Independent State of the Congo's territories, to its officials, who were then obliged to send it to their direct leaders. In September 1902, Governor General Émile Wangermée issued a decree prohibiting the hunting and capture of okapi in the state. Henceforth anyone not an agent of the state would require special permission from it to look for okapi, and the state claimed ownership over the entire known habitat of this elusive animal.[7]

By 1900, when Johnston visited the Congo, the atrocities committed by Leopold's agents in their relentless pursuit of rubber since the 1890s was receiving increasing publicity. African American historian George Washington Williams had visited the Congo and written a strong letter of protest as early as 1890, while Joseph Conrad's novella *Heart of Darkness* (1899) suggested that Leopold's allegedly philanthropic project in the Belgian Congo was going horrifyingly wrong. In the same year, the first African American missionary in the Congo, William Sheppard, experienced Leopold's regime's rubber-gathering tactics—systematic kidnappings, murder, and mutilations—firsthand. He and other missionaries began to publish articles critical of the regime.[8]

Rulers have given gifts of rare and exotic animals for diplomatic reasons to other countries since ancient times—famous twentieth-century examples include China's gift of panda bears to the United States, the United Kingdom, and the Soviet Union, among other countries, and Australia's gift of platypuses to the United States and the United Kingdom. In the same way, King Leopold II seized on the okapi as a means to win the favor of European governments. Gifting okapi presented one way of deflecting attention—from the rape of the land and butchery of the peoples of the Congo being enacted in his colony in the pursuit of rubber—to the more virtuous pursuit of scientific knowledge. The king's diplomacy was very much directed at maintaining his reputation as a philanthropist and lover of science and the arts in Europe but also importantly in the United States, the country that had first recognized his claim to the Congo, in 1884. The Independent State of the Congo donated numerous specimens of Congo fauna to European museums, including the British Natural History Museum, and to the American Museum of Natural History.[9]

By 1906, scholarly articles on the okapi had been published based on taxidermized specimens King Leopold II had gifted to the governments of France, Portugal, Spain, Sweden, and Italy. In Belgium, mounts were gifted to Antwerp's zoological society, the Museum of the Royal Zoological Society of Antwerp, with the majority of okapi remains going to the Tervuren museum (by 1908, thirty-seven had been acquired). The Duke of Brabant (later King Albert I) arranged for the gift of an okapi to the city of Munich, as a gesture to his future wife Elisabeth Gabriele Valérie Marie of Bavaria. Several other taxidermized okapis showed up in museums in England, Scotland, Ireland, Germany, Italy, and the first in the United States at Harvard in this period, their provenance not always certain (suggesting that they were secured without authorization).[10]

By 1907, complete skeletons had been gifted to the governments of Sweden, Switzerland, Portugal, France, and Spain and to the kings of Italy and Bavaria, while several were kept in Tervuren. As Julien Fraipont's monograph reveals, Leopold's canny officials often doubled the value of their okapi acquisitions, offering a taxidermized specimen to one government, and the same animal's skeleton to another.[11]

Although they were not naturalists and had many other duties to perform, officials of the Independent State of the Congo were of course best placed to find okapi, while local officials could advise expedition leaders on where to look for the animal. Further, on the orders of the colonial government, particular agents made concerted efforts to find okapi or source their remains from local Africans. These men are acknowledged by Fraipont in his monograph, the most important of them being Second Lieutenant Anzélius, chief of the post at Mawambi in the Ituri Forest. His district chief, Albert Sillye, first sent him to hunt okapi in the forest in the Ibina basin in 1902, and according to Fraipont he was one of the few Europeans to have shot an okapi himself and to have seen live and unskinned specimens. In 1902, Anzélius brought back the remains of six okapi that were then sent on to Tervuren, one of which was mounted. He provided the first precise information on the sex of slaughtered animals, which was critical to determining whether both males and females had horns.[12]

The Politics of Accessing Okapi

British explorers and collectors of okapi remains and the institutions that depended on them were reliant on the goodwill of the Belgian colonial authorities to give them access to its habitats. They were therefore reluctant to criticize the regime, despite the public campaign criticizing King Leopold II's motives and the atrocities perpetrated by his regime spearheaded by Edmund Morel's Congo reform campaign, launched in the United Kingdom in 1904. Protestant missionaries and Yorkshire Member of Parliament Herbert Samuel and Sir Charles Dilke of the Aborigines Protection Society had preceded Morel, but it was his campaign, along with Roger Casement's 1904 report, that turned the tide of public opinion against the king's extensive propaganda campaigns in Europe and the United States. The king's great "swindle" (Dilke's word) was raised in Belgium's parliament in 1906, beginning the debates that would ultimately result in the Belgian government buying the Independent State of the Congo from him (he was given no choice in the matter) in November 1908.[13]

The British Anti-Slavery Society had regarded King Leopold II as a hero of their cause after Arab slave traders were expelled from his colony in 1893. Johnston, duped like Stanley and many others, wrote a reluctant foreword to the first edition of Morel's important 1906 book *Red Rubber* in which Morel lays out the case against the Independent State of the Congo in forensic detail. Johnston recalled the lofty aspirations Leopold II had pretended to hold and his own enthusiasm for promoting these as part of a multinational program of European imperial intervention in Africa, admitting ruefully that "our own past Colonial history and that of other European nations in Africa was very far from being stainless." Johnston had personally fought (literally, hand-to-hand) against "Arab" slave traders in Central Africa. He was convinced that Leopold's efforts had helped suppress the trade in Central Africa (Leopold's agents had actually worked with slave traders like Tippu Tip at first).[14]

Johnston's condemnation of Leopold and his support for Morel's book was important, even if he equivocated in arguing that the Congo before Europeans arrived was far from idyllic and that there was a growing threat to the civilizing mission of Europe in Africa. However, those who later led expeditions into the Congo hoping to find okapi—like Percy Powell-Cotton, Hermann Schubotz, and Cuthbert Christy—still required official permission. They made sure to thank the Belgian authorities and local officials for assistance, offering few criticisms of their practices. Institutions like the Zoological Society of London and the British Natural History Museum also worked with the Belgian colonial authorities—the expeditions they organized to collect wildlife relied on colonial networks, connections, and infrastructure.[15]

The British Museum mounted an ethnographic expedition to the Congo in the final two years that Leopold II controlled the region led by Emil Torday and Melville William Hilton-Simpson. In a report on the mission titled "Happy Congo Natives," it was reported that the expedition spent four months in the " 'domaine prive' of the King of the Belgians" within the colony, and that "during the whole period of their stay" they "never saw a single act of brutality on the part of the state officials or the company's people." One hopes this was true. It is possible that colonial officials were instructed to refrain from engaging in any such behavior in front of members of the expedition, but it also might have been the case that the expedition dealt only with truly good men in the company's employ. Regardless, it seems regrettable that leaders of a British Museum expedition publicly defended the record of the Independent State of the Congo at this time.[16]

The American Museum of Natural History likewise collaborated with King Leopold II of the Belgians in the last years of his rule over the Congo colony.

Museum director Hermon C. Bumpus was much impressed by an exhibition of Congolese Africans he saw in St. Louis in 1904, which included four pygmies and aroused great public curiosity. Samuel P. Verner brought one of these men, Ota Benga, to New York in 1906, where he helped install some African mammal exhibits at the museum before being himself exhibited at the Bronx Zoo alongside chimpanzees. Verner worked for King Leopold II, from whom he received trade concessions, and became first president of a consortium of wealthy New Yorkers known as the American Congo Company. The consortium's focus was importing rubber from the Congo, and members included several patrons of the city's cultural institutions, including museum trustee J. P. Morgan Jr.[17]

In 1907, the museum entered into formal negotiations with Charles Liebrechts, secretary general of the Independent State of the Congo, regarding an Africa exhibition focused on the Congo. In that same year King Leopold II authorized his colonial agents to collect and send objects to the museum in New York, and the first of more than three thousand objects began to arrive over the summer. This was two years after publication of Mark Twain's critical pamphlet *King Leopold's Soliloquy*, and the same year that the US Senate resolved to pressure the king to reform. Internal debates ensued at the museum over the wisdom of collaborating with the king, and it was only in 1908—by which time it had been decided that he would have to hand over the colony to the Belgian government—that planning for a collecting expedition began under the new director, Henry Fairfield Osborn.[18]

Once in the Congo in 1909, expedition leader Herbert Lang did ask Protestant missionaries about the stories of terrible human rights abuses. He was told they were true. However, both he and his co-leader James Chapin came to believe the atrocities had been exaggerated or at any rate considerably curtailed after the Belgian government took over, and Prince Albert and government officials had toured the colony in 1909. They claimed to have seen no mistreatment of Africans or evidence of mutilations. When pressed by a reporter in 1910, Lang maintained that "the natives are perfectly happy, and . . . are in cordial relations with the representatives of the government."[19]

Scientists and conservationists working in politically difficult countries will always face hard choices regarding how to access the wild places: are they prepared to put up with tyranny to study and save wildlife? Those wanting to work in the Congo during the brutal regime of Mobutu in the country he renamed Zaïre faced the same dilemma. The human costs of such accommodations require careful consideration.

Exhibition and Illustration
Picturing Okapis

The first illustration of evidence for the existence of the okapi was the (uncredited) drawing of the okapi-skin bandoliers that Johnston had sent to the Zoological Society of London, which had been exhibited at a Society meeting by Sclater in November 1900. This illustration appeared in an article Sclater published in the *Proceedings of the Zoological Society* in February 1901. The arrangement makes the bandoliers look like two (somewhat peculiar) stripey legs. Each bandolier has tassels on one end and holes for fastening across the top of the other, together presenting a distinctly unorthodox documentation of a new species for a zoological journal.

The other crucial early evidence for what an okapi looked like was Johnston's remarkable artwork. While he had not actually seen a whole okapi, living or dead, he had discussed the animals in detail with Mbuti people and the Belgian officers in the Ituri forest and with African porters from the forest (who called okapi "ndumba") that he still had with him when he received the okapi remains in the Uganda Protectorate. He received two skulls and an entire skin (without hooves) whose condition enabled him to infer aspects of the body shape of the animal. While he had these with him, Johnston made drawings and "with the utmost care" painted two okapis.[20]

Johnston used the skulls and the descriptions given to him by the Africans he had questioned to guess the shape of the head of the animal. He was worried the skin (or pelt) would lose its amazing luster and noted to Sclater that he had not exaggerated the brightness of its colors in his painting. Further, he maintained his drawing was "correct in the tiniest details of the stripes." He had allowed for certain distortions resulting from the preparation of the skin, correcting these with input from his informants, chief of which were that he made the tail "a trifle longer" and allowed for some missing skin on the inner legs and belly. He concluded that also missing due to the skinning process were the penis sheath and testicles of the animal, the latter of which he depicted in his painting (he was wrong: it was a female animal).[21]

Johnston sent this watercolor painting to Sclater with a letter dated March 31, 1901, which Sclater showed and discussed at a Zoological Society of London meeting on May 7, 1901. Sclater published a full-page lithograph of Johnston's painting in the November *Proceedings of the Zoological Society of London*. There was clearly a coordinated effort going on between Sclater and Lankester over the sharing of Johnston's work, as the day after Sclater showed the painting at the meeting, it was exhibited by Lankester, together with the bandoliers, at the socially prestigious Royal Society conversazioni. The Royal Society held these glittering events in its Burlington House rooms, and the reports in *The Times* commence with long lists of aristocratic guests, peppered with earls, lords, knights, chief justices, major generals, admirals, and the like. Exhibits at the conversazioni ranged across the physical and natural sciences and included inventions and explanations of engineering feats such as the recent bridging of the Zambezi near Victoria Falls.[22]

Johnston's painting was the first depiction of a whole okapi to appear in Europe, and likely anywhere outside of Africa and was jealously guarded. When it was first shown at the Zoological Society of London meeting in May 1901, grumbled the *Field: The Country Gentleman's Newspaper*, it "was not permitted to be copied—an unusual restriction in scientific societies."[23]

Lankester received the skulls and the almost-entire skin from Johnston on June 17, 1901. These were exhibited at a packed meeting at the Zoological Society of London the very next evening, at which Johnston, just arrived from Uganda, was present. Having examined and described these remains at the museum, Lankester sent them to the renowned taxidermist James Rowland Ward at 167 Piccadilly (known as The Jungle) to mount, as he possessed expertise with African megafauna. He also supplied Ward with Johnston's painting.[24]

Ward liked to model from life, so this commission was "no easy matter," he remarks in his autobiography. Ward preferred to draw the living animal (later he worked from photographs), and ideally he received a carcass to allow him to study the muscles and skeleton. He conceived of a lifelike, expressive pose, then produced a model of this in clay. A skeleton was built of wood and metal, which he then covered with a plastic material to reproduce the muscles and flesh as modeled (in effect, he created "the form of a flayed animal" rather than "[filling] out the skin with stuffing"). However, despite the limited materials he had to work with in the case of this okapi, Ward had been mounting giraffes since the early 1890s for Walter Rothschild and others, including the specimens at the head of the staircase up to the east corridor in the main hall of the Natural History Museum. "Bearing in mind evidence of the giraffine affinities of the okapi," he notes in his autobiography, "I made . . . a very successful job."[25]

Ward's mount (see fig. 4.1) is a little more giraffid than Johnston's first painting, notably in the elongated neck (though held out more or less straight in line with the back), and the back, which is more sloped, as in a giraffe. However, presumably Ward was influenced by Johnston's painting, too. Ward's three-dimensional realization of an okapi—based on anecdotal descriptions, Johnston's

Figure 4.1. Rowland Ward's mount of Harry Johnston's okapi remains, photographed in a most incongruous domestic setting by J. Benjamin Stone in 1907 (© Trustees of the Natural History Museum).

painting, and Ward's expert work with the skin and skeleton—was to become an authoritative version of what okapi look like. Johnston himself made new drawings and another painting based on Ward's taxidermized okapi.

Appearances Deceive

That the first evidence of the existence of the okapi was limited to two bandoliers, two skulls, and a skin together with Johnston's painting had consequences for subsequent interpretations of the species. Based on Johnston and others' misidentification of the first almost-entire skin and skull as a male, it was at first wrongly assumed that neither males nor females have horns, and it was assumed that mature males are larger than females. The first version of Johnston's color painting of two okapi made its debut in a popular publication in his article in the *Graphic* on August 3, 1901. *Pall Mall Magazine* presumably did not have access to the image. In the feature it ran in the same month, *Pall Mall* includes a full-page artwork showing two okapi in the rainforest (see fig. 4.2). These literally match

Figure 4.2. Artwork commissioned to illustrate the okapi before images were available captioned "the head of a giraffe, the body of an antelope, the legs of a zebra" (*Pall Mall Magazine*, August 1901, 570).

the description in the caption ("the head of a giraffe, the body of an antelope, the legs of a zebra") and are almost comically incorrect, resembling two llamas with giraffid ears and horns and legs striped from hoof to shoulder.

Ward seems to have completed (or almost completed) his taxidermy work on Johnston's skull and skin by early August 1901, as a photograph of the front view of the head of this mounted specimen and a whole-body photograph from the side (both black-and-white) first appeared in an article in the *Sphere* on August 17, 1901, together with a drawing of a *Helladotherium* skeleton (supposedly an extinct ancestor of okapi) and photographs of a zebra and a giraffe (for comparison). Although the article claimed Ward's taxidermized mount would go on show at the Natural History Museum "within days," it seems to have only gone on exhibition in October.[26] These photographs of Ward's mount also featured in an article Lankester published in *Tatler* on August 21, 1901.[27]

Taxidermized mounts of okapi, beginning with Ward's setup of Johnston's skin, followed by others set up at Tervuren for various European rulers and governments, were crucial first because several were put on public exhibition and second because photographs of these taxidermized animals beginning with Ward's specimen (for better or worse) were the only available evidence in wider circulation showing what an okapi might look like. They featured, for example, in contemporary encyclopedias and compendiums of the world's fauna, the compilers of which were eager to include the new mammal.

Harry Johnston's drawings (no color) of Ward's mount (the full body from the side and a front and a side view of the head) appeared in an article he published in *McClure's Magazine* in September 1901. This article along with its accompanying drawings and Johnston's photograph of okapi habitat in the Semliki (Watalingi) Forest that was republished in the *Annual Report of the Smithsonian Institution* in 1902, were among the first notable American representations of the "new" mammal. The McClure's article was followed by a black-and-white photo of Ward's setup of Johnston's okapi skin and skull that appeared with a note in *Scientific American* in October 1901.[28]

An Image Evolves: Johnston after Ward after Johnston

Johnston painted a second version of his two okapi after having examined Ward's mounted specimen, and it is this version that adorns the frontispiece of his 1902 book *The Uganda Protectorate*. While on first glance they appear almost identical, a close, side-by-side examination of the two paintings reveals a few key differences.

The second version (see fig. 4.3) includes new incidental details like an additional hind hoof of the okapi in the background being visible, but a more notable

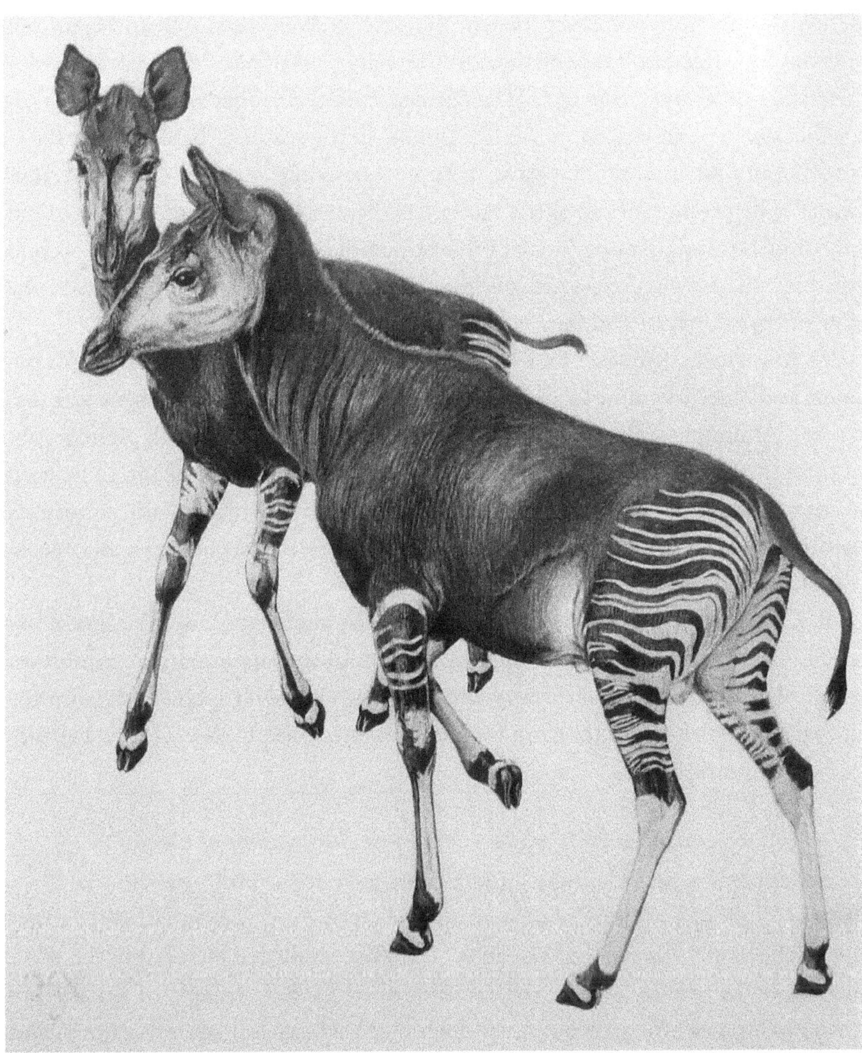

Figure 4.3. Harry Johnston's second, amended painting of the okapi (Harry Johnston, *The Uganda Protectorate*, vol. 1 [London: Hutchinson, 1902], frontispiece).

change is that the ears are shaped less like a donkey's and are less furry. The penis sheath of the male okapi in the foreground is also visible, unlike in the first version, though testes are shown in both versions (even though the taxidermized animal turned out to be a female). Another key difference is an adjustment to the carriage of the head and neck, which in the first painting more closely resembles an antelope like a kudu or a nyala but is more giraffe-like in the second (Johnston rightly did not copy the elongation of the neck in Ward's mount). In his second version, Johnston also made changes to the shape and slope of the back and overall shape of the torso, noting in *The Uganda Protectorate* that while he had given his first okapi "a very stout, horse-like build," in the second, "based on Mr Rowland Ward's restoration," he gave "a more giraffelike form to the mysterious okapi." The position of the nasal "bump" has been shifted upward on the skull in Johnston's second painting (it is not this noticeable in living okapi). In the second painting, the two ossicones are less like horns and more like lumps beneath the skin (incorrectly it turned out, as male skulls subsequently revealed).[29]

As Johnston points out in *The Uganda Protectorate*, until a photograph became available, it would be difficult to know which of his paintings was more correct. Although he illustrates the zoology section of this book with photographs of the animals he describes in it, there were no photographs of okapi, so he used his drawings with revisions made from Ward's mount. As it turns out, Johnston's paintings and drawings are remarkably accurate, given that he had never seen a live okapi.

In October 1901, presumably still disgruntled about not having access to existing imagery of the famous okapi, the *Field* published a note on the exhibition of the okapi at the Natural History Museum with its own (uncredited) commissioned black-and-white drawing of the museum's mount. It is clearly a depiction of Rowland Ward's setup.[30]

Scientific Exhibition and Technical Illustrations of Okapi

The okapi first appeared on the international stage at the fifth International Congress of Zoology in Berlin held August 12–16, 1901. At this international gathering of the stars of zoological research, *The Times* reported, "probably the most interesting feature . . . will be the exhibition, by Dr P. L. Sclater, of the skull of *Okapia Johnstoni*, the new quadruped which Sir Harry Johnston was the means of discovering in the Semliki forest of the Congo state," noting his specimen was "the only known one in the world."[31]

Sclater had only the larger, complete skull and one of the bandoliers to show, as the skin was being mounted by Ward for the Natural History Museum in London at the time. However, in announcing "one of the most remarkable discoveries in Zoology which has been made in recent years," he showed Johnston's first speculative painting of the complete animal. Control of images of even just this skull seems to have been tight, and there is no image accompanying Sclater's article in the conference's proceedings.[32]

The first scientific papers to include detailed drawings of okapi anatomy seem to be Forsyth Major's papers focused on okapi skulls published in May and June 1902. The second paper was especially important because it showed a skull with horns, the first to arrive in Europe. The most comprehensive early illustrated scientific paper was Lankester's August 1902 paper, showing the skulls and hair patterns of the okapi. It also included a drawing of the bandoliers from Sclater's earlier paper and a painting of two okapi in a rainforest scene based on Ward's reconstruction.[33]

That okapi had bony frontal horns was acknowledged by Lankester in the amendments he made to the 1902 version of his November 1901 Zoological Society of London paper, after remains of an adult okapi arrived in Brussels in early May 1902, but the information came too late for a revised color plate to be included. The paper does include Forsyth Major's figure of this skull as well as an unattributed color plate showing two views of Ward's mounting of Johnston's skull and skin (a subadult female okapi) with a rainforest background. Lankester tries to have it both ways by explaining that "the specimen is not an adult" and that therefore "the horns . . . are not yet developed" and that it cannot thus be "known whether the specimen here figured is male or female."[34]

The most comprehensive collection of early illustrations of the okapi is to be found in Fraipont's 1907 monograph. In addition to the numerous technical drawings of bones and skulls and seventy-six photographs of okapi bones and taxidermized okapis in the main body of the text, he includes 111 full-page plates at the end, including four attractive color plates with paintings of okapi in naturalistic backgrounds. The first two are likely to have been commissioned by Forsyth Major in 1902–3 while he was supposed to be writing the okapi monograph and are labeled "*Okapia Johnstoni*, Sclater ♂ (*Okapia Liebrechti* [*sic*] de Forsyth Major)" and "*Okapia Johnstoni*, Sclater ♀ (*Okapia Liebrechti* [*sic*] de Forsyth Major)," respectively. As the first plate (by P. J. Smit, who also did most of the lithographs of the bones and skulls) shows two males with horns (see fig. 4.4), it must have been completed after May 1902. The second plate shows a single female okapi with no horns and a notably longer neck than those on the previous plate (the artist's signature is illegible). The next two plates show, respectively, striking color illustrations of a horned male labelled *Okapia johnstoni* and a hornless female with a calf both also labelled *Okapia johnstoni*, the calf clearly based on a photograph taken in 1907. The artist is the same (signature illegible), and set these okapis against a forested background including palm trees and a water body.

Figure 4.4. Okapia liebrechtsi Maj. commissioned by Forsyth Major (Julien Fraipont, *Okapia* [Brussels: Independent State of the Congo, 1907], plate 1).

In the seven years following its discovery by the West, the okapi became well known to scientists and the public through visual representations based on scanty remains and Johnston's early drawings and paintings. These were joined by taxidermized okapi gifted to governments and rulers by King Leopold II as part of a propaganda campaign to obscure the real purpose of his colonial state in the Congo.

The eagerness of scientists to associate themselves with the sensational discovery, compounded by a shortage of remains for proper taxonomic deductions to be drawn, resulted in several misconceptions which only decades of further collecting, research, and eventually photography would help to correct. Most notably, there was confusion over whether there are two or more species, a smaller with horns in both sexes, and a larger without horns in either sex. It was uncertain whether there was sexual dimorphism between males and females (there is). A glance at Fraipont's plate XVIII in the closing section to his monograph will show that it was no trivial matter to decide on okapi taxonomy from limited remains: the four okapi skulls show notable variations in shape, at least to my untrained eyes. Further details around the slope of the back and carriage of the neck of okapi remained to be discovered (the neck is quite unlike the stacked tower of the neck of the giraffe), and the motley herd of fauxkapi produced by Europe's taxidermists did little to clarify this picture.

CHAPTER FIVE

Okapis Take Shape in the Western Imagination

It is in form rather outlandish and uncouth.

—The Times, *May 1901*

[It has] the head of a giraffe, the body of an antelope, the legs of a zebra.

—*Frank E. Beddard, August 1901*

Reporting on Boyd Alexander's hunt for okapi in 1906, an article in the journal *Nature* noted that very few specimens were publicly available; there was just one exhibited in the British Natural History Museum in London, one in Walter Rothschild's museum at Tring, another two in the Congo Museum in Tervuren near Brussels, and a few others in Europe, notably in Rome. "It is a great pity," the article complains, "that the Belgian Government does not take immediate steps to publish coloured figures of its specimens in order to aid in solving the question as to whether there is more than one species (or race) of okapi." Other points, for example, whether male's horns resemble deer antlers in protruding from the skin could have been clarified by illustrations of such specimens being made available.[1]

A clearer picture of the appearance of the okapi emerged only gradually in the West, beginning with a motley assortment of taxidermized individual okapis that were disseminated widely through photographs of them mounted in museums and private collections. The first photograph of an okapi (a calf) was published in 1907, and photographs of live captive okapis circulated in print in the interwar years. The first photographs of wild okapis and film footage of captive okapis appeared in the 1930s.

Taxidermy: Okapi Get Stuffed

While the ancient Egyptians showed a positive mania for mummifying the local fauna, the Victorians favored taxonomy and the mounting of impressive horns and antlers. There are parallels in the necrogeographies, however, from sites of killing or capture (ancient Egypt demanded tribute such as slaves and wild animals from Nubia which extended south along the Nile into what is now Sudan; the Kordofan region of what was Anglo-Egyptian Sudan [now Sudan] was a source of important specimens for the British collections in colonial times) to sites of embalming (Egyptian temples and European taxonomy studios) and exhibition (temples, tombs, and museums). In both cases, the animals were hollowed out, with their outer structures preserved.[2]

European taxidermists like Rowland Ward and Edward Gerrard and Sons worked with the often very detailed measurements and contextual information recorded in situ by hunter-naturalists like Percy Powell-Cotton that were taken as soon as possible after the animal was killed. They combined general knowledge of the species with precise information on particular specimens to provide as lifelike a representation of the dead animal as possible. Hunters strove to avoid damaging the hides and body parts of animals required for taxidermy, and evidence of the fatal wounds was artfully concealed. Expert taxidermists drew on natural history and anatomical knowledge as well as artistic flair and expertise with chemicals and the materials of taxidermy to fabricate a physically accurate and behaviorally convincing three-dimensional representation of the animal. Ward was working with very paltry evidence for his first okapi, however, and while it was one of his most awaited and possibly influential creations, it is not one of his finest.[3]

Fraipont includes photographs of sixteen different entire okapi mounts in his monograph, photographs of the heads of four mounts, eleven of their hind or forelegs that show patterning (some from entire mounts already pictured), and one photograph of a live calf. All are black-and-white images, and as is to be expected with different individuals of varying ages and gender, there is much variation in appearance.

Notable variation is apparent in how taxidermists interpreted the appearance of this animal as yet unseen in Europe. Some followed Ward's first mount, sloping the neck slightly upward following the line of the back and elongating the neck—for example, the mount gifted to the king of Italy on show in Rome, that is also rather spindly for an adult female (no. 488 in Fraipont's monograph *Okapia*) or the mount on show in Madrid (no. 536 in *Okapia*). The Lisbon mount (no. 484 in *Okapia*) is positively llama-like with a most elongated neck. The mount in the natural history museum of Genoa is distinctly equine, while others are more giraffid, notably two gifted to the French government (nos. 503 and 534 in *Okapia*). The taxidermized heads are variable in quality, ranging from the slightly odd (nos. 479, 484, and 488 in *Okapia*) to the wonderfully lifelike, notably the heads of an old female and a calf, both mounted at Tervuren (nos. 536 and 706 in *Okapia*).[4] (These can be viewed in the online supplementary materials.)

Photographs of mounts presented to the Natural History Museum in London by Harry Johnston, Percy Powell-Cotton, and Boyd Alexander and those at Tring and the Royal Scottish Museum in Edinburgh appear in the Natural History Museum's 1910 atlas. The British Natural History Museum's collection played it safe by including okapi mounts with necks at three different angles to the body. The neck on the mount of a young male presented by Alexander in 1907 takes an unusual downward slant, Powell-Cotton's larger male mount's neck from the same year takes a slightly downward angle, and Ward's first mount slopes gently upward as does the mount in the Royal Scottish Museum, Edinburgh ("sex doubtful," despite horns and penis sheath). Rothschild's Tring mount ("sex doubtful" and locality "unknown") is, like many of the mounts, a little uncertain about the shoulders, but the upward angle of the neck is more lifelike.[5]

World War I interrupted the exchange in okapis. A collecting mission that commenced in 1912 under the direction of the British collector Cuthbert Christy and that was authorized by the Belgian government amassed thirty-two cases of materials that became stranded because of the war. The Belgian minister for the colonies requested that these be deposited temporarily in the Natural History Museum in London, a request the trustees granted, sportingly agreeing that while the cases had to be opened so the specimens could be disinfected, "any new species in the collection were not to be described unless the consent in writing of the Belgian authorities had been obtained."

Two specimens of okapi were included in these materials sent to the museum, and the Belgian government requested that "the finer of the two skins" be mounted "for presentation to Lord Kitchener" (who never received them: he was drowned when *HMS Hampshire* struck a mine on June 5, 1916). The Belgian government expected the second skin and the two skeletons to be delivered to the Museum of Belgian Congo after the war, and it offered the Natural History Museum specimens of domestic animals of the Congo as a means of encouraging the British follow through on this (the Trustees approved all of this). In 1919, the transfer of okapi remains resumed.[6]

In November 2022, I visited the Natural History Museum's off-site storage location for the first time. It is located in a nondescript building with no identifying signage in South London. Museum scientist Natalie Cooper was taking students on a study tour and invited me along. Unlocking a red metal door on the ground floor, she led us downstairs across a cavernous industrial space and through a human-sized red metal door set within a huge (giraffe- and whale-accommodating) set of metal doors into a narrow space blocked by similar doors. Opening the small door, Natalie led us into a hangar-sized space crammed with double-storied rows of taxidermized mammal remains arranged on wooden shelves fastened to frames of metal poles. We walked past shelves of carnivores, wild pigs, and every imaginable bovid and cervid from giant kudu antelope to tiny mouse deer. Alongside one row there marched a parade of elephants (mounted skeletons and stuffed jumbos), and on the end of another stood a fan of giraffe necks and heads. The smell was singular.

I found the museum's four okapi mounts together in an impromptu family group (see fig. 5.1). Powell-Cotton's large male to the left was looking a bit tattered and cracked, but the taxidermy work on the head and ears were wonderfully lifelike, down to the details of veins squiggling beneath the skin and the long eyelashes. To his right, the poor old Alexander-Gosling expedition mount (Gosling died in Africa, Alexander survived to make the donation in their name) had a bandaged muzzle, tattered ear, and hairless patches down his long neck, which sweeps dramatically down and to the right. Next to him stood a large female from the Uélé district donated by Ward in 1931, which towers above the rest. The detail is very fine and includes a distinctively notched left ear, a notable improvement on Ward's first attempt. Her neck is inclined downward like the neck of the Powell-Cotton mount.

To the far right of the group stands Ward's first mount, the skin of the young female okapi sent by Harry Johnston (who thought it was a male) still bearing the okapi's distinctive red-brown luster, the legs beautifully striped, and the cheeks on the face still creamy. However, the right ear has disintegrated to twisted foil, and the left cheek is cracked from mouth to neck.

Seeing this last mount was uncanny for me, having read so many accounts of its appearance and how it was acquired. What first struck me was her now so obviously hybrid nature, as a fabricated object enveloped in bodily remains.

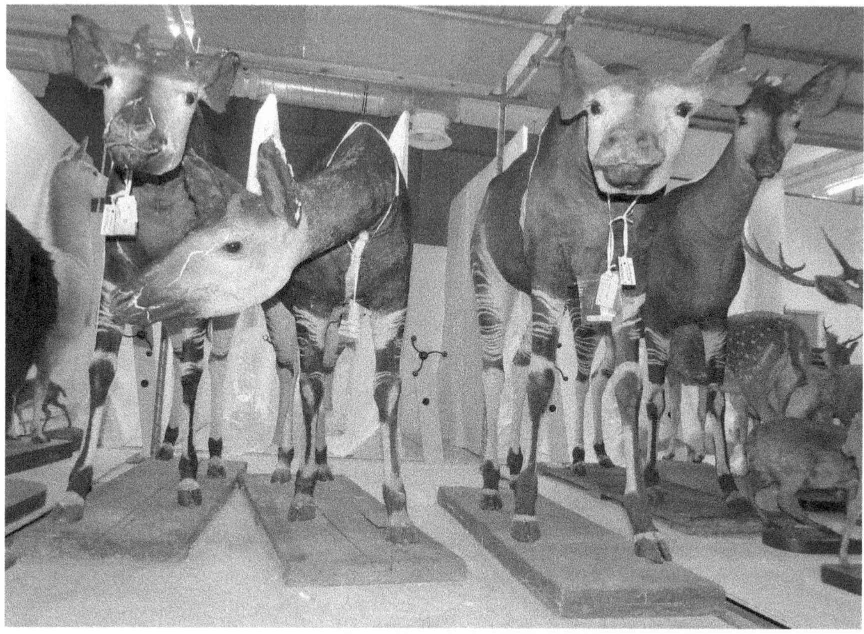

Figure 5.1. Okapi mounts in the Natural History Museum's storage facility. *Left to right*: Percy Powell-Cotton's large male okapi; the Alexander-Gosling Expedition mount of a male; a large female donated by Rowland Ward; Ward's first okapi mount out of remains sent by Johnston in 1901, misidentified as a male (© Simon Pooley, by permission of the Trustees of the Natural History Museum).

Then, the tags naming Karl Eriksson, Johnston, and Ray Lankester made me think of these long-dead and mostly forgotten men. Protected in a plastic sleeve, a red handwritten tag states: "Type of: *Okapia erikssoni* Lankester," commemorating the ghost of a taxonomically extinct species. The wooden panel on the heavy wooden stand for the mount, however, reads "The Okapi, *Okapia Johnstoni* (female), near Semliki Forest near Lake Albert Edward, presented by Sir Harry Johnston, K.C.B., 1901."

Observing this beleaguered little group, once treasures of the museum visited by thousands but now disintegrating in obscurity in an off-site storage facility, I felt a pang of melancholy. These venerable, hybrid creations of the taxidermist's art, strangely lifelike and yet cracking at their ever more visible seams like aging cyborgs, have been all but forgotten. I recalled the scene in the film *Blade Runner* when Roy Batty's replicant ("skinjob") talks of all he has seen, memories to be lost in time like tears in rain. I also thought of these four okapi, individual animals that had been killed so long ago in the rainforests of the Congo, their skins and bones removed, preserved, sent downriver and overland, then shipped to distant Europe.

Surprisingly, to me at least, there were at the time of writing no longer any okapi on show in the Natural History Museum. This is a pity, as the story of their discovery is intimately connected with the history of the museum. Displays featuring interpretation of male and female skulls that included the teeth and ossicones would be valuable, considering the importance of the teeth for identifying okapi as giraffids and the early debates over okapi ossicones, especially if compared to a giraffe skull. And so I hope this changes, but making that happen won't be straightforward.

Despite rehabilitation work in 2024 on the okapi mounts in storage ahead of a planned move of the entire collection to Reading, outside London (I noticed the bandage had disappeared from the reconstructed muzzle of the Alexander-Gosling mount when I returned for a second visit), these okapi are not in good condition. They would require careful positioning if displayed as lifelike (though why not exhibit one as a famous historical artifact?). The two skeletons—stored along the opposite wall of the storeroom from the mounts—are headless, and would require reuniting with their skulls, presumably stored on site at the museum in South Kensington.

While the taxidermized mounts show off the beauty of the okapi's skin and its outer appearance, and are more difficult to preserve, the skeletons are hardier, and valuable for revealing the extraordinary dimensions of the species. Okapi appear narrow and streamlined when viewed from the front (it's possible this shape allows them to ease their way through the dense forests of their favored habitats), whereas when you examine them from the side, it is their depth and power that impresses. Okapi skulls, viewed from the front, seem impossibly narrow, and from above, positively birdlike. From the side, they are as massive as a giraffe's, with powerful foreheads and jawbones. Displaying the skeletons in their collection seems the likelier option for any okapi exhibit in the Natural History Museum at present.

Alternatively, new okapi remains could be taxidermized, but that seems unlikely. The long relationship that existed between London Zoo and the Natural History Museum whereby animals that died at the zoo were automatically offered to the museum first seems to have ended. The male okapi Mbuti died at London Zoo in early 2024, but as far as I can discover, was not offered to the museum. New okapi to taxidermize are (perhaps fortunately) hard to come by, and the museum no longer has its own taxidermists. So, even if a dead okapi became available, the museum would have to outsource the taxidermy work.

Another place I had expected to see okapi on display was at the Powell-Cotton Museum in Quex Park, in Kent. When I asked to see their okapi materials in August 2022, the museum and archive staff were surprised by how many remains they discovered they had in storage. While recording that Percy had acquired two okapi himself on expedition in the Congo, and had given one to the Natural History Museum, the website for the museum claims that the "only Okapi specimen in the Powell-Cotton Museum" is the skull of an adult male given to Percy by a French colonial officer. The collection staff unearthed this skull, but also okapi dung, still unidentified leaves eaten by okapi collected by Percy in 1906, belts, and twenty-three bandoliers stored in a large wooden box (also containing sixteen buffalo tails).[7]

Ersatz and Actual Okapis
Photographs of Okapi, Dead and Alive

Photographs of taxidermized okapis provided an extra layer to the semblance of reality offered by these products of highly skilled surmise. However, they stood stark against plain backgrounds in photographs such as those in Fraipont's and Lankester's books and in the one used for the okapi card featuring Powell-Cotton's mount in the Natural History Museum's ten-card pack representing large ungulates. In museums, taxidermized okapis were mounted on stands or else they were encased in glass museum cabinets, particularly in Britain where museums have—with the exception of the Powell-Cotton Museum—eschewed dioramas. The skill of the taxidermist and a high-resolution photograph of the mount shot in such a way as to minimize the signs of artifice in the mount contributed to producing a facsimile of "the real okapi."

It was a testament to the rarity and elusiveness of the okapi that despite relentless searching from 1902, the first photograph of a live okapi (a calf) to emerge from the rainforest appeared in print in 1907 (see fig. 5.2). Alexander, who collected skins and bones of okapi in 1906, includes only a photograph of the taxidermized mount of the specimen he donated to the Natural History Museum in his book *From the Niger to the Nile* (1907). Powell-Cotton similarly failed to glimpse, let alone photograph, a living okapi, but the fine mount taxidermized by Ward from the skin and skeleton he had obtained, was a great success at the Royal Society conversazione at Burlington Gardens on May 8, 1907. According to Ward, this was the finest skin to have arrived in England, and Richard Lydekker pointed out that it was the first male okapi brought into the country. A large illustration showing two views of this taxidermized okapi in an equatorial forest was published in the *Illustrated London*

News on December 31, 1910, and subsequently reproduced in several Belgian newspapers. This is by far the most atmospheric image of okapi in this period.

This illustration is an artful collage made up of a photograph of an okapi drinking place taken by Powell-Cotton as a backdrop, with two photographs of Ward's setup of his okapi taken from different angles superimposed on this scenic background. It was published to accompany the news, which turned out to be false, that Herbert Lang had captured a mother and calf okapi and was sending them back to the United States. The collage appears to show an okapi standing on a sandbank about to drink, with another standing in shallow water, facing its companion.

Images of the okapi remains Powell-Cotton secured do not appear in the journal articles he published about his Congo expeditions, although images of the skull and bones of the okapi and of the skin being dried on a rack do accompany a popular article he wrote in 1907 for the *Wide World Magazine*. He also showed these images and a photograph of his Natural History Museum mount in a talk to the African Society.

The first photograph of a live (calf) okapi, taken, according to Lankester, by a "Signor Ribotti" in Bambili on the Welles (Uélé) River in the Independent State of the Congo, was shown by Lankester at a British Association meeting in Leicester on August 5, 1907. According to the *Illustrated London News*, in an article published a month later, this was the first time a European (Ribotti) had seen a living okapi, a claim that was subsequently disputed. The same article reports that Lankester had dispelled doubts circulating that "okapi" was not the "native name" of the animal. Apparently, he had shown the photograph and a taxidermized okapi to pygmy people brought to London from the Ituri forest by Colonel James Harrison, and they had recognized the animal and used the name "okapi" to talk about it.[8]

Just as there were controversies over who had discovered or seen an okapi alive, so this first photographic evidence was the subject of conflicting stories about its creator. Lankester said he was given the photograph by the naturalist and politician Marquis Giacomo Doria of Genoa, who said he had obtained it from a Lieutenant Ribotti (who had sent okapi remains to the natural history museum in Genoa in early 1905). However, Fraipont states that in April 1907, Africans brought a live eight-day-old okapi to the Angu post in the Rubi zone and that the deputy head of culture for the Independent State of the Congo, a Mr. Lamboray, visited Angu and photographed this young animal when it was one month old (it died soon after).[9]

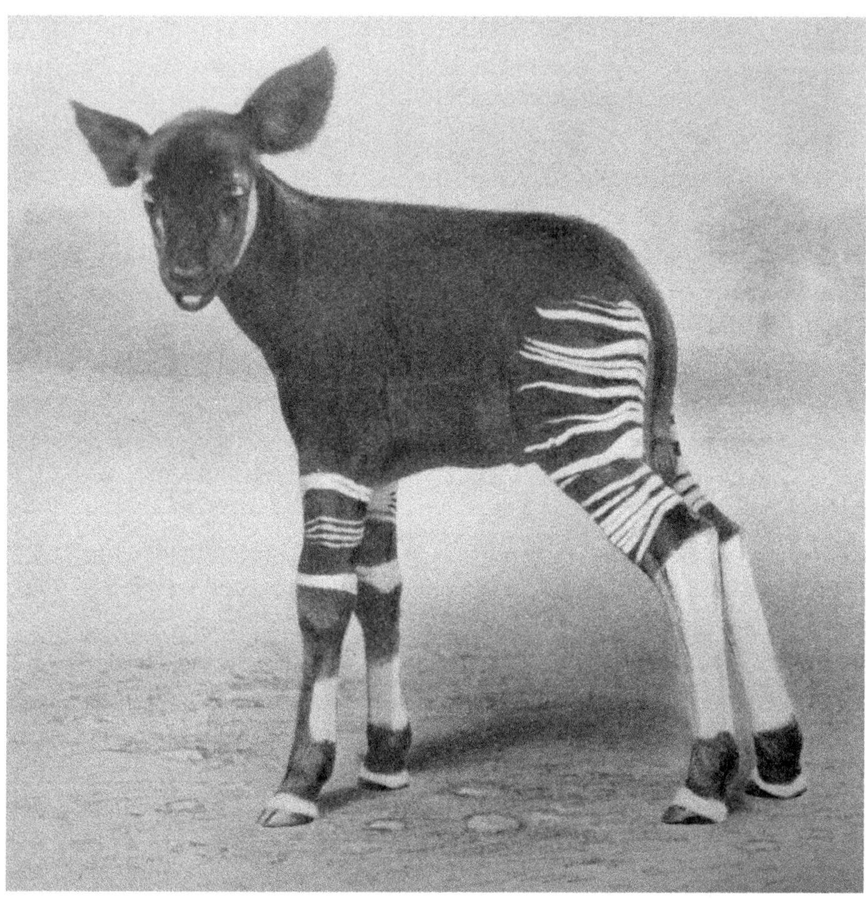

Figure 5.2. The first photograph of a living okapi published in the West, a one-month-old calf photographed at Bambili by "Monsieur Ribotti," according to Lankester, although Julien Fraipont suggests it was taken by M. Lamboray at Angu (Julien Fraipont, *Okapia* [Brussels: Independent State of the Congo, 1907], 96).

The German explorer Adolf Friedrich, Duke of Mecklenburg-Schwerin, whose party spent time with the Mbuti in the Semliki valley in 1908 searching for okapi, obtained only a skull, five skins, and some belts. His book includes photographs of these and a painting of an okapi in the rainforest by Wilhelm Kuhnert. The duke sent these body parts to Berlin as required in this period by the German empire, claiming they were the first okapi remains to reach Germany. The skins were not of sufficient quality to be taxidermized.[10]

Friedrich made a journey from the Congo to the Nile from 1910 to 1911 during which one of his party, Hermann Schubotz, acquired two okapi in mid-1911 through the services of a local hunter. Schubotz's black-and-white photograph of the body of one of these was reproduced in the German magazine *Die Woche* in January 1912 and is possibly the first photograph of a wild adult okapi published in the West (as claimed in the caption). Of good quality, it shows a female okapi lying prone on her right side on a bare patch of earth against a background of dense vegetation. Her head and long neck are outstretched toward the lower left corner of the photograph, ears erect, her torso extending away to the right, with her left foreleg crossed daintily over her right.[11]

The book describing the duke's 1910–11 expedition (published in German in 1912 and in English in 1914) also includes this photograph. In addition, the book features a photograph of a skin being stretched, a reproduction of a painting by the party's artist, Ernst M. Heims, and a photograph of a very fine mount of a male okapi donated by Schubotz to the Senckenberg Naturmuseum in Frankfurt (another went to Hamburg). By 1914, the German museums in Bremen, Frankfurt, Hamburg, Magdeburg, and Stuttgart would all have okapi specimens.[12]

No further photographs of an entire okapi appeared until photos emerged from Herbert Lang's American Museum of Natural History's expedition to the Congo, which departed in 1909 (although Christy had shot and photographed an okapi in 1913, his book was published in 1924). Until 1915, Lang had only managed to glimpse one dying okapi on his many hunts with local Africans. His 1918 expedition report, published after he had returned from the Congo, includes six photographs of okapi, living and dead. The first shows the renowned African hunter Abawe sitting next to a huge male okapi posed side-on to the camera, prone on its stomach with front and hind legs folded in, but its long neck raised, supporting a massive head with large horns. The caption notes that this "record bull okapi" stood "over five feet at the withers," and was destined to be the "the main feature of the habitat group to be installed . . . in the African Hall of the American Museum of Natural History." Lang notes that the head shows few giraffid characteristics, and "the lips are not in the least prehensile as has generally been stated."[13]

The second photograph shows the head, neck, and trunk of a propped-up, obviously dead male okapi against a forest background. It is unironically captioned with "True to life and in its native environment: This is an exceptionally fine study of a still fresh okapi bull." The misrepresentation is obvious in the final photo in the article, which shows a side-view of a badly battered large bull sprawled on the forest floor, its head supported by a Congolese man (we are not told if he trapped it or indeed anything about him—he is not Abawe).[14]

Figure 5.3. The okapi calf (looking emaciated) captured by Azande hunter Abawe for Herbert Lang in 1915. It died of starvation not long after this photograph was taken (Herbert Lang, "In Quest of the Rare Okapi," *Zoological Society Bulletin* 21, no. 3 [1918]: 1608).

More sensationally, in 1915 the Azande hunter Abawe caught an okapi calf for Lang, who photographed it before it starved to death. Three black-and-white photographs are included in his 1918 report, one showing the calf standing with its left side toward the camera, looking at the photographer. The second is another side view that underscores (as Lang notes) the only slightly inclined neck, as shown in Ward's first mount. The third shows the calf sprawled on the ground, "all legs and feet," and looking, frankly, gaunt (see fig. 5.3).[15]

After the war, the Congo became a major supplier of African fauna to European zoos and museums, including okapi. The first live okapi to arrive in Europe (and apparently, the first to leave Africa) was Buta (most captive okapi were named), who was captured and sent to the Antwerp zoo in 1919. There is a black-and-white photograph of Willy L'Hoëst, the son of the zoo's director, in shoes and long socks, jacket, tie, and boater hat, welcoming the new acquisition to the zoo. Willy is holding up something out of shot (probably food) next to the semiadult okapi, which is looking up toward his hand. Photographs of Buta at the zoo were widely circulated, including a photograph of a keeper feeding him.[16]

In the interwar years, various individuals were commissioned to hunt for prized animals in the Belgian Congo or obtained permission from the Belgian government to do so. Christy's memoir of his Central African expedition, *Big Game and Pygmies*, was in part made "in quest of the okapi" (as noted in the subtitle), and includes photographs of two okapis he shot, along with five photographs of one shot by a Mr. Reid, shown with and without Mr. Reid. One photograph is a closeup of an okapi head that shows the polished bone tips of the horns—Christy claimed Reid's was "the record head," with five-inch horns.[17]

It seems rather pathetic that even these relatively small horns for a ruminant were subject to comparison and competition between sports hunters vying for an entry in *Rowland Ward's Records of Big Game*. This was the standard work for the recognition of trophy animals, which was established in 1892 and is still going today.[18]

A pioneering figure for capturing and supplying live okapi was the Norbertine monk Franz Joseph Hutsebaut, who was stationed at the Catholic mission in Buta (in the Rubi zone, today the Bas-Uélé province). Hutsebaut managed to capture and tame fifteen okapis, which were sent on to zoos overseas. When King Albert I visited the mission in 1928, Hutsebaut showed him a young okapi named Congo which he eventually accompanied to Antwerp Zoo. There are photographs of Hutsebaut with okapi in this period, including a large male named Kadanga, notable for his unusually positioned horns and the few stripes on his hindquarters. Hutsebaut stands in the background in hat, shirtsleeves, and braces, while an African keeper, barefoot and dressed in a kepi and colonial white shorts and short-sleeved shirt with piping, stands to the right of Kadanga.[19]

Beginning around this time, as captive okapi began to arrive in Europe, photographs (albeit in zoos) became less rare. The arrival of each new okapi was generally headline news and accompanied by photographs of the individual, at least in the country where it arrived. When Congo, the second okapi to reach Europe arrived in Antwerp, pictures were widely circulated. The *Illustrated London News* featured photographs of Congo alone as a calf, and being bottle-fed by Hutsebaut in a white hat with pipe, alongside a photograph of an adult okapi in the Belgian Congo with an African keeper.[20]

Photographs of wild okapis on the other hand—at least live ones—remained incredibly rare. In 1931, the *Illustrated London News*, boasting that it could "claim to have played a pioneer part in introducing the okapi to the world through the medium of illustration," published a front-page photograph of an adult female okapi "photographed for the first time in its natural haunts" in the Ituri Forest. She is shown from behind, her head turned to the side and so fully visible against a background of foliage (see fig. 5.4). The story includes a description of the hunt for the picture by the photographer Cornelius Bezuidenhout, a member of Lord Howard de Walden's British Museum expedition.

Bezuidenhout claimed to have disguised himself "in the skin of a forest hog" in order to capture this okapi image from close range. After initial attempts to approach okapi disguised in this way failed, he took the additional step of disguising his smell with the aroma of certain crushed leaves. With the aid of pygmy trackers, he again got close to an okapi but narrowly escaped being kicked. Finally, after further misadventures, he managed to approach "a female with a calm temperament" and got his photos. The story includes three photographs of this okapi, one of a captive okapi, and a photo of a calf at the Buta mission "making friends with Lady May Cambridge" who appears wearing dark glasses and pith helmet.[21]

The first live okapi to arrive in Britain reached London in July 1935, a gift from the Duke of Brabant (newly crowned King Leopold III of the Belgians) to the Prince of Wales (who became King Edward VIII). Named Congo, he was featured in *The Times* and elsewhere. Photographs showed him in captivity in the Congo, in Antwerp, and then at London Zoo (the prince gifted him to the zoo). He became a huge zoo attraction over the summer, notably on a particularly hot August bank holiday that year. Congo sadly did not live long, but a photograph of Zoological Society of London's second okapi, named Buta, who was delivered in 1937, featured in the zoo's official guide beginning in 1938.[22]

The Italian American Attilio Gatti, who undertook two expeditions to try to capture and photograph okapi along with other rare mammals in the mid- to late 1930s, published photographs of okapi in his books *Great Mother Forest* (1936) and *South of the Sahara* (1946). These included adult okapi captured in pit traps, and several photographs of a live calf captured for him through the efforts of the Mbuti man Makulu-kulu.[23]

By the 1930s, then, okapi could be seen in published photographs and in a few European and American zoos, and so they were no longer as mysterious as they had once been. In addition to photographs, the first film footage of captive okapis was shown.

Many of these early zoo okapi died relatively young, however, and some European okapi died during the Second World War, when it was not possible to import more. In 1946, London Zoo's male okapi, Buta, chosen from among three okapi available in the zoo in Antwerp in summer 1937 to be gifted to King George VI

Figure 5.4. Allegedly the first okapi photographed in the wild, in the Ituri forest, by Cornelius Bezuidenhout (*Illustrated London News*, July 11, 1931, © Mary Evans Picture Library).

by King Leopold III of the Belgians, was one of only two okapi remaining alive in captivity in Europe. In the postwar period, once again, okapi were therefore difficult to see in the flesh outside of the Belgian Congo.[24]

The editorial to the first issue of *Zoo Life*, an illustrated bulletin launched by the Zoological Society of London in 1946, announces that "the main features of each issue will be an article on some animal of special interest, because of its rarity or novelty, with a photographic illustration." The cover of the first edition features a painting of a Markhor (*Capra falconeri*, a wild goat), but the lead story is on the okapi and is accompanied by a large photo of Buta, a "loose coloured art plate" of whom appears in the second issue.[25]

The circulation of okapi imagery in the media did not necessarily mean that Westerners flocked to zoos to see okapi in the flesh, however. While Buta and Congo had attracted large crowds on their arrival in Antwerp and London, as had the mounted specimens in London and Tervuren previously, the Paris Zoological Park found at least initially that most visitors ignored its okapi. Learning from this, Basel's zoo staged the arrival of its first okapi, Bambe, in 1949, "on a grand scale," ensuring it was photographed and filmed. There is a wonderful photograph of a smiling Jean de Medina (Congolese director of the okapi capture station in Epulu in the Belgian Congo) at the zoo in December 1957. He stands in an enclosure with one arm resting on Nanuk, the zoo's second okapi (a male). Partially visible to the right of the shot is the zoo's new female, Bibi.[26]

As a result of the availability of photographs and films, artworks depicting okapi were now usually accurate—for instance, those on the stamps issued by Antwerp Zoo for fundraising in 1961 (six rare species, including okapi) and for the zoo's 150th anniversary in 1992.[27] There is a wonderful photographic record of all of the okapis kept at Antwerp Zoo between 1919 and 1970 in the publication "Seventy Years Okapi," including one of Besobe (who answered to Bela), a male okapi with a distinctive right horn with a twisted tip who was the model for the okapi emblem adopted for publications of the Royal Zoological Society of Antwerp. A quiet animal, Besobe was never aggressive with his mate Dasegela, and sired four calves. The photo of their calf Mafuta shows the fluffy coat characteristic of a juvenile, and the longer white hairs (longer than the dark hairs) of infant okapi.[28]

Photographs of wild okapi remain rare. While camera trapping—photographing wildlife, especially rare or shy and nocturnal animals, using automated time-lapse or motion-triggered cameras left unattended in the field—has become a mainstay for wildlife research, especially from circa 2005/06 when digital cameras became mainstream, the first okapi caught on camera in this way was still newsworthy and only happened in 2008.[29]

Reel Okapis

The famous husband-and-wife wildlife filmmakers Martin and Osa Johnson failed to film okapi in the Ituri forest in 1930 or even photograph them, despite using flash-trap cameras left where fresh tracks had been found in the forest. They did buy some badly cured okapi skins from Mbuti pygmies they filmed.[30]

Another husband-and-wife team, the Pearsons, claimed their 1938 film *Jungle Adventure* was the first to show the okapi in its native habitat (I have not located this footage).[31] Footage shot during the American Museum of Natural History's William D. Campbell African expedition in 1938 shows an okapi pen, and includes a still of a dead okapi and one of the cleaning of its skeleton and preparation of the skin.[32]

Gatti includes about twenty seconds of an okapi browsing (after a staged stalk with expedition members led by pygmies) in his 1940 film *Jungle Yachts in the Belgian Congo*. Belgian-born wildlife filmmaker Armand Denis visited the American anthropologist Patrick Putnam in the Ituri forest in 1946 hoping to film wild okapi but saw only captives, though he accompanied de Medina on capture operations. Denis filmed some captured okapi, including a male he helped transport to captivity from a pit trap, but he was not satisfied with his work, noting he'd not managed to film one using its long blue tongue to pluck leaves off trees.[33]

American, British, and Belgian filmmakers who obtained permission to film in the Congo were essentially the captives of their guides, advisors, and facilitators, and avoided criticizing the colony or its operations, including mentioning, it seems, the high death tolls of okapi capture and export operations. A propaganda film directed by André Cauvin titled *Bwana Kitoko* funded by the Belgian colonial office recorded King Baudouin's visit to the Congo in 1955. It includes an extended sequence on capturing okapi that was staged using captive animals. It only portrays the successes of the Epulu capture center.[34]

Another significant film with a royal connection was *Les seigneurs de la forêt* (1958), directed by Henry Barndt and Heinz Sielmann, released in English as *Masters of the Congo Jungle* (1960) with sonorous voice-overs by William Warfield and Orson Welles. It was motivated and part funded by the abdicated King Leopold III (who had been forced to give way to Baudouin in 1951 following controversies dating back to his surrender to the Nazis in World War II), who maintained a keen interest in the region's wildlife and ethnography. The film was shot over two years and includes a thirty-seven-second sequence of an apparently wild okapi male walking through a rainforest and plucking leaves with his long tongue.[35]

Of later films, a 1991 documentary titled, contra Joseph Conrad, *Heart of Brightness* by the Kenya-born wildlife filmmaker Alan Root merits brief description. If anyone could capture an elusive animal on film, it was Root, a dedicated naturalist and innovative filmmaker. In the late 1980s, he began work on an okapi documentary. He knew how difficult it would be, but he was encouraged because in 1986 an American couple, John and Terese Hart, had begun working with Mbuti pygmies in the Ituri forest to trap, radio-collar, and track wild okapi. At the time, American conservation NGOs were helping to revive okapi research in the region. In addition to the Wildlife Conservation Society's funded tracking study, Gilman International Conservation (which subsequently became the Okapi Conservation Project), were reactivating the old okapi capture station established by the Belgians in Epulu.

Because Root knew his chances of filming wild okapi were slim, he decided to instead make a film about the Harts' research and their life with their three children and the Mbuti in the rainforest. For the Harts' capture operation, two hundred narrow pits were dug along rainforest paths, and carefully covered with a lattice of sticks disguised with large *mangongo* leaves and a scattering of leaf litter (Europeans not infrequently fell into these pits to the Mbuti's great amusement). A number of okapi were caught, collared, and tracked, and as a result, scientific knowledge of their movements and territory size grew. The NGOs' capture operation, run by Swiss couple Karl and Rosmarie Ruf, also had several okapi that Root could film in enclosures in the forest.

The film (aired on BBC television in 1991) provides a wonderful record of the Harts' time with the Mbuti, celebrates the pygmies' encyclopedic knowledge of the forest, and reveals the rich biodiversity of "the heart of brightness." However, there is little footage of okapi. Most of it shows okapis in pits being fitted with collars or being freed from the pits. As Root writes in his autobiography, "None of us was ever able to observe [okapi] in the wild and all our behavioural information came from the captive okapi."[36]

Media and Popular Responses to Okapi in Britain

A perusal of articles published in the British newspaper of record, *The Times*, over the course of the twentieth century reveals how okapis entered mainstream print culture in England. The larger project of exploring how people responded to the discovery and exhibition of okapi and how they entered the public consciousness in Africa and especially the Democratic Republic of Congo and in Europe and the United States is one for the future.

In the English print media, reactions to the discovery and presentation of the okapi run the gamut from the serious and scientific to the jocular and utilitarian. A May 1901 *Times* article on "the new mammal" commences with a reference to a *Punch* magazine drawing of an overfed footman "grumbling over the constant reiteration of beef and mutton, and crying out for the invention of a 'new hannimal.' If the Jeames of the early '60s is alive today, his wish has been gratified," the writer jokes, because Johnston had discovered just such a new mammal.[37]

In the course of a long discussion of the okapi's taxonomical relationships, and a detailed description of its appearance (based on Johnston's painting), the article also reports that "its flesh is said to be excellent eating, and there is no reason why an attempt should not be made to domesticate it." Another *Times* article observes that it was unsurprising okapi were "excellent eating," as the species' relative the giraffe was considered delicious too (explorer of the Nile Sir Samuel Baker was cited as an authority on roasted giraffe).[38]

Although the okapi was generally agreed to be of striking appearance, this did not necessarily mean it was admired. Describing Johnston's first painting (see front cover), a writer in *The Times* opines that "its general appearance is not very graceful" and that "indeed, in spite of its vivid and varied colouring, it is in form rather outlandish and uncouth." Another writer who had seen the first mounted specimen on view at the Natural History Museum refers to it as "one of the very oddest forms of animal life in a country teeming with strange and singular creatures." This writer characterizes it as "a blending of three different mammals—an antelope, giraffe, and a zebra," arguing that it has a "shape and body colouring resembling hartebeest, a giraffe-like head, and zebra stripes on the legs" and that nonetheless it is "beyond all doubt a true species" and "not a hybrid or a 'sport' of any kind." However, where the eye of the giraffe is "large, tender, and melting," the eye of the okapi is "small, dark, protruding, and unlovely." The writer claims that it "can scarcely be called a handsome contribution to the fauna of Africa," as it lacks the magnificence of eland, kudu, or sable, or the stately beauty of the giraffe. It is a rather primitive beast, similar to the tapir, "a singular and bizarre instance of arrested development."[39]

The okapi was known to readers of *The Times* by spring 1901 and to visitors to the Natural History Museum in London as of October 1901. A November 12, 1902, *Punch* cartoon features a portly Ray Lankester in top hat and tails astride an okapi resembling Ward's mount of Johnston's okapi (see fig. 1.2). The male okapi Congo arrived at London Zoo in 1935 and had his photograph published in British newspapers, but he didn't survive long. Buta arrived in 1937 and fared rather better (he died in 1950), and photographs and notes on his progress at the zoo were published in *The Times* and other British newspapers.

By the interwar years, the okapi had become a symbol of rarity and timidity in British popular culture. In 1938, visitors to London Zoo were informed that "the elusive okapi can always be seen by ringing the bell for the keeper. A strip of glass in the wooden partition allows visitors to get a view of the animal without disturbing his privacy unduly." Somewhat impatiently, the article adds that "after 18 months in captivity he is still in excellent condition, but he shows the temperamental shyness of his species in what appears to be an exaggerated individual form." Articles published in the *Star* and the *Evening Standard* during World War II similarly remark on his timidity and nervous disposition. Much later, the English writer and humorist Alan Coren compared two nervous diners to "timorous okapi."[40]

In an article celebrating "the fiftieth anniversary of the introduction of that daily obstacle course of the wit that has become a national institution, *The Times* crossword puzzle," Philip Howard refers to "the okapi, invaluable five-letter word beginning and ending with a vowel," describing the animal itself as "the elusive okapi." The obituary for Edmund Akenhead, long-time editor of *The Times* crossword (1965–83), documents his rigor in instructing crossword compilers on how to create clues, quoting him as telling them that "the dictionaries do not define okapi as a giraffe. The fact that it is related to the giraffe does not make it one."[41]

In describing the middle class as a "doomed institution" in a November 1966 article, the Jamaica-born art critic Edward Lucie-Smith claims "the solid citizen will soon be rarer than the okapi." Not everyone saw rarity as demanding conservationists' attentions, however: political writer and columnist Matthew Parris recalled seeing okapis at the zoo in Antwerp and thinking how miserable they looked in "the Belgian drizzle." "The okapis," he claims, "realised that it was all over bar the shouting." What was the point of struggling on "if you were a moth-eaten traveler down a dead-end evolutionary track?" Okapi should "fight for the right to extinction" rather than suffer "survival, with ignominy."

Parris returns to okapi analogies in a parliamentary political sketch titled "Labour's Okapi Gallops towards a Darwinian Dead-End." He then proceeds to commit a natural history howler in describing okapi as having "the markings of a faded zebra, the body of a stunted giraffe and the long tongue of an anteater . . . perfectly adapted to lick termites off high branches in the Congo rainforest." Parris was straining to make an analogy between an animal doomed to extinction and the UK's then-secretary of state for health, a socialist and in Parris's opinion an endangered species in Tony Blair's new Labour Party cabinet. Belgian ambassador Lode Willems wrote a letter back assuring Parris that the okapi was doing well in the Antwerp zoo, "with 30 born there over the past 26 years."[42]

Andy Warhol Introduces Okapis to Western Art

I was surprised to discover the long association of okapis with New York (through zoos and museum displays and collections). I was truly astonished to discover that a priest of pop art, Andy Warhol, had created artworks of endangered species—including okapi. As a young man more enamored of abstract expressionism, I had made the usual assumptions about pop art and Warhol's blankness and lack of depth. Certainly, his famous images from the early 1960s didn't suggest any interest in the natural world.

In 1973, ten years after Warhol had made his big artistic breakthroughs and moved his studio to East 47th Street, the Endangered Species Act came into effect in the United States. Another decade later, Warhol produced ten large psychedelic silkscreen prints of endangered animals, including three African mammals but no okapi. This portfolio was commissioned by art dealers Frayda and Ronald Feldman following conversations with Warhol about ecological issues.

Three years later, in 1986 Warhol produced sixteen more prints (silkscreen over collage) for a book with Kurt Benirschke, an animal pathologist and former director of research at the San Diego Zoo. Titled *Vanishing Animals*, it was a call to arms to save the world's endangered species. Quoting Chief Seattle and E. O. Wilson, the introduction warns that extinctions were accelerating due to human influence and that natural evolution could not keep up. The book highlights the plight of "some of the less well known endangered animals," including the okapi. Although it is number six on the list, the okapi features on the cover, a reproduction of a line drawing of a mother and calf from Warhol's colored print.[43]

Each of the sixteen species receives a chapter featuring Warhol's artwork, and photographs of the animals, with text aimed at the lay reader written by Benirschke. He begins his okapi chapter by remarking that even erudite friends were unaware of its existence. He briefly describes its discovery (mistaking Sclater's son for his father—Philip was never "curator of the Capetown Museum"), noting it was first assumed to be an equid but that on arrival of the first skin and skulls in London it was "quickly renamed for various complex reasons, not all necessarily logical." (Johnston, however, as we've seen, had already identified it as a giraffid and not an equid based on its teeth and not, as Benirschke claims, on its cloven hooves, as hooves were missing from Johnston's skin.)[44]

The text is informative regarding efforts to transport okapi to the West and keep them in zoos (San Diego Zoo had five at the time). Warhol's print shows an okapi mother and calf, their outlines and markings sketched with white lines, tinted also with blue and green on a black ground, partially framed with bars of red, pink, yellow, green and blue. We see the mother from the front, her head and neck turned to face left (her right), her calf side-on in front of her, its head turned to look backwards, also facing left. The book also features two photographs, one of an okapi suckling from its mother and one of a calf wearing a harness to prevent injury from its mother's tongue. Warhol's position on the okapi and endangered species conservation remains opaque, which is entirely in character. But he certainly gifted this little-known species its fifteen minutes of fame in the 1980s.[45]

Despite their former prominence in museum displays and short-lived fame in the art world, however, it seems that okapi have slipped back into a rainforest-like obscurity, elusive as they ever were. They have disappeared in plain sight, from famous venues that once displayed them like the British Natural History Museum, and generally from the public consciousness in the West, slipping away more than a hundred years after they were named, categorized, and first displayed in the capitol cities of Europe.

CHAPTER SIX

Okapis in African Art, Ancient and Modern

The okapi was recognized—perhaps even by the ancient Egyptians—and was part of a far older imperial network. Okapis were depicted on a frieze of a procession of delegates paying homage to Xerxes in the early fifth century BCE. . . . This okapi had traveled three thousand miles from central Africa to the Mediterranean world.

—Sandra Swart, 2020

Inspired perhaps by the striking appearance of the okapi, and the sensation surrounding its discovery by the West during a golden period for Egyptian archaeology, a curious legend emerged alleging that it was known to the Ancient Egyptians. Any account of okapi in African art needs to take account of this enduring suggestion that it appears in Ancient Egyptian artworks, also because of the associated idea that okapi actually inhabited a once more fertile Nile valley. Two lesser-known stories allege that okapi were known to the Achaemenids and appear in early Hebrew texts, but the evidence for this is even more slender.

In the period in which African artifacts and artworks were collected by Europeans in the Congo basin region, strips of okapi hide appear on significant objects, but okapi are not themselves represented. Anne Eisner and her husband Patrick Putnam collected artworks including bark paintings in the Ituri Forest and the wider region (c. 1946–58), and this collection provides a means to investigate whether local Africans depicted okapi in their art, figuratively or abstractly through patterns referring to their striking patterning.[1]

Ancient Egyptian Okapis?

In the early 1800s, Henry Salt, the British consul general in Egypt, who arranged for a giraffe to be sent to King George IV, had a lucrative side hustle in the Egyptian antiquities trade. The British Museum were his eager buyers. European powers—notably France, Germany (Prussia), and Britain—had been involved in discovering and removing Egypt's antiquities for decades (some through purchases, some through gifts, many through looting). This extraction accelerated after the brilliant young French Egyptologist Gaston Maspero took over the director-generalship of Egypt's antiquities service in 1881, as he actively encouraged foreigners to excavate in Egypt. He aimed to control rather than halt the antiquities trade.

Britain invaded Egypt in 1882, installing Tewfik as khedive but ruling through Sir Evelyn Baring (later Lord Cromer), who took control the following year. British control of Egypt enhanced the influence of British collectors and excavators and raised the profile of Egyptology in Britain. Well-heeled tourists took Cook's Tours up the Nile to see the ancient ruins. In the 1880s, Reginald Stuart Poole, keeper of coins and medals at the British Museum, urged the British public (and wealthy philanthropists) to save Egypt's antiquities (by which he meant find them and then safely remove them to the British Museum). Working with the formidable Amelia Edwards to save Egypt's heritage, in effect, from Egyptians (and other Europeans), he set up the Egypt Exploration Fund in 1882.

Ancient Egypt was a subject of great interest in Britain in this period. Biblical scholars, for example, hoped to find proof of Old Testament stories, a search that crossed over into science (of a sort). The eugenicist Francis Galton worked with the distinguished Egyptologist William Matthew Flinders Petrie on attempts to organize depictions of humans on Egyptian memorials and artworks into racial categories.

Just at the time that Johnston entered the rainforests of the Congo, Egyptology was going through a new period of discovery that lasted until 1914. Two chief inspectorates were created for the antiquities service, and with the help of Cromer, both ended up being headed by British archeologists, James Quibell and Howard Carter. At the same time, extensive natural history survey work was being conducted on the wildlife of Egypt, both extant and extinct. The British Museum was deeply involved in both sets of investigations.

All this may explain why a short note published in 1902 by the German Egyptologist Alfred Wiedemann, an associate professor at the University of Bonn, asserting that the newly discovered okapi had been known to the ancient Egyptians gained so much attention. Wiedemann was trying to solve a long-standing dispute in Egyptology: what species of animal was the beast sacred to the god Set-Typhon? (Set was identified with the Greek monster Typhon).[2]

In his note, Wiedemann compares a drawing based on Johnston's original painting of an okapi with two representations of the god Set, one from a monument of the time of Seti I (circa 1290–1279 BCE), and one from a bas-relief from the time of Hatsheput/Tuthmosis III (1479–1425 BCE). In the Middle Kingdom and the New Kingdom, the god Set was usually shown with the body of a man and head of an animal with long, squared-off ears and a curved snout. Weidemann argues all previous attempts to associate Set with a particular animal failed on particularities of their appearances. However, in his view, details of the head and body of the okapi fitted exactly. Weidemann notes the disappearance of larger fauna from the Nile valley and claims that okapi had once lived there but must have fled hunting pressure into the surrounding desert.[3]

Wiedemann's note featured in the list of publications included in the 1904 *Zoological Record*, probably because of the long response it provoked from M.-G. (Claude) Gaillard, assistant naturalist at the city of Lyon's natural history museum. Gaillard's article includes a detailed and reasonably accurate summary of the state of knowledge on okapi at the time and compares measurements of various species in Lyon's museum to the okapi, concluding that the okapi could not be the sacred animal of Set-Typhon. He explains that for many Egyptologists, Set-Typhon personified the desert and sterility unlike Osiris and Isis, who represented the Nile and the fertile land of its valley. The contrast with the rainforests the okapi was known to inhabit seemed relevant—how could it possibly have survived in desert conditions?[4]

Egyptologists had identified various desert animals as figurations of Set-Typhon, including the fox, the gerbil, the elephant shrew, the camel, and the giraffe. Gaillard allows the okapi was a plausible match for New Kingdom theriocephalic images (animal heads on human bodies). However, he notes that in earlier representations, like the tombs of Beni Hassan (Old Kingdom and Middle Kingdom), the sacred animal of Set is shown in its entirety and does not resemble an okapi. Gaillard agrees with Eugène Lefébure that it much more closely resembles a greyhound.[5]

Gaillard argues the earlier images are decisive because the later ones are stylized evolutions of the former. Even a comparison based only on the head raises problems for the okapi theory, because the frontal horns (known from Charles Forsyth Major's 1902 paper) are missing (the god is male) and the ears are squared off and not rounded like an okapi's. Given the remarkable exactness with which the ancient Egyptians portrayed the species of wildlife they lived alongside, Gaillard argues, it is unlikely they would have gotten such details wrong.[6]

Gaillard's paper should have put Wiedemann's passing suggestion to rest. Certainly, Fraipont says so in his monograph (1907), also citing the French architect and Egyptologist Amans Paul Hippolyte Boussac's article of 1907. However, the tying together of an amazing zoological discovery with the ongoing European fascination with Egyptology was such a striking story that the association of the okapi with Set and the idea that the ancient Egyptians knew about the animal persisted.[7]

American professor of Egyptology James Henry Breasted repeats it in his book *A History of the Ancient Egyptians* (1908), and Boussac was still trying to discredit the story in 1920. Thirty years later, Ludwig Keimer, a student of ancient Egyptian fauna and flora taught by the German explorer Georg Schweinfurth (who associated the Set animal with a kind of wild pig), claims in an article published in *Acta Tropica* that any resemblance to the okapi was coincidental as the animal had not lived in Egypt, not even in prehistoric times. The reason he had bothered to write the article, he states, was "solely because A. Wiedemann's assertion was accepted by great scholars and lives on in their works."[8]

In her 1959 monograph *Das Okapi*, Agatha Gijzen, then a zoologist at the Antwerp Zoo, summarized five recent publications on this supposed link with ancient Egypt. She notes Keimer's objections but remains noncommittal on the claims for such a link made by three other authors. She includes five line drawings to demonstrate supposed resemblances.

One drawing is a copy of a rock carving from the Libyan desert, resembling a rather short-bodied and long-necked okapi, another is a copy of a rock carving from In Habeter in the Sahara that clearly represents a giraffe but with a shortened neck and elongated head and ears. Another is a drawing of rock art from an unnamed location in South Africa bearing a supposed resemblance to an extinct short-necked giraffid (no rock paintings old enough to have depicted a living short-necked giraffe have been found). These drawings are printed alongside an example of an ancient Egyptian representation of Set. In their book on the okapi, Susan Lindsey, Mary Green, and Cynthia Bennett say only that "according to some Egyptologists, certain representations of the god Seth-Typhon resemble an okapi or giraffe."[9]

As is the case with many myths about animals, this story—like Herodotus's story of the myth of the plover that cleans the Nile crocodile's teeth—is just too entertaining, it seems, to be debunked.

Achaemenid and Abrahamic Okapis?

A number of sources claim that a live okapi was given in tribute to the great king of the Achaemenid dynasty (559–331 BCE), possibly Darius I, by Africans variously described as Ethiopians, Kushites, and Nubians. This was memorialized, the sources state, in a carving on the eastern staircase of the Apadana, the audience hall of the king's palace in Persepolis, in what is now Iran. It could also represent a badly carved giraffe with its neck shrunk to fit the available space, or as others have argued, a nilgai (*Boselaphus tragocamelus*), which is found in Pakistan and India (although it is unclear why Africans would bring one as tribute). Even if aspects of the body and neck of the carved animal are giraffid and the head fairly credible, the body has no stripes, the horns are in the wrong position, and it has a mane (adult okapi do not, whereas nilgai do). The debate is unresolved.[10]

One scholar tried to link the okapi to Abrahamic religious texts: S. M. Perlmann asked Walter Rothschild to task librarians of his family's notable Hebrew library to prove that the okapi is "the same animal called Tachash in the Bible." However, there is no record of an answer to Perlmann's request, though he did manage to publish a short piece on his theory in the illustrated London-based journal *Zoologist* in 1908.[11]

South African environmental historian Sandra Swart explains that the okapi was the symbol of the International Society of Cryptozoology (1982–98), because "the okapi is more than an animal; it is a symbol of an even rarer and move elusive thing: the embodiment of other possible worlds." If that is so, then the tales of Egyptian and eastern Mediterranean okapis are instances of a failure to progress beyond discovery of myths that indicate the possible existence of undiscovered creatures, that is, a failure to demythify "the content of received information," which is the necessary corollary of cryptozoologist Bernard Heuvelmans's twofold mission to demonstrate "a true scepticism, that which opposes both an *a priori* incredulity, and a naive willingness to believe" when confronted with such myths.[12]

Harry Johnston was first inspired to search for the okapi by stories of unknown animals in the Congo rainforests, possibly the unicorn. He was convinced of its existence by what Heuvelmans terms "parataxa" (fragmentary evidence, verbal or physical, for a possible new living animal) in the form of anecdotes and then bandoliers. However, he soon progressed to more complete evidence, using this to dispense with fanciful versions of a "rainforest horse" or "living fossil equid." Heuvelmans therefore claimed its discovery as a triumph for cryptozoology, and the okapi was adopted as the icon of the shortlived International Society of Cryptology.[13]

The okapi is surely remarkable and singular enough even if it wasn't either a unicorn or known to the ancients and inscribed on their monuments?

Local African Depictions and Uses of Okapi, Early 1900s

In *African Reflections*, a thoughtful survey of art mostly by the Mangbetu people of northeastern Democratic Republic of Congo, Enid Schildkrout and Curtis Keim explain the complexity of attributing artworks to specific ethnic groups or localities in the region owing to both precolonial movements of peoples and cultural interactions of African societies in the region and Arabic influence through the slave trade. Encounters with Europeans beginning in around 1870 also influenced Mangbetu styles.[14]

Schweinfurth presented a romantic image of the Mangbetu as a strong, centralized, and regionally dominant kingdom with a sophisticated court and artistic culture. Later European collectors influenced by his account attributed regional styles to the Mangbetu. They were impressed by the figurative art of the region and regarded it as evidence of a higher artistic development. But they confused artwork reflecting court fashions of the turn of the century with a "timeless" African style, assuming African art was not made for its own sake but rather in the service of deeper and enduring ritual or spiritual traditions.[15]

Herbert Lang and James Chapin led the American Museum of Natural History's Congo expedition (1909–15), and their notes show that, like other European collectors, they influenced the artifacts they collected. Numerous artworks were made for them, particularly figurative objects. What of depictions of wildlife, and the okapi?

Schildkrout and Keim's book draws on the American Museum of Natural History's incredible hoard of objects from the Congo basin region. This includes about 3,500 objects from the Congo of which eight hundred came from the Uélé region gifted to the museum by King Leopold II of the Belgians in 1907 and around four thousand objects collected by Lang and Chapin. Lang's Anthropology notebook 1 records acquiring *bandaka* swords with shoulder straps made of okapi skin in October 1909, apparently the first okapi artifacts Lang acquired.[16]

African Reflections includes photographs of four objects made from okapi skin. These are a hat, an ivory harp with an okapi-skin-covered soundbox, a belt with attachments intended to protect a child from witchcraft, and a ninety-three inches (238 centimeters) long chief's ivory horn with bindings of okapi hide, for use in battle. The authors note the convention that all okapi hides were to be handed to the chief or king, especially in areas outside of its natural range. Okapi were rare and regarded as symbols of aristocratic power. Okapi skin belts were also, according to Schildkrout and Keim, worn by dancers. The museum's website includes photographs of a few additional objects that incorporate okapi skin, including ivory horns, a belt, and a shoulder strap for a rifle.[17]

Significantly, only one depiction of the animal itself, a carving of a young okapi in profile incised into an ivory horn by an unnamed Azande artist (possibly a man named Saza) is among the expedition's collection. This horn is not reproduced in *African Reflections* but can be found in the digitized collection on the museum's website. It seems possible that the artist created the ivory horn for Lang, who everyone knew was trying desperately to find an okapi. It may even be a carving of the okapi calf eventually captured with the help of the Azande hunter Abawe in the final months of the expedition. The horn was acquired in 1915, which was the year an okapi was finally caught alive.[18]

Certainly, the okapi was among the significant animals whose body parts were used by locals to create artworks in what is now northeastern Democratic Republic of Congo. Their skins could symbolize social power and prestige, and they were incorporated into objects intended to have magical powers as well as used to beautify musical instruments and to adorn as belts, straps, and hats. There is scant evidence of artworks depicting okapi figuratively in this period, however. Overall, few animals are depicted on the artworks collected by Lang, aside from carvings on a few ivory horns and other objects, including a drum carved to resemble a cow. Others searching for okapi in the first half of the twentieth century found only artifacts made from okapi skin.[19]

It appears that it was not only figurative art depicting humans that had been influenced by the desires of Western visitors to the region. Westerners were evidently more in need of images depicting okapis than the locals were.

Primitivism, Anne Eisner, and Bark Cloth Painting

A survey of okapis in African art after 1945 is beyond the scope of this book, but one particularly intriguing story revolves around bark cloth painting, its long history among the peoples of the region that features aesthetic cross-fertilization between pygmy and farmer groups, and the influence of Western collectors. Where it seems clear that Lang had influenced Azande artists to introduce figurative elements into their work—he had even tried tutoring them in drawing animals—mutual influence between Mbuti and Western artists is harder to determine.

An examination of American painter Anne Eisner and the Mbuti women creating bark cloth artworks in Epulu in the Ituri Forest offers insights. Eisner was a New York artist, active from the 1930s and part of the dynamic 1950s art scene when New York became the epicenter of modernist painting. Jackson Pollock and other Abstract expressionists pursued abstraction but also dream symbolism, and artists were also interested in what has been called primitivism in this period.

Primitivism in art is a vexed subject: was it a deep engagement with other cultures and sacred visions or, as Adam Kuper and others suggest, more a means for European artists born in the 1880s to distinguish themselves from existing Western bourgeois ideas about fine art? Picasso claimed that, unlike his peers, including Braque and Matisse, who, he maintained, just saw African carvings as sculptures, he understood the spiritual dimensions of the African carvings that became so collectable in Paris in the 1890s. However, on other occasions, he also described those carvings he had in his studio as witnesses to his creativity rather than models for generating it.

Robert Goldwater argues that the masks and carvings depicted in the paintings of the postimpressionists like Gauguin are allusions, and he distinguishes between European ideas about primitivism in art, and the primitive—the primitive being, he claims, the subject of anthropologists rather than artists and art critics.[20]

Whatever the case, French dealers in "primitive art" had set up shop in New York by the 1910s, and in 1914, Alfred Stieglitz's gallery exhibited African wood carvings that it described as the "roots of modern art," following up with a show juxtaposing works by Picasso and Braque with African sculptures. In 1923, the Brooklyn Museum put on an exhibition titled *Primitive Negro Art, Chiefly from the Belgian Congo*, and the Museum of Modern Art put on a show titled *African Negro Art* in 1935. Claude Lévi-Strauss haunted antique shops in Greenwich Village looking for African and American masks and sculptures after he arrived in New York City in 1939. By the 1940s, New York had replaced Paris as the major market for primitive art, and Nelson D. Rockefeller was a major collector. Anne Eisner would have been aware of African art and ideas about primitivism long before she moved to Africa herself.[21]

In 1945 when she was on Martha's Vineyard, Eisner met and fell in love with Patrick Putnam, by then a long-time Congo resident. She first saw him swimming in the surf, and was impressed by this "giant of a man" who was "raw-hide thin, with a strong face and a devilish glint in his clear blue eyes." She was introduced and soon learned about his life in the Belgian Congo, later remarking in her book about her life with him that he was "easy to like," a "romantic figure, full of strange tales of strange places," and in the heady island summer atmosphere they resolved to marry and travel to Africa. To her friends' consternation, she traveled to West Africa with Putnam in 1946, and from there they drove

Figure 6.1. Anne Eisner bottle-feeding an okapi at Camp Putnam (Anne Eisner Putnam, *Eight Years with Congo Pigmies* [London: Hutchison, 1954], opposite 49).

down to the Ituri forest in northeastern Belgian Congo together. She married him in 1948 (she was his third American wife), despite discovering he had Mbuti wives, too.[22]

Although Putnam had originally been drawn to the Congo as an anthropologist, he ended up establishing a settlement known as Camp Putnam in Epulu, where he ran a medical clinic, the only one in the region. Eisner helped run the accommodations she and Putnam maintained for visitors like anthropologist Colin Turnbull, whom Putnam had encouraged to study the local pygmy peoples. Eisner also helped nurture the menagerie of wild animals Putnam kept and bottle-fed an okapi calf (see fig. 6.1).[23]

Eisner dealt with her frustrations around sharing Putnam with his other wives by escaping for days at a time to paint and share daily life with Mbuti pygmies in their camps. She became as well acquainted as Putnam with their lives and customs, compiled a collection of their folktales, and was an important (though little acknowledged) influence on Turnbull's anthropological work. She stayed on in Epulu after Putnam's death in 1953 and tried to keep the camp running, but returned to the United States permanently in 1958.[24]

Camp Putnam was in financial dire straits when she arrived, so she and Patrick began collecting artworks across the Congo, with a view to selling them as a way to support the camp. They did not encourage African artists to create more tourist-friendly art featuring representations of African peoples and wildlife, though they did collect curios created specifically for tourists. They classified the artifacts they collected as either imitations of genuine artworks and implements created for sale; or "genuine" museum pieces. They were particularly interested in bark paintings, and they encouraged local women in Epulu who made these.[25]

Bark cloth paintings are created by stripping bark from various trees and creepers and softening them in water or over a fire. They are then hammered out (Turnbull published a photograph of an Mbuti man doing this with a hafted ivory tusk hammer). These cloths might be dyed with mud, or what Turnbull refers to as red *nkula* paste, before being painted with black or blue vegetable pigments applied with reeds, twigs, or fingers. Women do all the decorating. The cloths are worn by men and women and are used to catch babies as they are born and to carry babies.[26]

The patterning is abstract and striking and not directly representational. The rectangular or square strips of cloth feature repetitive geometric shapes, for example, dots, triangles, or lozenges. Suzanne Blier suggests they may provide cartographies of the rainforest and life in the forest, though it is hard to tell if they refer to specific places, settlements, and incidents or are more generic in their references.

A few of these paintings feature stripe motifs that may represent okapi pelt patterns. Bark cloths decorated using okapi motifs or those of other powerful animals like leopards were typically worn by men. Although it might be expected that okapi would be widely represented in bark cloth painting undertaken in Epulu given the focus on okapi capture there, this doesn't seem to have been the case in any obvious way.

Neither Eisner nor local women's artworks include figurative representations of this iconic animal. The extent to which Eisner influenced local women is difficult to establish. She was certainly influenced by local women artists, and bark cloth motifs appear in her work from this period. The motifs that may represent okapis in regional bark cloth paintings predate Eisner's visit.

Blier speculates that the depiction of motifs related to ungulates like antelope and okapi in bark cloth paintings produced at the turn of the nineteenth century may reflect increased pressure on pygmies to supply bushmeat to villagers at this time following food shortages that the colonial imposition of commercial rather than food crop growing brought. A more systematic historical study of bark cloth paintings could enable more precise statements on the influence of external events on African art traditions.[27]

Eisner was instrumental in changing how pygmy peoples were studied by Westerners. In her influence on Turnbull, through her book, and through Turnbull's famous book *The Forest People* (1961), she helped encourage physical anthropologists to move away from focusing on measuring pygmies and studying them as a "primitive race" of humans, and toward developing an aesthetic appreciation of their arts and culture. Her appreciation of their lifestyles and visual arts and Turnbull's accounts of their music, customs, and philosophies opened up a more general interest in these peoples as artists, social beings, and thinkers.[28]

Eisner was aware of the difficulties of representing her Mbuti friends to an American public, and she waged a sometimes bitter and only partially successful campaign against the cliches perpetrated by her ghost writer Allan Keller in their book about her life at Camp Putnam, *Eight Years with Congo Pygmies*, published in 1955. While we learn nothing about Mbuti depictions of okapi from this book, we do learn about their deep knowledge of the forest, including okapi behavior.

Eisner's paintings do not strike me as romanticized depictions of forest life, although they have primitivist elements. While her paintings of the 1950s are influenced by the colors and flat fields of color or pattern of bark paintings, they remain strongly figurative. She develops her own, abstract response to the Ituri forest in her final paintings of the 1960s, some of which borrow the striped patterning appearing on bark cloth paintings. Although she nursed okapi at Epulu, she never depicted them. Her art is focused almost entirely on the human inhabitants of the rainforest rather than the wildlife of the region.

Considering all this, a photograph of the contemporary Mbuti artist Wendo's depiction of an okapi on a bark painting with a traditional abstract patterned background is intriguing (see fig. 6.2). Alas, more information about Wendo's intentions is unavailable (Chris Hamley bought these artworks from the artist outside the headquarters of the Okapi Wildlife Reserve in 2023).

Conclusion

Despite the best attempts of Western scholars to claim that the okapi was known far and wide across Africa and perhaps even beyond in what is now the

Figure 6.2. Mbuti artist Wendo with his bark cloth paintings, one (*displayed on bottom left of photo above, and enlarged below*) featuring an image of an okapi (© Chris Hamley).

Middle East, it is likely that with the exception of Africans living in the rainforest region of the Congo basin inhabited by okapi in what is now the Democratic Republic of Congo (DRC), very few Africans had seen an okapi or even an image of an okapi before colonial times. Although okapi skins were used to add potency to or to decorate ceremonial and other artifacts, the available evidence suggests that at least until Europeans arrived in search of representative artworks, Africans in the region did not create images of okapi. The one example collected by Herbert Lang may well have been created for him.

It would be interesting to know when Africans across the Belgian Congo would have first become familiar with the image of the okapi. Presumably this would have been through the publication of photographs in regional or national newspapers or magazines, or on postcards like the one on the cover of this book. Certainly, they would have become aware of them through watching propaganda films like *Bwana Kitoko* (1955) and *Les seigneurs de la forêt* (1958), and seeing their image on postage stamps. A stamp from 1955 commemorating the fifth international congress for African tourism in Elisabethville in the Belgian Congo reveals that the icon for the Belgian Congo's royal touring club was the okapi.[29]

Figurative representations of the okapi became well known across what is now the Democratic Republic of Congo after independence from Belgian rule, partly because of its international fame as a unique and endangered large mammal. It is on the official logo of the country's national conservation organization, the Institut Congolais pour la Conservation de la Nature. Park guards wear a shoulder roundel that includes an outline map of the country with an inset okapi head and shoulders.[30]

The okapi was featured on the country's postage stamps during the Mobutu era and has been used on stamps since it became the Democratic Republic of Congo. It appears on banknotes (e.g., the fifty-centime note issued in 1997, and the one-thousand-franc banknote issued in 2013), and its image is used on household products and by restaurants and companies including cocoa producers. The national radio network Radio Okapi, established in 2002, uses a cartoonish image of the head and neck of an okapi in its logo.[31]

CHAPTER SEVEN

Catching Okapi, 1901–15

News has just come to hand from Captain Boyd Alexander . . .
that he has procured a specimen of the okapi, and, what is more
important, has seen the animal alive. This, it is believed, is the first
instance of a white man ever having had the good fortune to come
across a living okapi.

—Daily Mail, *May 1906*

Sir, without wishing in the least to detract from the merits of
Captain Boyd Alexander's arduous journey across Africa, I venture
to point out to you that he is not . . . the first European who has
hunted the okapi in its native wilds. Dr J. J. David, a Swiss
naturalist . . . has already preformed this feat.

—*Philip Sclater, May 1906*

I spent some months at Makala, the post near which a native
hunter named Agukki shot the two specimens for Dr David.

—*Percy Powell-Cotton, September 1906*

The discovery of the okapi was followed by a scramble to secure physical evidence of its existence, but the greater prize was a live specimen. European and American scientific and hunting expeditions (the latter often masquerading as the former) to Central Africa now routinely included northeastern Congo on their itineraries. Westerners spent months, even years, trying to secure the elusive animal. This was a daunting proposition requiring substantial funds, diplomatic permissions, organizational skills, the capacity to cajole and negotiate with local officials and Africans, and large teams of African assistants.[1]

Africans were recruited to perform the day-to-day labor of running expeditions once they were in the rainforest, a difficult terrain and climate where outsiders fell victim to tropical diseases and other accidents, and in the first decades of their pursuit of the okapi, Westerners depended entirely on locals to find and capture the elusive giraffid. The extent to which they did so was admitted to varying degrees in their accounts of their exploits, the local African often remaining nameless, but on occasion particular African experts at finding and catching okapis were named. Either way, Africans are seen through the lenses of visiting Westerners. Their ethnic affiliations and social roles and even their names are often misunderstood or incompletely recorded.

I have recovered the names and images of as many of the mentioned experts as possible but must caution that the names used are reliant on Western historical sources and so are probably incorrectly spelled or wrong for other reasons; they are, however, what remain and are recoverable from the written records.

There were also a few men of so-called mixed racial origin who played an additional role as go-betweens gaining access and influence for Western expedition leaders with local Africans, or Africans recruited to assist expeditions. The most notable of whom in this first period of the quest for okapi was José Lopes.

Catch Me an Okapi

After Johnston discovered "the famous okapi—the crowning zoological discovery of the Nineteenth Century," as the *Standard* oddly expressed it (considering the first evidence was found in mid-1900), he found himself too busy to return to the Semliki (Watalingi) Forest to look for further evidence himself. Further, his health was failing (he had seizures related to blackwater fever on two occasions), and so he completed his tour of duty and returned to England in June 1901. Johnston returned to "applause" at a meeting of the Zoological Society of London "so full that it was difficult to get near the table on which two skulls of the much-discussed okapi . . . were exhibited." Johnston had left the field open for others to find an okapi, preferably alive.[2]

In March 1902, the Zoological Society of London council resolved to form an okapi committee "to consider the best steps to be taken for acquiring more information about the Okapi and the possibility of obtaining living specimens of it for the Society's Menagerie." The Committee for Research on the Okapi

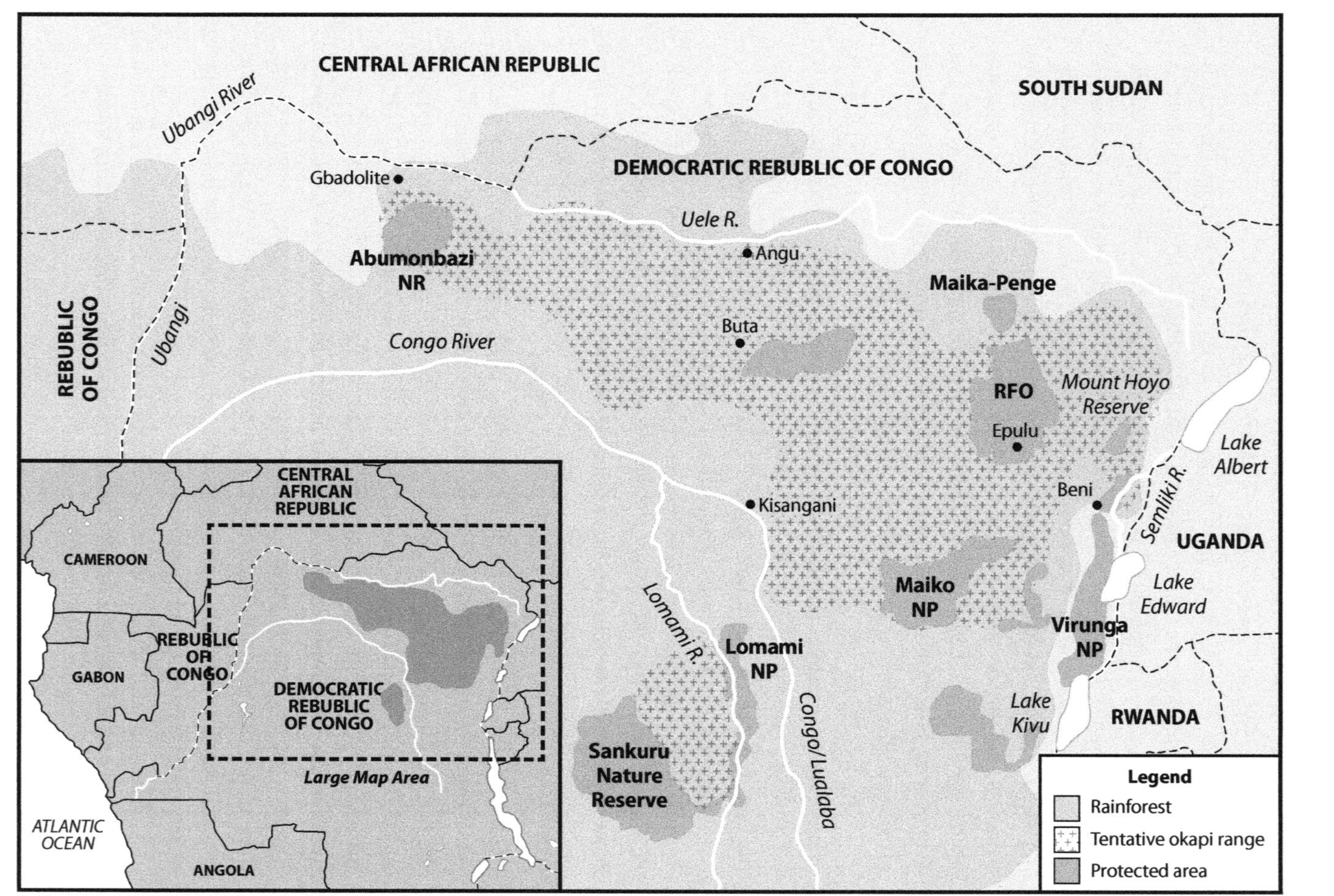

CENTRAL AFRICAN REPUBLIC
SOUTH SUDAN
DEMOCRATIC REBUBLIC OF CONGO
REBUBLIC OF CONGO
Ubangi River
Ubangi
Gbadolite
Uele R.
Angu
Abumonbazi NR
Maika-Penge
Buta
Congo River
RFO
Mount Hoyo Reserve
Epulu
Beni
Lake Albert
Semliki R.
UGANDA
Kisangani
Lake Edward
Lomami R.
Maiko NP
Virunga NP
Lomami NP
Congo/Lualaba
Lake Kivu
RWANDA
Sankuru Nature Reserve
CENTRAL AFRICAN REPUBLIC
CAMEROON
REBUBLIC OF CONGO
GABON
DEMOCRATIC REBUBLIC OF CONGO
Large Map Area
ATLANTIC OCEAN
ANGOLA
Legend
Rainforest
Tentative okapi range
Protected area

included Ray Lankester, Johnston, Major Charles Delmé-Radcliffe, zoologist John Samuel Budgett, Michael Oldfield Thomas from the Natural History Museum and Philip Sclater, the Zoological Society of London's secretary.[3]

One hundred pounds sterling was placed at the disposal of the committee, which they used to fund Budgett's expedition to the Uganda Protectorate in mid-1902. Budgett's purpose was primarily to study the life history of African fishes, but that included a planned detour into the Semliki Valley where he could explore the Semliki River and look for okapi. As he related in his talk to the Zoological Society about this expedition delivered in January 1903, however, Budgett would learn that there were no okapi in forests in British territory. Further, he reported, the Belgians were finding numerous okapi in the Welle (Uélé) region around Angu and Buta, much further west (see fig. 7.1 for locations).[4]

Budgett also learned that the Semliki River had only one species of the *Polypterus* fishes he was interested in, whereas the sources of the Nile further north had at least three. Furthermore, the Semliki Valley was an inconvenient place to camp due to a shortage of food, venturing into the rainforest would be difficult with his delicate equipment, and he maintained the grass was too high at the time of year to track large wild animals (it is unclear how this circumstance was relevant to forest-dwelling okapi). Therefore, he abandoned his plans to find okapi (never his focus) and chose to avoid the Semliki altogether, traveling north to join the upper reaches of the Nile instead. Budgett conscientiously refunded the hundred pounds with a note pointing out that there was in any case "not much left to be found out about the animal," given that the Belgians were sending specimens to Europe.[5]

The Zoological Society of London's okapi committee's plans had come to naught, but as Sclater correctly replied to Budgett, "You are quite in error in supposing that the subject of the okapi is exhausted. All that has been ascertained from the specimens sent to Brussels is that the adult male carries horns, but we still want further specimens in England and information of all sorts about the habits and life of this animal."[6]

Jingoism and Priority

The widely reported claims to priority in seeing and shooting okapi mostly left out those of Belgian officials working in the Congo because, as Auguste Lameere

Figure 7.1. (opposite) Map showing estimated current okapi distribution and key protected areas. Country names are the current ones. For scale, the Congo/Lualaba River is 2,720 miles (4,370 kilometers) long and Kisangani is nearly 1,000 miles upstream from the Atlantic (map by Jade Myers).

dryly notes in the conclusion to his 1903 paper on okapi, it wasn't that the okapi was unknown to these officials but that they were "unknown to science." He reports the exclamation of a Belgian officer on returning from the Congo and being shown the taxidermized okapi at Tervuren: "That! I have been eating it for ten years!"

The authorities of the independent state authorized, "with a generosity which has not always been repaid, four foreign missions for the purpose of hunting the Okapi," Fraipont testily remarks in his 1907 monograph. Subsequent expeditions which collected okapi remains up until the outbreak of World War I interrupted collecting efforts included those of Duke Adolf Friedrich von Mecklenburg-Schwerin, the Italian Captain Emilio Piola, and the Austrian explorer Rudolf Grauer. It was never made clear to the Belgian authorities exactly what had been found by these expeditions, and most of the materials ended up in the museums of other countries.[7]

Harrison Hunts an Okapi

Colonel James Harrison was an independently wealthy former army officer and keen trophy hunter. He came from a family of landed gentry and magistrates in Yorkshire, England. James set off on his first hunting expedition to the Congo by traveling down the Nile via Khartoum in Anglo-Egyptian Sudan and Lado (on the Nile, where it separated the Belgian-controlled Lado Enclave from Britain's Uganda Protectorate) in early 1904. His main intention was to find "ochapi." When he arrived at Poko, in the Bomokandi zone of the Uélé district, on March 11, 1904, he was told that there were no okapi in the vicinity, though he saw the skin of one on the commandant's veranda. After a week of hunting in the area, he returned to Poko to meet Chief Okengi, an important Azande chief, who presented him with an okapi belt.[8]

Harrison next traveled west to Enguetra on the Likati River, where he was told okapi were present and where he saw many locals wearing okapi skin belts. He lodged with a Chief Bakia, who suggested he wait in camp while a soldier and a group of locals went out to search for an okapi for him. On April 6, he went out in pursuit of an okapi, experiencing "one of the hardest days I ever had," as he trudged through a semiflooded landscape, "going mile after mile in a half doubled-up shape, always creeping under jungle." He saw okapi spoor and managed to glimpse one, he said, but he didn't see how he would get a chance to shoot one. After a night of "terrific rain and thunder," he decided to "chuck the trip and make a push to catch the next boat, for it's quite hopeless." He sailed for home on May 3, 1904.[9]

Harrison set off again on December 29, 1904, traveling to the Congo via Khartoum. Reuters reported that he was heading for the Ituri district "[hoping] to be able to secure an okapi in the forest region" and that he had also "obtained leave in Brussels to bring home some pygmies" (providing they came voluntarily). He shot several elephants, selling the ivory to pay for his trip, and collected rare species, but found no okapi. This journey was remembered for his bringing back of six pygmies to England, which he records in his published booklet *Life among the Pygmies*.[10]

In this booklet, Harrison describes seeing one okapi in the forest near Enguetra in early 1904. He crossed the Ituri River on Monday, February 27, 1905, where he met people described as pygmies for the first time. On March 3, he went on his first hunt with Mbuti pygmies. In his book, he claims that the Mbuti's method of hunting in large numbers with dogs and much noise meant "there was no chance of an okapi within miles." Despite "hearing of an Italian who had shot one the previous year," he was of the opinion that "it was an absolute impossibility for any white man to shoot an okapi."[11]

Harrison made enquiries about this kill by an Italian and was told that in fact an African hunter had killed the animal. According to Italian taxonomist and zoo curator Spartaco Gippoliti, however, locals had captured the female okapi in the Enguetra forest in December 1904 and the skin and skeleton had been acquired by Lieutenant Antonio Millo Ribotti. Ribotti sent these okapi remains to the Natural History Museum of Genoa.[12]

Harrison bought two skins of okapi killed by locals on March 11, 1905, but he reports that the problem in buying such remains was that as soon as these animals were killed, they were cut up and so became "quite spoilt." The European hunter, he concludes, would have to be "right on the spot" at the moment of the kill to prevent this despoilation, which would have required him to wait for at least a month "for the little people [to kill] one, and then [pack] off to the place without delay." Harrison had only eleven days in the Ituri forest and so came away with only a photograph of the two skins and presumably the (somewhat damaged) skins themselves.[13]

Harrison's brief account of okapi hunting provided a kind of template for okapi expeditions, where the Western hunter's skills are shown to be useless when it comes to pursuing the elusive animal in the difficult terrain of the rainforest. The hunter relies on pygmies to find okapi yet complains that their hunting techniques scare away living okapi and that they damage the bodies (and thus trophies) of any okapi they successfully trap. Westerners end up demonstrating a curious mixture of admiration for pygmy expertise, frustration with the results of their hunting, and a grudging dependence on them.

Harrison writes appreciatively about the pygmies he met, but he names no individuals in connection with his pursuit of the okapi (or any other game). After the final diary entry for his 1905 expedition to the Congo, Harrison scribbles a few notes about the okapi: it "goes about singly never in herds. If startled does not go far before stopping. It has the same habits as a pig—rooting up mud with its feet and then eating it. It also eats the young shoots of trees and bush." No source is provided for this mostly accurate information.[14]

Harrison returned to the Congo with the six Mbuti he had taken to England in December 1907. He spent about ten days in the Ituri but saw no okapi, and by early February 1908, he was ill with malaria. In his diary, he recorded only: "Dr Davis—Swiss. Gun bearer shot okapi in 94." Harrison crossed the Semliki on the way to the east coast on March 9.[15]

Harrison made two further expeditions to the Ituri in early 1909 and in 1910, but saw only "quite fresh okapi spoor." These were the last of his expeditions. He married the American socialite and trophy hunter Mary Stetson Clarke in 1910 and lived with her in Brandesburton Hall in Yorkshire with its trophy rooms and fifty glass cases of taxidermized animals until his death in 1923.[16]

The Alexander-Gosling Expedition

Boyd Alexander grew up in Kent in England, where he spent much of his boyhood studying and collecting birds. His father was a military man, and Boyd was educated at Sandhurst. He enlisted as a young man, more to travel and study ornithology than from any enthusiasm for soldiering. He made several expeditions to Africa and adjacent islands to describe avifauna, publishing reports on all of these in the journal *Ibis*. Alexander's major expedition was a three-year journey memorialized in his two-volume book *From the Niger to the Nile*, for which he was awarded the Founder's Medal of the Royal Geographic Society.[17]

In February 1904, Lieutenant Boyd Alexander, his brother Claud, Captain George B. Gosling, a Mr. Percy A. Talbot, and José Lopes set out on an expedition whose initial aim was an exploration of the then little-known Lake Chad region. Starting at the mouth of the Niger, they traveled inland to Lake Chad. Here Claud died of enteric fever in October. Talbot turned back after they had surveyed the lake. Boyd, Gosling, Lopes, and African expedition members including John the cook, Lowi, Quasso (formerly Claud's personal retainer), Mustafa the Arabic interpreter, and six armed Hausa men commanded by Agoma Lafia, then turned south, following the Chari River upstream from the lake through what is today Chad and Central African Republic. They continued until they reached what was then the Independent State of the Congo.

On January 25, 1906, after a strenuous journey up the Uélé River during which they were beset with rapids and battered by rocks in the wide riverbed, they reached the small Belgian post at Angu in the western part of Equatoria Province. Here they heard "rumours of the existence of the okapi in the neighbourhood." They had already seen a few hunters wearing striped okapi bandoliers. This was the only place near the Uélé where okapi were rumored to be found, and they spent three weeks trying to catch one. Gosling, Alexander, and Lopes competed "to be the one to capture the mysterious okapi."[18]

Alexander made his headquarters in a "Mobatti" (Mbuti) village called Lobi, where he was impressed by the skill of local hunters with traps, pits, and nets. They brought him numerous rainforest animals, which he and his skinner Quasso skinned and made up. He was less pleased with the rigors of hunting in the forest, his initial awe and wonder soon giving way to a "haunting fear that rests on the forest like a spell." He experienced the darkness "like a live thing struggling in the meshes of the trees." According to Alexander, even locals found the okapi a mysterious and restless creature which was hard to track. He saw none.[19]

Gosling kept notes on his search for okapi, called by the Bangala name *n'dumba* in the region. He recorded observations by local African hunters about okapi, including their habits, routines, favored habitat and diet (discussed in chapter 11). Gosling saw okapi tracks in glades and clearings, and got very near to some, but didn't see one.[20]

Lopes, meanwhile, was hunting okapi some three days' journey south of Alexander in the vicinity of the village of Beritio. Guided by a local hunter, he followed a single okapi for three mornings, noting it always followed a particular course away from a stream. Based on this knowledge, he dug a capture pit "as the natives do" with local help. They caught and killed an okapi in this pit the following day. He sent some meat with a note to Alexander, who shortly after receiving it was met by an exhausted, feverish, and disappointed Gosling.[21]

"Cheer up Goose," Alexander told him, "I am going to give you okapi cutlets for luncheon!" Soon, "like Roman emperors, we were feasting on the rarest animal in the world, fully appreciating the fact that we were the first white men to eat the meat of the okapi. It was very tender and tasted like beef" (of course, Belgian colonial officials had already eaten okapi before this).[22]

Excited about acquiring a good okapi skin, Alexander and Lopes waited three nervous days to dry it amid frequent heavy rain. They then traveled on to meet Gosling at Angu, buying a partial okapi skin from a chief en route for "a great deal of cloth, for the striped part is much prized by the natives." Locals used it for bandoliers and chair covers, and okapi skins were regarded by local Africans as emblems of rank on account their attractive markings and rarity, and the difficulty of acquiring them. En route to the Belgian station at Niangara (or Nyangera), a chief negotiated with Alexander to buy a strip of his okapi skin, following their boat for two days in the hope of acquiring it.[23]

They were all feverish after their exertions in the rainforest, which were exacerbated by the 124-mile-long (200 kilometer) journey to Niangara from Angu. In June 1906, Gosling died of blackwater fever at Niangara and was buried there. According to Alexander, "It was undoubtedly in the pursuit of the okapi, upon the capture of which he had set his heart, that he got his death." Alexander and Lopes eventually made it to the Nile and caught a boat to Khartoum, having covered about five thousand miles (eight thousand kilometers) on foot and by boat, and arrived in England in February 1907. The progress of the Alexander-Gosling expedition, including their hunts for okapi, had been widely reported in British newspapers.[24]

This expedition confirmed the elusiveness of the animal. Despite hunting it for three weeks in known okapi habitat, neither Alexander nor Gosling saw one alive or even saw an entire corpse. Lopes may have seen the okapi he trapped with the aid of locals while it was still alive, but he could not prevent them from butchering it on the spot. He was fortunate in being at the scene, which allowed him to retrieve a good skin.

The ways this sensational event was reported are instructive. *Scientific American*, for example, got the Alexander-Gosling okapi story quite wrong, claiming that "Capt. Gosling on three occasions observed the [okapi] busily feeding, and was able to approach quite close to it without frightening it away, so that he was able to follow its movements and habits among the swamps with perfect ease. Capt. Gosling is the first white man to see the okapi alive." The *Daily Mail*, on the other hand, claimed that Alexander "has seen the animal alive." Alexander had pointed out that neither man had seen one.[25]

The Times notes that while several expeditions had sought living okapi, the credit, "which is sufficiently remarkable to be for ever noted in the annals of zoology, . . . must belong to the Alexander-Gosling expedition." The article reproduces Alexander's description of the specimen, which he clarified "was obtained by José Lopez [*sic*], his Portuguese collector." Also included are Gosling's notes on what he had learned about okapi "in the forest contained by the rivers Welle [Uélé], Lubuati, and Rubi." However, the big news was (allegedly) that "no white man has hitherto been able to report having had the good fortune to come across a living Okapi."[26]

This story marked the beginning of claims and counterclaims in the British press about which "white man" first saw an okapi. Sclater responded by arguing that the Swiss geologist and naturalist J. J. David, based at Beni on the Semliki, was really the first European to have seen an okapi. David had sent a skin to a Professor Burckhardt at the University of Basel, and described shooting it in an article in *Basler Nachrichten* on May 22, 1904. David did procure an okapi at an early date. Curators David Marques and Loïc Costeur at the Natural History Museum of Basel kindly confirmed that okapi remains, including the skull and mandible of an okapi shot on November 23, 1903, "on the left bank of the river Loya, a tributary of the Ituri," were donated to the museum by him in around 1906. Similar remains were donated for an okapi shot in January 1904, at "Nord-Kivu Beni." David provided further fragmentary okapi remains described as "küchenreste" ("kitchen leftovers"), and in 1908 he donated the full skeleton and skin of an okapi shot on June 17, 1905, at Makala on the Lindi River. The skeleton was mounted and is still at the museum. Who shot the okapi is another matter.[27]

The *Times* story also prompted commentary on the status and agency of Lopes. In a review of Alexander's book, Harry Johnston argues that as a native of the Cape Verde islands, Lopes (see fig. 7.2) was more African than European ("not quite correctly styled . . . a Portuguese"), though Johnston continues that "José Lopes, according to his employer's account, played . . . a splendid part in resourcefulness, courage, and adroitness." Meanwhile, Lankester in one of his articles on the okapi refers to Lopes as Alexander's "half-breed servant." Both imply that Lopes was not "the first white man" to see an okapi, because he was not "white." A letter to *The Times* states that the okapi "was caught . . . by Alexander's Cape Verde boy, José Lopez [*sic*]. . . . Lopez was thus the first man born outside the Continent of Africa to see a live okapi. He claimed to have Portuguese blood in him."[28]

These remarks, their deplorable racism aside, raises a question about the agency of—and whether credit was due to—the assistants of expedition leaders. This is clear from the *Daily Telegraph*'s article on Alexander's return from his three-year expedition titled "Alone across Africa." Despite traveling with a team of Hausa bearers, Quasso who became a skilled skinner, Lopes, and his "excellent Senegalese cook" John, only "the four Englishmen who entered Africa" are mentioned in this article. After one returned and two had died, this left Alexander somehow "alone." The article does admit the existence of Lopes, described as "Alexander's Portuguese Collector."[29]

Figure 7.2. José Lopes, who found an okapi, later mounted in the Natural History Museum in London (Boyd Alexander, *From the Niger to the Nile*, 2 vols. [London: Edward Arnold, 1907], 2: 235).

What of Alexander's representation of Lopes? In the introduction to his book, Alexander allocates one sentence to Talbot but devotes nearly a full page to "my collector José Lopes, my right-hand man for usefulness." He describes the importance of Lopes's skills not only as collector, transport manager, boat handler, expert skinner (there is a photo of him skinning birds), and crack shot, but also as interpreter and go-between with the Hausa men that Lopes had hired for Alexander in West Africa. Lopes was able to "get further into the confidence of the rulers and chiefs we came across, and so we obtained a great deal more information of interest than would otherwise have been possible."[30]

Alexander reveals that Lopes had been his "faithful servant for ten years. . . . I found him as a little boy, working on his father's trading-boat that plied between the Cape Verde Islands." (It is unclear under what conditions Alexander took over this boy.) Lopes features at various points throughout Alexander's account and is credited for catching their only okapi. Despite this, many contemporary newspaper reports of the sighting of the okapi fail to mention Lopes. The gift of the taxidermized okapi skin to the Natural History Museum is credited to the Alexander-Gosling expedition, with no mention of Lopes.[31]

Missing from many accounts of scientific discovery is the role of Indigenous peoples, locals and intermediaries like Lopes, not to mention the many helpful local colonial officials, missionaries, and settlers. To recover such hidden histories, it is helpful to pay attention to who and what are marginalized in reports and exhibitions of new discoveries in the West. The role of Africans is the focus of chapter 13, while the focus here is on go-betweens.

It has been argued that the homogenization of the modern world that resulted from the global ascendancy of the West in the nineteenth century saw the decline and disappearance of the go-between as key player in making interactions between different cultures possible. However, as Simon Schaffer and others show, go-betweens played a vital part in the construction of the modern world, notably in the domain of scientific knowledge. Without diminishing his roles as collector and traveler, we can see Lopes as one such go-between. While Western explorers often acknowledged the contributions of individuals like Lopes, these go-betweens were often screened out or trivialized in media reports and omitted from scientific publications.[32]

Major Percy Powell-Cotton, a Sportsman Frustrated

Percy Horace Gordon Powell-Cotton grew up in London and in Margate in Kent, southeast England, helping his father modernize Quex House on their Kent estate, which became the family home. He joined the Third (later Fifth) Northumberland Fusiliers in 1885, retiring from military life with the title of major in 1901. Between 1887 and 1939 he conducted nearly thirty hunting and collecting expeditions in Asia and Africa, becoming a renowned trophy hunter. He set up his remarkable personal collection of taxidermized specimens by arranging them in dioramas with painted backdrops at Quex Estate (in what is now the Powell-Cotton Museum). On his 1902–03 expedition through Kenya and the Uganda Protectorate, during which he shot two giraffe of a new subspecies named *Giraffa camelopardalis cottoni* which were later exhibited in the Natural History Museum in London, he heard about the okapi and decided to return and hunt one.[33]

Percy set out for the Lado enclave on the White Nile on November 2, 1904. Such journeys were expensive and required considerable planning and diplomacy, and his diaries provide a good snapshot of the logistics involved. He sent his equipment (except his tent and cartridges) by post (!) to Lado on the Nile, filling 367 eleven-pound parcels (the maximum for parcel post) and three vans. He traveled by ship to Alexandria and then by train to Luxor and Shellal, where he boarded the steamer *Ibis*, beginning the first of three river voyages to Lado. Once there, it took nine days to unpack his goods and rearrange them into loads for his bearers. They set off overland on December 15, 1904.[34]

Powell-Cotton's main aim in the area around Lado was to shoot a northern white rhino, then known by a single specimen only, and large elephants for their ivory, succeeding in both endeavors by early 1905. He next traveled south past Lake Albert, then west to cross the Ituri into the rainforest. He remained in the Congo forests from late June to October 1905, hunting between Irumu, Mawambi, and Beni. He spent time with pygmies in the Ituri forest, observing that only men wore okapi skin belts (the items referred to as "bandoliers" by Johnston, Sclater, Alexander, and others). The substantial collection of these that he made (loaned to the Natural History Museum to be photographed for its 1910 atlas) are still hidden away in a packing case at the Powell-Cotton Museum. He had no luck finding a live okapi, however.[35]

In October, Percy traveled down to Mombasa on the east coast to meet his fiancée, Hannah Slater. They were married in Nairobi in the cathedral (which Hannah called "the little tin church of Nairobi"). Percy had lured her out, she notes, with promises of "Christmas among the Mountains of the Moon" (the Rwenzori). After Christmas in the mountains, these two "vagabond spirits" (her words) returned together to the Ituri forest in January 1906.[36]

They stayed in the forest for more than six months, despite suffering intermittent fevers and stomach complaints. Hannah, who did not materialize into the awful tyrant that Percy's Swahili caravan feared a European woman would be, mostly enjoyed "life under canvas." Their days ended, according to Percy's terse diary entries, with "jaw" (conversation), "tiff" (snack, from "tiffin"), "laze" (obvious), "tub" (a bath), and "din" (dinner). This was certainly an unusually challenging and extended honeymoon.[37]

In March 1906, the Powell-Cottons traveled to the station in Makala, an important rubber collection center on the banks of the Lindi. Percy describes it as "a typical forest station, with an avenue, lined with pineapple and bananas, leading up to the post, which crowned a little hill. It consisted of a well-built brick house, another of wattle and daub, a mess-hut and a number of rubber and other stores, besides the usual village for the soldiers and personnel." They had come here to hunt okapi (known as *kangi* in the area), because this was where "a native hunter had secured the two specimens brought to Europe by Dr David."[38]

At the station, they found Ahkuki (Cotton-Powell's spelling, rendered Agukki by others), "the native renowned for his skill in okapi hunting," and sent him off in search of tracks. News reached the post a few days later that Agukki had shot a fine male okapi at about midday on March 21, 1906. Cotton-Powell was disappointed to miss shooting it himself but immediately sent men to the place where it fell to retrieve the carcass. He worked most of the night and the next day to preserve this skin. He records that it "smelt of faint vinegar; no mane but ridge there, skin quite ¾ in thick; horns were loose and came off skull when boiled . . . kept tongue skin; meat excellent steaks; took 8 men at a time to carry without stomach."[39] The skin and skeleton of this okapi was sent to Rowland Ward in London to mount, which Powell-Cotton donated to the Natural History Museum in 1907.

Powell-Cotton continued hunting in the hope of shooting an okapi himself, but despite days spent following fresh tracks, usually found by Agukki, he never saw one. "The chances of a sportsman coming across these extremely timid denizens of the forest undergrowth," he reports, "seem to me to be very remote." The Powell-Cottons spent over three months in the area, during which time Agukki wounded another okapi (on May 5) but didn't recover it. In 1907, after twenty-seven months in Africa and several more bouts of fever, they returned to England.[40]

In a letter to the *The Times* written from Kasindi in the Congo responding to letters from Gosling and Sclater, Powell-Cotton acknowledges that the Mbuti had "aided me in the chase for several months, and as hunting is the sole aim and ambition of their lives, there is little they do not know about forest game." He relayed information on the social and maternal behavior of okapi, and identified the plant Alexander said okapi liked to eat (*Megaphrynium macrostachyum*), while noting that the Mbuti denied that okapi ate it (which is correct).[41]

Powell-Cotton offers more specific information regarding local expertise on the matter of David shooting an okapi. Agukki told him he had shot two okapi for David. David had then (as Powell-Cotton had) gone directly to the spot "and superintended the preservation of the specimens." Powell-Cotton further claims that despite his "very careful inquiries," he had been "unable to find conclusive evidence that any European has killed [an okapi] himself."[42]

He learned that the Independent State of the Congo's first specimen, procured by its agent Lieutenant Anzélius, had been killed by Africans near Mawambi. The first European to see a live okapi, he reports, was a young Swiss official of the Independent State of the Congo named Jeannet, who saw one while supervising road work between Avakubi and Irumi in spring 1905. Alerted to its presence by an African soldier, Jeannet ordered him to shoot it. The skin was sent to Brussels.[43]

Clearly, determining which European had first seen or shot an okapi in the forests of the Congo (to me a senseless undertaking, but revealing of what motivated pursuit of okapi at the time) was difficult. Certainly, there was great interest in ascertaining to whom the credit should go, and the various claims made revealed not only racism but also an element of jingoism on the part of the various European claimants.

These early accounts of European attempts to find okapi reveal that, while not very numerous, they were also not particularly rare in some regions of the Congo rainforest. The problem was locating them. Until the 1910s, only a handful of outsiders had seen one alive, all directed to the animal by local Africans. Even for locals, okapi were hard to hunt down (rather than catch in pit traps), although some (like Agukki) were specialist okapi hunters, but almost all knowledge of the behavior of wild okapi was learned from local Africans.

Von Mecklenburg-Schwerin's Congo to Niger and the Nile Expedition (1910–11)

The German empire was founded in 1871, and many colonial African territories were formally recognized beginning in 1884, including German colonies in Togoland Protectorate (now Togo), Kamerun (Cameroon), and German East Africa (Rwanda, Burundi, and Tanzania). Numerous German hunters, explorers, and collectors were active across Central and East Africa. After Berlin's Natural History Museum moved to its new site in 1889, all objects collected by colonial officials or during expeditions funded by the German empire, had to be sent there. However, the spoils also went to the museums of prosperous cities like Frankfurt and Hamburg, whose zoological societies and wealthy burghers funded collecting expeditions.[44]

Adolf Friedrich, Duke of Mecklenburg-Schwerin, made an unsuccessful attempt to hunt okapi in the Ituri forest during an expedition he led to Central Africa from 1907 to 1908. According to an expedition member, the dry season had made it possible to move through the forest, but that meant there were no okapi tracks to be seen. They found only remains in 1908, which they sent to the museum in Berlin. The skins they found were in too poor condition to be taxidermized, but they also bought a skeleton, which could be displayed (the museum's 1919 guidebook lists a skin and a skeleton from Friedrich).[45]

In 1910, the duke again assembled a team of scientists that included the zoologist Dr. (Johann Gustav) Hermann Schubotz, and Ernst Heims, an artist who had previously worked in Kamerun. They sailed up the Congo and Ubangi (or Oubangui) rivers, before splitting into different groups, one heading northward to Kamerun and then Lake Chad and one heading east through the Congo to the Nile. The sections in the volumes recording the expedition (in which the duke appears, photographed in a white military tunic festooned with medals) are written by different expedition members. The first is narrated by Captain Walter von Wiese und Kaiserswaldau. He notes there were no okapi around Libenge on the Ubangi, but apparently okapi were found further north, especially south of Yakoma (at the fork of the Ubangi and Uélé), near a place he refers to as Linyati (I can find no such place, but perhaps he means the Likati River).[46]

Schubotz was an enigmatic individual whose career progressed from natural history through diplomacy to broadcasting, including a period in exile in German South West Africa during Nazi rule. An assistant at the Institute of Zoology of the Royal Friedrich Wilhelm University of Berlin from 1905 to 1907, he was seconded to join Friedrich's first expedition, and on the second, he led (and wrote up) the expedition that traveled eastward across the Congo.[47]

Schubotz's party traveled through the Uélé district to "the home of the okapi" and then across the Mangbettu country to Khartoum on the Nile. His main aims were to explore the fauna of the northern regions of the great equatorial rainforest and of the adjoining plains of the Sudan. He knew of the Alexander-Gosling expedition's collection, "at the head [of which was] the okapi." He was determined to secure okapi for museums back in Germany; indeed, he felt obligated to because the expedition had secured significant funding from museums in Frankfurt and Hamburg.[48]

Schubotz headed for Angu, a small but important rubber station on the south bank of the Uélé River, which a Commandant van der Cruyssen had told him was located in good okapi country. He knew that the "only photograph of a living okapi had been taken in this neighbourhood." He had also read about Angu in Alexander's book *From the Niger to the Nile*. In his narrative, Schubotz laments the fate of the members of that expedition, including the subsequent death of Boyd Alexander, who had recently been killed in an altercation with locals in the Sahel near Darfur.[49]

On May 20, 1911, Schubotz's expedition reached Angu. Here, Schubotz told the African soldiers (the Belgian chief of post was absent) of his desire to shoot an okapi, among other forest animals, and offered them rewards for specimens.

When he showed the men Alexander's photographs of dead okapi, they informed him it was well known in the area and that their name for it was "ndumbe."[50]

Etumba Mingi (or Etumbamingi), the young chief of a neighboring village, soon arrived to offer his services as an okapi hunter. In his account of his expedition, Schubotz describes the chief as "a tall, sinewy Mobatti youth who was famous as a hunter as well as a warrior." Etumba Mingi brought a large basket with handles made from okapi and elephant skin as proof of his skills. He said he could find okapi, but "this kind of hunting was an arduous undertaking, quite unsuited to white men." It would require a pursuit over many days through dense bush and over swamps, because, according to him, okapi constantly wandered through swamps and dense jungle. Furthermore, he said, Europeans were too noisy: while they could shoot "stupid animals" like elephant and buffalo, the "shy and wary *ndumbe*" was another matter.[51]

Schubotz was inclined to agree, having tried unsuccessfully to hunt okapi during the 1907–8 expedition with Friedrich. However, he was "not prepared to give up so easily my great desire to be the first European to shoot this rare animal." Therefore, he sent Etumba Mingi off to try and find okapi while making further plans to shoot one himself.[52]

The chief of post (one Andersson) confirmed to Schubotz once he returned that the solitary, wandering okapi was a very difficult target, asserting that studying its mode of life would be impossible. Like Harrison, Alexander, Gosling, and Powell-Cotton before him, Schubotz undertook numerous "very fatiguing" hunting expeditions accompanied by an experienced local hunter named Koki, before admitting defeat. This was no fault of Koki, a former soldier, whom Schubotz described as "a model of strength and intelligence."[53]

Andersson, like so many colonial officers in Belgian employ described in these travels of hunters and collectors, was very obliging in facilitating Schubotz's hunting. He helped him source bearers and introduced him to a chief named Koloka, from a "Mobatti" village two days' march south of Angu, where the men lived mostly by hunting. After two weeks of hunting during which he was joined again by Koki and during which he found that in the dense jungle his high-powered rifle and binoculars were useless, Schubotz succumbed to fevers: "So, like all my predecessors, I was obliged to fall back upon the skill of the native hunters."[54]

After six weeks, news reached Schubotz that Etumba Mingi had shot an okapi with his high-caliber elephant gun after tracking it for nine days. Schubotz rushed to secure the skin and skeleton, accompanied by a skinner and porters. He hoped to get a good photograph while the okapi was still in good condition, remarking that most of his predecessors had only heard about the kill of an okapi from the Africans who brought in the skin. Their photographs showed butchered animals and so did not reveal the animal's form accurately. Unfortunately, after a six-hour march Schubotz's party realized they were lost, and Schubotz's men dispersed to find the kill while he returned to camp. Fortunately, the skin of an adult female arrived in excellent condition, but he never saw the corpse.[55]

Much encouraged, Schubotz induced Etumba Mingi ("not a little proud of his fine booty") to kill a second okapi, and eight days later, Etumba Mingi succeeded. This time Schubotz managed to arrive before the okapi was dismembered and "see the strange beast that few Europeans had had the opportunity to see, even in death." Having rushed to meet them before daylight faded, he describes how an hour before sunset "the head of the advancing procession came into view. An animal, about the size and weight of a horse, was suspended by the legs from a young tree, carried at each end by fifteen staggering, panting negroes, urged on by scolding soldiers." He admired the "powerful dark-brown body, the disproportionately long neck, the black donkey's ears above the grey face with its long protruding tongue . . . the beautiful, slender, black and white striped legs." None of the taxidermized specimens had captured the okapi's "fine muscular form," he declares in his account. Schubotz had his men chop a clearing in which he took photographs. One is included in the book of the expedition, and Heims's painting of two okapi adorning the cover of volume 2 was based on these photos. Also included is a photo of "the hunter" (see fig. 7.3) standing next to the prepared skin of one of the okapis.[56]

News of this exploit was first published in the German publication *Die Woche* in January 1912, accompanied by a full-page photograph of the okapi captioned "the first photographic image of a freshly killed okapi." The caption celebrates Schubotz's "finding of two specimens of the extremely rare okapi for German zoological museums." Schubotz himself notes in his article that while King Leopold II had bestowed "princely gifts" of okapi on France, Italy, and Spain, only the court museum of Munich had received an okapi and that the other German museums had "searched in vain for the precious object."[57]

The first okapi shot by Etumba Mingi was taxidermized and donated to the Naturmuseum Senckenberg in Frankfurt-on-Main (volume 2 of the expedition books includes a photograph of it). The skeleton of the other went to the Hamburg zoological museum. The natural history museum in Berlin had to go without an okapi mount, probably because it had been given a skin and a skeleton from Friedrich's first expedition. Having generously funded the second expedition, the other museums received more materials, including Schubotz's okapi

Figure 7.3. This is very likely Etumba Mingi, the hunter who shot two okapi for Schubotz. Schubotz's caption reads "hunter with the prepared skin of the okapi" (Adolf Friedrich von Mecklenburg-Schwerin, *From the Congo to the Niger and Nile,* vol. 2 [Philadelphia: John Winston, 1914], after 14).

remains. Furthermore, the wealthy burghers of those cities could sponsor the expensive mounts. As to Schubotz, although he was appointed a professor of zoology by 1916, he began a new career as a diplomat after World War I.[58]

Cuthbert Christy Collects

Cuthbert Christy, who took a medical degree at Edinburgh, was gored to death by a buffalo in the Congo at the age of sixty-eight. By then he had conducted extensive tours of duty in South America, the West Indies, Africa, the Middle East, and India working in tropical medicine (notably sleeping sickness) and had contributed significant collections of zoological specimens to the Natural History Museum in London. One of his most extensive and significant periods of scientific exploration took place from 1912 to 1914 in the Congo, on behalf of the Belgian government. His book about this trip, *Big Game and Pygmies,* was published in 1924.[59]

Christy wanted to be the first white man to shoot an okapi, "perhaps the most coveted of all African sporting trophies," and ultimately, he claimed, he succeeded. He was quite specific in saying he aimed to be the first white man to shoot an okapi *and* bring back its skin and skeleton to Europe for a museum because he had heard that in 1912 an English elephant hunter had been floating down the Ituri a few miles from Mawambi when he encountered an okapi on the riverbank and shot it, but neither the skin nor the skeleton were preserved.[60]

Christy knew at the outset that "to be successful" he first had to "make the acquaintance and gain the confidence of the real forest natives—the Bambute Pygmies." However, they were "naturally difficult to deal with," and African villagers who had relationships with them were "jealous of any interference."[61]

Christy spent several weeks traveling along the forest edges, making forays into the forest, and trying to contact these people with no success. While traveling through the Ituri Forest, the chief of the section at Avakubi "gave" him a "live Pygmy" (a former slave of an Arab). Christy spent a month hunting with him, learning "a good deal . . . of forest natural history." However, the man knew no other pygmy people.[62]

The chief of post at Mawambi, almost sixty miles (one hundred kilometers) east of Avakubi along the river, introduced him to the chief of a village with connections to pygmy people, whom Christy eventually met and befriended. Taking along a small pocket tent, he roamed the forest with them, tracking animals. In his book, he describes their tracking skills and the stealth with which they moved and communicated. "Without the assistance of the Pygmies," he remarks, "the sportsman in these forests is helpless. He can neither find any animals nor get back to his starting point."[63]

In 1913, Christy finally managed to shoot a young male okapi (his book includes a photograph), which had been tracked for him by two Mbuti men (shown with the dead okapi, in his photograph). A few weeks later, a Commandant Hedemark of what Christy describes as "the Congo Service," visited him and shot a large adult male okapi he encountered by chance near camp. Christy shot a very large elderly female okapi in 1914, near Moera on the Semliki side of the Ituri forest. Later that year, the chief prospector for a mining company, one A. E. H. Reid, was returning to camp near Mawambi when he heard a large animal break away and fired off a shot: he discovered, to his surprise, that it was a fine old male okapi (Christy's book includes photographs).[64]

Christy claimed to be the first European to shoot okapi in the wild and to bring back to Europe complete skins and skeletons as well as preserved viscera. This had required eighteen months of searching in the Ituri-Aruwimi forests, during which time he had glimpsed the animals alive on only a handful of occasions. Besides spending months trekking through dense forest and waist-deep swamps with a heavy firearm, he had also spent much time learning bushcraft from the Mbuti. He remarks that it is paramount to remain silent and to interpret every sound, "for the hunter, like the animals, is almost wholly dependent upon his ears."[65]

The difficulties of preserving the hard-won skins of okapi is also well described by Christy. Having shot a male okapi after hours of tracking, he and his hunting party followed the wounded animal and found it with an hour of daylight remaining. Covering it with leafy branches, they returned to the Ituri, marking their way as they followed an elephant path to where they had left their canoe. From there, they paddled down through rapids to their island camp a mile below, had their first meal since the previous night, and snatched three hours of sleep.

They then poled back upriver and walked back to the kill site, where they took measurements and cleared the bush so Christy could photograph the animal. This done, they removed the skin and soaked it in preservative. They then carried it to the river and spread it on a platform of sticks to make use of the remaining sunlight before paddling it to camp that evening. The drying of the skin the next day was slowed by a lack of sunlight and "clouds of brown butterflies, which clustered upon it in spite of its having been dipped in a strong solution of arsenical soap." The weather turned wet, so Christy sat up all night with his African hunters drying the skin in the smoke of a campfire.[66]

In his book, Christy drew on his experiences during these extended periods hunting okapi with Mbuti people to provide detailed information on their habits, including their favorite haunts in the rainforest, keen senses, and shy and solitary nature (discussed in chapter 11). He was forced to leave the Congo in August 1914 after Britain declared war on Germany. The news came as a great shock: "Never shall I forget with what surprise I read the contents of a letter, wrapped in leaves and fixed in the end of a cleft stick, brought me by a native runner from the *Chef de Poste* of the nearest Belgian station many miles away. . . . All my plans had to be abandoned."[67]

Herbert Lang: An Okapi for the Bronx Zoo

Americans were just as interested in acquiring okapi specimens as Europeans, and the New York Zoological Society and American Museum of Natural History were particularly determined. They wanted live specimens for the new Bronx

Zoological Park that had opened in November 1899, and okapi remains and other material for an okapi display in the proposed African hall of the museum.

Capturing okapi was a key goal of the museum's 1909–15 Congo expedition led by Herbert Lang, a German zoologist who emigrated to the United States in 1903 and worked at the museum. A formal portrait from the time shows him to be a slightly nervous-looking man in pince-nez glasses with wispy moustaches, but he proved a meticulous organizer, a disciplined leader, and a prodigious field worker. Lang was assisted by a young American ornithologist named James Chapin.

The expedition sailed to Brussels to secure the necessary permissions in May 1909 and then to Africa. They traveled over twelve hundred miles (nearly two thousand kilometers) up the Congo River from the coast to Stanleyville (Kisangani). From there they marched to Avakubi, and then on to Medje, making their headquarters there. On the advice of a local officer, a Lieutenant E. Boynton, they traveled south into Chief Banda's territory, where they found local men drying okapi meat over a fire.[68]

By late 1910, Lang reported that they had collected the necessary material for an okapi group for the museum's hall of African mammals, including watercolor paintings, photographs, vegetation samples from trees, lianas, bushes, leaf molds, soil samples, and even okapi dung, and that he had made a study of the "little-known life history." What they did not have was a live okapi.[69]

Having walked "more than a thousand miles in the tracks of the Okapi," Lang concludes that "a great wariness and nocturnal habits efficiently protect it from being successfully stalked by white men." Anyone who had seen one alive or shot one, he maintains, could only have come across it accidentally. He was convinced that okapi had escaped European detection for so long because they frequent "the most unhealthy [territory] in the world": the vastness and monotony of the forest, the appalling heat and humidity, and violent storms, not to mention the risk of falling victim to cannibals, deterred most Westerners from entering its habitat.[70]

Lang therefore relied on African hunters to catch an okapi. He and Chapin trained eighteen Africans as field assistants for their collecting activities. After two months, they acquired a specimen of a dead okapi and were able to take casts of the head. Lang only managed to glimpse one dying okapi on his many hunts with Chief Banda's people, complaining that their methods of trapping killed or badly injured the animals. He learned much about trapping techniques from the Medje hunter Aposho, but he secured only remains, no live okapi. The expedition was scheduled for completion in 1913, but it was extended as Lang continued his hunt for a live okapi.

Lang and Chapin received news of the outbreak of war in Europe while at Avakubi in 1914. Chapin completed packing up the collections and transported them to the coast, from where he sailed to Liverpool on a British steamer, slipping past the German naval blockade. Lang in the meantime continued his search for an okapi into 1915.

Lang only acquired a live specimen after—on the advice of a Commandant M. Siffer—he joined the annual hunting expedition of a powerful Azande chief. Chief Akenge's son Abawe was a renowned hunter who would collect many rare animals for Lang, including killing the "record bull okapi" (see fig. 7.4) that was sent back and taxidermized for an okapi display in the hall of African mammals in the American Museum of Natural History.

Abawe finally located a mother okapi with calf near Niapu, and in Lang's company they got so close that the mother bolted, leaving her calf behind. When the hunter seized the calf, everyone expected a struggle, but it only licked the face of its captor and sucked the fingers held out to it. Abawe seized the calf (see fig. 5.3), which proved a "most endearing creature," sleeping at night in a hut

Figure 7.4. The famous Azande hunter Abawe with the "record bull" okapi he killed for Herbert Lang (*Zoological Society Bulletin* 21, no. 3 [1918]: frontispiece).

"with his head on the breast of his guardian." The calf was "tame as a lamb and enjoyed being patted and stroked." Although only a week old, it could walk, run, and jump. In all, it was "a most endearing creature . . . the pet of all."[71]

Lang soon ran out of condensed milk, however, and couldn't get more, as low water had prevented steamers traveling up the Ubangi and Aruwimi rivers. Lang tried a mixture of rice and water, but the calf weakened and died within ten days. Lang was disappointed but convinced that "under proper conditions Okapi could be brought to civilized countries." In a short article on an okapi calf raised by one Mrs. Landeghem, Lang included a heartbreaking photograph of this calf Abawe had caught for him, shown "bleating for his mother." This is one of the few references to okapi vocalizations.[72]

As Lang had a German passport he couldn't board Allied ships, so he sailed to Luanda in Angola and from there to Lisbon, as Portugal remained neutral. From Lisbon he returned to the United States with the remainder of the collection. Lang and Chapin brought back thousands of zoological and ethnographic objects, including artworks and items of clothing and adornment made from okapi skins, as well as copious field notes, drawings, and photographs.

Lang subsequently became a curator of mammals at the museum and helped mount specimens for the Akeley Hall of African Mammals that included two taxidermized okapi. This fine pair of okapi mounts can still be seen in the art deco diorama display at the museum in the African hall, which is Carl Akeley's legacy, with its painted backdrop (look out for the chipmunk!) and elaborately recreated foliage from samples taken by Lang in the Congo.[73]

Conclusion

The outbreak of World War I put an end to the first period of Western efforts to collect okapi, dead or alive. The accounts of the expeditions offered here reveal something of the interactions between Western collectors, local colonial officials, and local Africans drawn into these attempts. While it is clear that local African peoples hunted and trapped okapi prior to Western interest in the animal, and assisted Westerners in their attempts to find okapi, it is difficult to assess what kinds of impacts the Western demand for okapi after 1901 had on indigenous peoples in okapi habitats in the northeastern Congo, and whether this affected their relations to okapi.

As for the okapi, we cannot know what they made of these novel and concentrated efforts of Westerners to track them down and capture or kill them. With their great ears and highly developed hearing, they would certainly have heard them blundering through the rainforest. While we have evidence that okapi circled human dwellings in their wanderings, including an elderly couple who sent messages using slit gongs, and would have been familiar with the wonderful polyphonic singing of pygmy peoples, we can only wonder what they made of Percy Powell-Cotton playing "The Laughing Policeman" to delighted Mbuti on his Edison Standard Phonograph.[74]

CHAPTER EIGHT

Pursuing Okapi in the Interwar Years

I knew that the okapi was a zoological freak, having the neck of a
giraffe, the head of the prehistoric *Samotherium*, . . . the body of an
antelope and striped legs much like a zebra's. What I didn't know,
was that the okapi family was to assume complete control of my
personal life and push me around for the next two years.

—*Ellen Gatti, 1950*

The interwar years were the final years in which the pursuit of okapi was, at least
in the eyes of the outside world, almost entirely in the hands of the amateur explorer and collector, or charismatic and singular individuals like Brother Franz
Joseph Hutsebaut. By 1938, the Belgians had established their own capture operations, and there was less need for the likes of Attilio Gatti to capture okapis
for zoos and circuses in the West. One or two well-connected wild animal
capture specialists persisted, but the export of okapis became regulated, and captive okapis could be procured from the capture station at Epulu.[1]

Gatti merits particular attention, because of his important observations of a
live okapi calf in captivity in particular. From observation of the calf together
with what he learned from Africans, he amassed a fund of okapi knowledge virtually unparalleled at the time, which he disseminated through the press, his
books, and films. Some of his ideas remained influential for decades after because
so few methodical observations of wild okapi were made or reported. His story
also brings to light details of how okapi hunters interacted with those who spon-
sored them and with the colonial conservation authorities in this period. Of the

few others pursuing okapi between the wars, only Delia Akeley's story is told here.

Delia Akeley's Okapi Family

For years, Delia Akeley had accompanied her husband, the famous taxidermist Carl Akeley, on his African expeditions, collecting megafauna for his Hall of African Mammals at the American Museum of Natural History (she shot the larger of the taxidermized fighting African elephants on display). However, she and her husband became estranged, and in 1924, nothing daunted, she set out on her own expedition to Central Africa.

Delia had often had to manage her husband's large caravan of African porters and assistants during his frequent fevers. She was confident medicating herself, given her iron constitution and her experience in treating his many tropical afflictions. Further, she "had the firm conviction that if a woman went alone, without armed escort, and lived in the villages, she could make friends with the women and secure authentic and valuable information concerning their tribal customs and habits."[2]

In the American summer of 1924, Delia sailed to Africa to find "uncivilized" Africans as yet unaffected by Western influence. She had observed a "craving for the barbaric" and the "primitive" in the America of her time, seeing its influence in jazz, fashions, jewelry, makeup, and "negro spirituals." She believed her compatriots were entranced by a deluded imitation of the actual "primitive," which she sought in equatorial Africa. She resolved to "view the natives . . . as naturally as I viewed the specimens I was collecting," setting aside romantic Western notions in doing so. The results are in equal measure interesting, remarkable, and difficult to read due to the prejudices of her time.[3]

Akeley funded her ethnographic expedition with a commission from the Brooklyn Museum to collect specimens of rare African antelope, though she didn't get a permit for okapi. She therefore first spent several months shooting antelope in British Kenya before traveling to the great lakes of Central Africa. Here, she was advised to seek out the "primitive" pygmy peoples living in the Ituri forest in the Belgian Congo.

Akeley describes frustrating weeks of trying to find pygmies who were not already familiar with being photographed and questioned by outsiders. Finally, she found what she regarded as an unspoiled village where she met pygmy women who were visiting to barter elephant meat and an okapi skin for the villagers' palm wine and vegetables.[4] She introduced herself and then took up residence in the camp of these tolerant people, who included her in their subsequent activities.[5]

In her book *Jungle Portraits* (1930), Akeley reports seeing okapi in the wild during this three-month period. She was able to because her Mbuti hosts allowed her to accompany them on their hunts, despite her inadvertently sabotaging several of these. On one occasion, Akeley was on an elephant hunt with her hosts when they unexpectedly came across a little group of okapi "standing on the sandy bed of a shallow stream of water." Her encounter was remarkable because what she witnessed, very unusually, was "an okapi family"—"male" (unlikely to be the father of that calf), mother, and baby.[6]

While Akeley was transported by this vision (painful biting insects notwithstanding), her companions were intent on getting close enough to kill an okapi for food. Something startled the mother okapi, and she encouraged her calf to get out of the water, so they fled just in time. The male was felled by poison-tipped pygmy spears. Akeley describes this okapi as larger and darker in color than those she had seen preserved in European museums. She did not manage to preserve the skin.

Reflecting on the incident, she notes that she regretted "killing or seeing an animal killed for any other purpose than food or science." In her opinion, Africans had a much better right to kill such animals than white men did, and, she adds, they were unlikely to slaughter them all. She did not see another okapi, but her book features a photograph of two okapi calves who had been brought to the Norbertine mission at Buta, where they survived for several months.[7]

The Rainforest Safaris of Attilio and Ellen Gatti

Italians with an interest in African exploration would have been aware of the existence of the okapi soon after it was discovered, as Italian museums acquired a number of specimens very early in the century. This was partly owing to a formal agreement of 1903 according to which the Kingdom of Italy would supply military personnel to the Independent State of the Congo, and Italian colonists would be encouraged to settle the fertile Kivu region. An okapi was gifted to the king of Italy by the Independent State at this time and was mounted and displayed in Rome. As it turned out, an investigative mission to the Congo led by Eduardo Baccari witnessed atrocities in the colony and advised against colonization or the formal supply of military personnel. Nevertheless, for several years Italian soldiers did serve in the Congo and sent back their natural history collections to Italian museums.

Lieutenant Antonio Ribotti sent a skin and skeleton to the natural history museum in Genoa in 1905, and in late 1907, the collection of Captain Emilio Piola was donated to the natural history museum of Parma University, including two mounted okapis. Further okapi remains were sent to Turin in 1909 or 1910, probably donated by Prince Vittorio Emanuele, the Count of Turin. The count deposited the bulk of his collection from an African hunting expedition (1908–10) in the Museo di Storia Naturale of the University of Florence, including a taxidermized okapi.[8]

Attilio Gatti was a flamboyant explorer, filmmaker, and author who completed thirteen expeditions in Africa. Born in Vonghera, Lombardy, he began traveling to Africa in 1922, settling in the United States in 1930 after getting into financial difficulties on his seventh expedition. A showman who played up the safari-leader stereotype, he retained a sense of humor about himself. In the United States, he became a minor television personality and was good at publicizing his travels (with the help of his American wife, Ellen), for example, his expedition to the Rwenzori mountains in a nine-ton jungle yacht built by International Harvester. He made two lengthy expeditions in search of the okapi and other rare fauna in the 1930s, and one of his great coups was to capture and observe live "okwapis" as he insisted on calling them.[9]

Gatti's Eighth Africa Expedition

For his first okapi expedition (1934–35), Gatti received permits from the Belgian authorities to capture and export two okapis to London Zoo. It is curious that Gatti didn't persuade Bronx Zoo director William Hornaday to support him: Hornaday had offered the Buta mission twenty-five hundred dollars for an okapi, plus five hundred dollars for transport. Perhaps it was because Hornaday didn't believe it could be done without Belgian government support for shipping and care. Running in large measure on bluff and bravado as usual, Gatti only managed to finalize a deal with London Zoo for payment on delivery of the okapi nine months after arriving in the Congo, in October 1934.[10]

Gatti spent two years in the Ituri forest on this expedition, accompanied by his second wife, Ellen. She was a New York copywriter and publicity director for "Mr. James P. Bond, purveyor of culture to the elite." She was a woman who claimed to have "shunned sports or exercise" and who "detested the outdoor world." In her snappily written memoir of her travels with Gatti, she recalls that before meeting "Tillo" in May 1931, she took a dim view of explorers, seeing them as "cases of arrested development."

Ellen was therefore astonished to be swept off her feet by Gatti when he arrived at her office as a potential lecturer and almost immediately agreed to accompany him to Africa. After lengthy planning and marriage in Italy, they departed for East Africa, landing in Mombasa in January 1934. They traveled through Kenya and Uganda to Rwanda before heading north into the Ituri forest in the Belgian Congo.[11]

Once in the Ituri rainforest, the Gattis (see fig. 8.1) established a camp they called Tzamboho, northwest of Beni, from which base Gatti set about searching for okapi with the help of pygmy and African hunters. Gatti describes the usual disappointments and near misses of seeking an okapi, seeing only footprints. After many months, a large male okapi was caught.[12]

Figure 8.1. Attilio and Ellen Gatti in camp (Attilio Gatti, *Great Mother Forest* [London: Hodder and Stoughton, 1936], frontispiece).

THE CAPTURES (AND ESCAPES) OF BEAUTIFUL

This first okapi the Gatti expedition captured was trapped in a *zemu*, or pit trap. Gatti hastened to the scene with a large party of helpers (on this occasion, one hundred men). However, the trapped male (which he named Beautiful) was much larger, more powerful, and determined to escape than he had anticipated. Gatti's plans to anesthetize Beautiful and carry him back to camp on a hammock proved impractical. He discovered it was impossible to administer either of the two German anesthetics he had brought, one to be given by mouth, the other to be injected into the cervical vein. Gatti himself nearly succumbed to the chloroform he tried to apply to the okapi, despite his gas mask. His ever-present pith helmet may have been an encumbrance, though Ellen blamed the "antique" mask.[13]

After fruitless hours spent trying to chloroform or inject the animal and with daylight fading, Gatti instructed the men to construct a sturdy palisade around the pit. They then dug a ramp into the pit to allow the okapi to emerge. This work took all night, with Gatti having to do much of the final digging, as the men were afraid to approach okapi.

Once Beautiful was freed from the pit next morning, he surprised them with his forceful determination to escape. Beautiful had observed where Gatti climbed over the palisade to retrieve his camera and rammed it at that point, snapping two sixteen-inch (forty centimeter) thick poles in the process. Inspecting the breach, he turned and gave it a good kick before charging again. The commander rushed back to the palisade, waving his arms to make Beautiful swerve, but he would not. He smashed into the poles and sent Gatti flying some ten feet (three meters) through the air (by this time most of his helpers were hiding up trees and shouting advice down to him). Beautiful squeezed between two splintered tree trunks, shook himself, and galloped off into the forest.

Gatti concluded after this that the okapi has a dual nature, at times docile and sensitive but at others liable under stress to turn into a "stupid, brutal force intent only upon its one obstinate determination to get away at any cost." Clearly, while Gatti could appreciate an okapi as a sensitive, beautiful creature, when it tried (understandably) to thwart his efforts to capture it and escape the yelling crowd of onlookers, he pronounced it stupid and brutal! In this case, natural history observations reveal more about the observer than the observed.

Seeing Gatti's disappointment after the okapi escaped into the rainforest, two of his men, Lumalese and Kunabo, tracked the animal to a nearby waterhole, where it was fastidiously washing off the thick coat of mud it had acquired in the capture pit. Here, they managed to noose and secure it again with ropes. Another struggle ensued before the okapi was tied fast to sturdy trees. At this moment, a violent storm arrived, and all Gatti's helpers retreated to shelter. The okapi, meanwhile, seemed to lose the will to fight, lying down on the forest floor. Gatti decided not to keep the animal tied up in a heavy storm after a long night of struggles and reluctantly cut it free.

Realizing he was free, Beautiful sprang to his feet, "his face resuming that implacable, savage expression." Entrapped by thick vegetation, his only apparent way out was past Gatti, who had staggered back only a few paces. To Gatti's astonishment, however, the okapi veered away and "crashed through the vegetation as if it didn't exist, passing by chance or by fear, or, who knows, perhaps by gratitude, a good six feet from me, disappearing as if sucked in by the green wall."[14]

Gatti came to two conclusions: he should rather try to capture and tame a young okapi, and he should bring along an African hunter in addition to the pygmies he had employed, who seemed afraid of the animal. Carrying out such a plan required a change of tactics, as they would be unlikely to capture a young okapi in a pit trap. This he concluded because he had been told that young okapi remain in a hiding place chosen by their mother for their first months, and when they emerge, they follow their mothers. So, if a pit was used, the mother rather than their offspring would be the animal trapped.

THE CAPTURE OF TOTO

Gatti then employed eight patrols, each comprising one African hunter along with pygmy people, who went out in search of a female with calf. A key figure throughout this period of hunting okapi was a chief of the Mbuti forest people who Gatti called Makulu-kulu (probably not his real name). Gatti claimed Makulu-kulu (see fig. 8.2) prayed to a god Muungu to help him catch an okapi for Gatti in return for a reward of salt, money, a pipe, and a blanket.[15]

Makulu-kulu found the tracks of a female okapi that doubled back on the same day as he observed them, and so he judged it to be a mother who was returning to the place she had hidden her calf. The men tracked down the hiding place of her calf and then, according to Gatti, employed his suggestion to mimic the sounds of a falling tree (one of the few things that frighten okapi). The mother fled, and the calf, on emerging from its thicket, was seized, fastened securely inside a large canvas bag, and carried back to camp.

This calf was nicknamed Toto by the Africans in Gatti's camp and cared for by Gatti himself. He and a local man named Kaluèse accompanied the calf at all times when it was outside its palisade. The problem was what to feed the little

Figure 8.2. Having made his offerings to Muungu, Makulu-kulu, according to Gatti, assures him that an okapi will soon be caught (Attilio Gatti, *Great Mother Forest* [London: Hodder and Stoughton, 1936], opposite 196).

orphan. Initially, they force-fed Toto goat's milk, though he soon fed by himself, three times a day. Photographs of Gatti bottle-feeding Toto later featured in *The Times*.[16]

After a week, Toto began turning up his nose at the goat's milk, however. Gatti bought and installed a cow in camp together with its calf, but after it nearly gored Toto, they had to be separated. (Gatti had to buy the cow outside of the forest and transport it to his camp, as cows tended to die from sleeping sickness in the forest.) Gatti trained Toto to eat some salad leaves (which Ellen grew in a camp garden), to drink water from a bucket (Toto splayed his front legs like a giraffe when drinking), and to lick salt off the ground. Toto was a quick study. The calf was affectionate and tame with Gatti (see fig. 8.3), who became very attached to it.

Gatti observed that Toto seemed happy until sundown, when he guessed the calf was used to the return of its mother. Toto missed being groomed and having physical company through the night, which apparently none of the humans (unlike those in Herbert Lang's party) were prepared to offer. Gatti tethered a goat in with Toto at night, but he objected to its "stupid baaing." As Gatti ad-

Figure 8.3. Gatti established a relationship with the okapi calf Toto (Attilio Gatti, *Great Mother Forest* [London: Hodder and Stoughton, 1936], opposite 221).

mitted, his lettuce leaves, caresses, words of sympathy, and a blanket were "small consolation and quite a poor substitute" for a mother's care.[17]

Where Lang only managed to keep his calf alive for ten days, Gatti managed to keep Toto alive for fifteen days after capture. After Toto died, Gatti performed an autopsy and declared the cause of death to be "an intestinal disease of long standing, evidently contracted before the capture." He suspected that its mother's instincts and natural remedies in clay mineral licks and leaves might have saved Toto but also that the goat's milk given to the calf in his first days of capture had made things worse. A buffalo calf fed milk from one of the same goats also died. It was later discovered that okapis' milk had special properties that the milk of none of these other animals could provide.[18]

FURTHER EFFORTS, A DEATH, AND PERMIT PROBLEMS

Gatti continued to try to catch okapi for a further six months. Although he knew it would be harder to deal with an adult, he resorted again to capture pits. However, when they caught an adult female he was reluctant, given his prior experience with Beautiful, to keep her in a palisade or lead her back to camp, as either of these approaches could result in injuries or even death. He feared, rightly as it turned out, that the Belgian authorities would not be very understanding if the okapi died. The capture party went as far as tying this okapi to a large stretcher, wrapping her in blankets on a bed of leaves, and carrying her through the dense forest. But it was muddy and difficult going, and the shouting of the men and the constant bumping into trees and branches meant progress was very slow and stressful for the animal.[19]

When a violent storm broke out, the men abandoned the stretcher and hid under large trees. Gatti bribed some to construct a makeshift shelter over the animal and returned to camp to organize food for the men. He knew he should have released the okapi but admitted that "after so much labour and struggle, I could not bring myself to do it—a decision which I have bitterly regretted ever since." While he was away getting supplies, the okapi "fell into that state of hopeless discouragement, that nervous breakdown, which is . . . the greatest danger for an Okwapi in captivity," Gatti laments in *Great Mother Forest*. By one o'clock, she was dead.

Gatti admitted that this was a very serious loss for him, as he was limited in the number of okapi he was permitted to capture. The Buta mission, he complained, was "under no limitation whatever, and it didn't matter very much if a live animal arriving in Europe was the sole survivor of six or seven Okwapi which had died in pits, or in transport to the pits, or at the Mission itself." He also knew that many young okapi had died at the mission or en route to it.[20]

As callous as Gatti's observation was, it was also accurate. Like Lang, Gatti argues in *Great Mother Forest* that it was possible to raise and transport an okapi calf but claims that the reason most died was because of the "innate cruelty [of the pygmies] in dealing with animals," resulting in many dying before they reached Buta. Anne Eisner Putnam would later put it another way, arguing that their "instinct was too strong," by which she meant the Mbuti's urge to kill a captured animal so it could be eaten overpowered any rewards offered for its safe delivery.[21]

Whatever Gatti thought of his chances of raising an okapi and transporting it safely to England, the commissioner of the province decided that the death of two okapi was enough and told him his permit had expired. Despite his protestations, including the argument that for elephants the law allowed for a proportion of losses for every animal captured, the authorities would not budge. This was particularly exasperating, Gatti claimed, because just a month previously the regulations had been modified to allow hunting permits for scientific purposes and for the first time okapi had been included on the list of animals that could be captured or hunted. There was no license fee but there was a tax of three thousand francs per specimen. However, permission was required from the Museum of Belgian Congo, and as the governor-general explained to Gatti, no permit would be granted unless "facilities or gifts of a similar value [were] first assured to that Museum."[22]

The problem for Gatti now was that okapi were still falling into the pits dug on his orders, and he was forced to assist in the release of two more adults, and pay pygmies to fill in all the pits. Gatti complains in *Great Mother Forest* that this was "a very sad end, after so much work and sacrifice and expense. And one which, in full conscience, I felt to be neither just nor deserved."[23]

What made it worse was that a young Italian friend of his, Silvio Rota, had been passing through a village en route to visit him when two okapi calves were brought in by villagers to be cooked and eaten. Knowing of Gatti's quest and oblivious to the permit regulations, Rota bought them and was driving to Gatti's camp when the governor of the area stopped him and investigated the crate the okapi were being transported in. He tried to prosecute Gatti for arranging this, but the court accepted Gatti's explanation of events (although he was still fined five hundred Belgian francs). Meanwhile, the okapis had died in the care of the local police.[24]

Gatti salvaged two consolations from his okapi episode. First, he claimed to have discovered a great deal about okapi behavior. Second, based on his examinations of okapi, living and dead (including at the Buta capture facility), he concluded that he had identified "an unknown, well-distinct race of okwapi." He proposed to name it *Ocuapia kibalensis* after a forest region east of the Epulu and Lenda rivers, arguing that his new okapi occupied a distinct geographic range, separated from that of *Okapia johnstoni* by the deep and impassable Epulu (although okapi are now believed to be strong swimmers).[25]

Gatti argued that nobody had noticed this distinct species before because all okapi remains to reach Europe aside from Harry Johnston's had been collected west of this region. Gatti based his identification of a new race of okapi primarily on differences in the skulls (variations in skulls are apparent in Fraipont's monograph), but he also made claims about the differing temperaments of these "races," confirmed, according to him, by different attitudes to okapi displayed by Africans in the territories of the proposed races (his arguments on temperament are considered in chapter 11). To Gatti's frustration, experts at the British Natural History Museum would dismiss his claims (Lankester had noted high variability in okapi skulls in 1910), arguing that much more detailed inquiries and comparisons would be required before a new species could be confirmed.[26]

Gatti's Second Congo (Ninth African) Expedition

Gatti returned to the Congo from 1938 to 1940, aiming to capture two okapi and to photograph and film okapi in the wild. During this expedition, he

Figure 8.4. Pygmy men covering a pit trap (Attilio Gatti, *South of the Sahara* [New York: Robert M. McBride, 1946], opposite 97).

claimed to have taken the first photograph of a wild okapi in a forest south of the Stanleyville–Irumu road, near Mambasa. (The vagueness of captions to photographs of okapi in *Great Mother Forest* leaves it unclear whether they were taken "in the wild" or not.) However, Cornelius Bezuidenhout claimed to have already done so in 1931, the photographs published in the *London Illustrated News*. Regardless, if it truly showed a wild okapi, then Gatti's photograph was still a remarkable achievement.[27]

Gatti again resorted to the help of local Africans, notably a pygmy chief he calls Kotu-Kotu and another named Abdullah. He succeeded in capturing one okapi in a pit (see fig. 8.4), but it was killed by a leopard over night before they discovered it. They caught another fine male that survived in captivity initially but soon died of intestinal parasites.

Gatti once again infringed on his license, claiming to have been frustrated by the authorities' refusal to extend his permit to capture two okapi. According to the Belgian hunting officer G. L. E. Swaluë, however, Gatti only had a permit to film okapi.[28]

Those Left Out

To the histories of the okapi-related exploits of James Harrison, Percy Powell-Cotton, Boyd Alexander, George Gosling, Hermann Schubotz, Cuthbert Christy, Herbert Lang, Delia Akeley, and Attilio and Ellen Gatti, should be added the names of the local officials who facilitated their pursuits—men like Meura, Eriksson, Boynton, Andersson, van der Cruyssen, and Siffer—sometimes at the request of their superiors, sometimes on their own initiative. This assistance was invaluable, though it could also complicate relations with locals who were angered by the behavior of colonial agents and their African assistants, particularly members of the notorious Force Publique, the Belgian Congo's army. Lang had been forced to dispense with the armed escort provided for him by the Belgian colonial government for this reason, despite his fear of cannibals.[29]

Traditional leaders like the chiefs Akenge, Abdullah, Banda, and Kotu-Kotu were essential to the success of okapi collecting expeditions, authorizing activities in the areas they controlled, linking up collectors with local experts, providing security, arranging guides, and supplying labor, notably bearers. Also vital were the often invisible go-betweens like José Lopes, who were part of their expedition, and yet through shared African heritage, linguistic and diplomatic skills, and force of personality could cross cultural barriers and work with locals in ways closed to most Westerners.

The rainforest and the okapi were of course central to all these pursuits. The difficult terrain, obscuring vegetation, and the physical discomfort experienced by Westerners due to the heat, humidity, and frequent storms all made hunting okapi difficult, dangerous, and dispiriting. Tropical diseases had considerable agency in curtailing the activities of would-be collectors, a number of whom died in the rainforest. Advanced hunting technology like binoculars and high-powered rifles capable of felling large megafauna from a comfortable distance on the open savanna was much less effective in the dense forest. African hunters preferred to hunt okapi without the encumbrance of large Europeans lugging heavy gear and crashing about noisily in the forest.

The okapi was an elusive and difficult animal to find: okapi generally live alone, and were said to restlessly wander the trackless forest in ways that were not well understood. As attested by their large ears, they have acute hearing, as well as good vision and no doubt a keen sense of smell, and are easily startled. They are well camouflaged in the dappled light and complex vegetation of the rainforest, and move through it effortlessly and fast.

Most important for the acquisition of okapi, therefore, were the African hunters and trackers who knew the habits and whereabouts of the animal and how to catch or kill it. Without the skills, knowledge, and experience of men like Agukki, Aposho, Etumba Mingi, Koki, Abawe, Lumalese, Kunabo, Makulu-kulu, and many others not named (for example, pygmy hunters who appear in some photographs taken of or by Powell-Cotton and Christy), none of these Westerners would have achieved their aims in the rainforest.

The mass hunts of some pygmy peoples were not effective in catching okapi, as hunters like Christy soon learned. The technique of using pit traps was a much better bet, though if the Western hunter was not present when the trapped okapi was discovered, it was likely to be killed and butchered immediately in a way that ruined the specimen.

To find and shoot okapi, specialists like Agukki were required, and they usually hunted alone, or with an experienced partner. Catching live okapi was usually achieved through supervised pit trapping, though sometimes locals randomly came across and brought in calves. Exceptionally skilled hunters like Abawe and Makulu-kulu could track down live calves via their mothers.

Recovering the names and recognizing the agency of local Africans in finding okapi is not always straightforward, but finding out more about these men's histories and legacies, work I hope to inspire in others, is even more difficult. As David van Reybrouck notes in his remarkable history of the Democratic Republic of Congo, the histories have been written by Europeans. While some oral histories have been recovered, there are few sources recording Africans' experiences from their points of view or relating their personal histories.[30] For pygmy peoples, it is not clear that history is understood or recorded in a way that fits easily with Western notions of linear history and the importance of the individual. As I discuss in chapter 15, however, historical events and personalities have been incorporated into their traditions and oral histories.

What survives are mostly the recollections of Westerners with particular agendas and colored by the prejudices of their times and backgrounds. The local Africans they named usually appear in their diaries or occasionally in the books and articles they wrote about their travels. They seldom appear in newspaper reports about expeditions and hunting for okapi, and are absent from scientific and geographic journal articles. In some cases, European hunters (like J. J. David) took the credit for shooting animals they had acquired from African hunters.

Conclusion

Attilio Gatti's two journeys to the Congo in the 1930s were the last of the major Western expeditions to find okapi. After the Second World War, expeditions to collect okapi (aside from journeys to fetch okapi from the capture station in Epulu) were not as necessary as they had been before. Capture and export operations, begun after World War I, had by 1938 been formalized by the Belgian government and would be set up on a professional footing in 1947. The pointless questions regarding which white man had first seen an okapi or shot an okapi, or had caught and exported an okapi, ceased to matter.

Reflecting on what I have recovered on the first four decades of Western attempts to find okapi, including what seems irrecoverable or only partially so with regard to the contributions of Africans, it seems worthwhile to also acknowledge the contributions of the now largely forgotten European and American men and women I have portrayed so far. These have also been largely expunged from scientific knowledge and exhibitions of the physical remains and natural history of the okapi in the West.

While acknowledging my dislike of the racial prejudices of their times as expressed in the writings of these first Westerners to seek the okapi, and their shabby treatment of African retainers on occasion, along with my unease at their curious mixture of tenderness and callousness toward wild animals, I must confess to growing rather fond of some of these intrepid eccentrics, in particular Attilio and Ellen Gatti, Percy and Hannah Powell-Cotton, Boyd Alexander, and Harry Johnston (perhaps the most interesting, perplexing, and self-contradictory of them).

CHAPTER NINE

Capture, Transport, and Survival of Okapi after 1918

The Zoo, with some 45,000 visitors . . . was among the most popular of all [London's] attractions [during the bank holiday weekend]. Every available elephant carried full loads of young passengers. Congo, the okapi, was the most sought-after animal of all, as the newest and strangest inhabitant. Though he appeared a little shy, he obliged the visitors by coming out into his paddock.

—The Times, *August 1935*

Why, given the terrible record of keeping okapis alive en route to the West, and in its zoos, were they nevertheless exported? There are several reasons, including the diplomatic and economic value of okapi, the prestige of exhibiting one, and the claims for advancing scientific knowledge, which preceded the conservation arguments for capture and export that became prominent starting in the 1960s.[1]

First Arrivals in the West

The first live okapi to arrive alive in Europe was a female named Buta who was delivered to the Antwerp Zoo in Belgium in 1919. The second two were destined for Antwerp and London in 1928, but London's (Congo I) died en route. Britain was first in line after Belgium for several reasons. Relations between Britain and Belgium had improved during the First World War (England had previously been the center of opposition to King Leopold II's activities in the Congo), Englishmen had played a significant role in "discovering" and describing the okapi, and the royal families of Belgium and Great Britain were extremely closely related.

Figure 9.1. Congo, the first okapi to arrive in the United Kingdom, being fed carrots by his keeper Ernie Bowman (known as "the professor") in his (bare) paddock at London Zoo in August 1935 (© F. W. Bond/Zoological Society of London).

King Leopold I of the Belgians was uncle (by blood, not marriage) to both Victoria and her consort, Albert.[2]

The first okapi to arrive safely in England was gifted to Edward the Prince of Wales by the Duke of Brabant, later King Leopold III of the Belgians. Congo II, a large male who arrived in July 1935, was housed in a shed fronted by an enclosed and treeless paddock. He seemed healthy (see fig. 9.1), and after recovering from his journey he was allowed to enter his paddock, where large crowds gathered to view him. He was moved to the antelope house for winter but died in November due to "chronic liver disease."[3]

Starting in the late 1920s, then, live okapis were supplied to Western European zoos. The first five went to Antwerp, then in the 1930s, Paris, Rome, London, and New York all received okapis. After the Second World War, other European zoos received okapis, including those in Copenhagen, Paris, Amsterdam, and Basel. Almost all of these animals passed through Antwerp's zoo (established in 1843) until colonial rule ended in the Congo. The zoo established itself as in effect the national zoo of Belgium, cultivated royal links, and developed strong relationships with the colonial authorities and collectors in the Belgian Congo. The zoo had a similar relationship with the Royal Zoological Society of Antwerp, as London Zoo had with the Zoological Society of London.

New York's Bronx Zoo was the first US zoo to receive an okapi, also named Congo, in August 1937. ("The only living okapi in captivity" exhibited in the 1920s in the United States by Al G. Barnes was a donkey-zebra hybrid.) The zoo also received the second okapi to reach America, in June 1949, unfortunately, for breeding purposes, another male, named Biloto. Congo died without issue in September 1952, but Biloto went on to sire five calves, three of whom lived to adulthood.

Brookfield Zoo in Chicago acquired four okapi in the 1950s, a male (Aribi), a female (Museka) from Epulu in the Belgian Congo in November 1956, then an unnamed male and female from Epulu in September 1959. Other early arrivals from Epulu included a male named Bunia who went to St. Louis Zoo in 1956, and a male and female, Kopo and Viviane, who arrived at Dallas Zoo in September 1960 and went on to produce seven offspring over the next decade.

The first okapi born in the United States, Mr. G, was born to Aribi and Museka at Brookfield Zoo in September 1959, but ended up being transferred to Ringling Brothers Circus in 1960. This sorry tale had its origins in 1954, when Ringling Brothers and Barnum and Bailey Circus decided to enlarge the menagerie that it hoped to exhibit in a separate tent in 1955. It dispatched its agent, McCormick Porter Steele, to Africa to source an elephant, a hippopotamus, pairs of rhinos and zebras, Africans for an ethnographic exhibit, and an okapi. Steele failed to obtain any Africans or the elephant but was otherwise successful. He bought a four-year-old okapi male from Jean de Medina at Epulu in December 1954. This was Aribi, a great prize as the only other living okapi in the United States at the time was the Bronx Zoo's Biloto.

Aribi was flown from Stanleyville to quarantine in Hamburg, Germany, by Sabena Belgian World Airlines. From Hamburg, he was flown to the US Department of Agriculture station at Clifton, New Jersey, for a second stage of quarantine. In February 1955, adverts heralding the okapi's arrival were published in New York newspapers, and McCormick Smith published an article titled "African Jungle Freak." Unfortunately for the circus, the department refused to release Aribi to a traveling circus, as an okapi could transmit rinderpest and hoof and mouth disease. It insisted Aribi could only be housed in a secure, well-managed facility. Henry Ringling North therefore asked his friend Robert Bean to house Aribi on loan at Brookfield Zoo. Brookfield then imported a mate, Museka, the understanding being that any offspring would go to the circus. The prohibition against a circus owning cloven-hoofed ruminants applied only to imported animals.

By 1960, the circus was only exhibiting its menagerie at Madison Square Garden, and so in March, Mr. G was flown to New York. He was shown for two seasons and was moved about among other locations in between. Following a delay in unloading him for the spring 1962 season at Madison Square Garden, Mr. G stepped from his crate and dropped dead. He was the last okapi exhibited by circuses, his remains going to the American Museum of Natural History. In 1966, Aribi was sent to Busch Gardens in Tampa, Florida, where he survived until August 1978.[4]

Although Aribi was able to live for thirty-plus years in captivity, the death rates of this rare and elusive animal after it was captured and exported tended to be shockingly high. An okapi stud book created by the Okapi Management Website records the fates of exported okapi, the data showing that few of the early arrivals survived for long. Of the okapi born between 1918 and 1939 recorded in the stud book, only Antwerp's Tele and New York's Congo and London's Buta made it to double figures (Tele and Congo died at sixteen, while Buta died at thirteen). Half of the recorded okapi born up to 1950 didn't make it out of the Congo, dying in captivity in Stanleyville or Leopoldville.[5]

Given that until the 1950s okapis were virtually all sourced from the wild and that captive breeding of okapis became the norm for zoos only in the 1960s, it is hard to make an argument that zoos contributed significantly to conserving wild okapi until the late twentieth century. In fact, the question arises as to why zoos continued to request okapi for so long and why the colonial authorities continued to ship okapi out of the Congo, given the high death rates.

Why Capture Okapi? Curiosity, Currency, and Conservation

Okapi were highly desirable, prestige animals for zoos and museums. They were sent to kings, princes, national zoological gardens, and Ringling-Barnum's circus. The survival of okapi in the wild was apparently not a priority for many of these individuals and institutions, for King Leopold's colonial government or the Belgian one that succeeded it.

Further, selling okapi was lucrative. In the interwar period an okapi could fetch as much as 100–150,000 Belgian francs (approximately US$430,000 in 2022). Sales continued until the 1950s, when it was decided that trading in endangered species was not consistent with the Antwerp Zoo's mission.[6]

According to *The Times* in 1939, although the Zoological Society of London conducted a census and made a valuation of its animal collection every year,

Buta—the second okapi to arrive at London Zoo, in 1937—had no value listed because he belonged to the king. However, the yearly report notes it would in any case "be impossible . . . to put a reasonable price to such a rare creature" (though Felix the Indian rhinoceros was valued at two thousand pounds sterling). A story run in the *Evening Standard* in 1943 referred to Buta as the "£1,000 okapi," and "the most valuable animal in the Zoo," but in an interview in 1945, head keeper F. G. Perry valued him at two thousand pounds.[7]

In 1952, the female okapi Zendy (who had been at London Zoo since June 1949) was valued at two thousand pounds (about US$105,000 in 2024), more than either a gorilla or African or Asian elephant. In January 1952, Zendy was still the most valuable animal in the zoo, though as *The Times* notes, at this time the okapi was "beyond price," as she was "in fact, irreplaceable except through the courtesy of the Government of the Belgian Congo."[8]

Until the 1960s, the capture and export of exotic wildlife to zoos was also justified on the grounds that such wildlife had recreational or entertainment value, served an educational purpose, and supported scientific study. Beginning in the 1960s and 1970s, however, in situ and ex situ conservation became routinely self-proclaimed missions for zoos.[9]

Although King Leopold II and the Belgian government after him used okapi as diplomatic currency, as a way to take the spotlight off their toxic colonial reputations, in the interwar years the Belgian authorities also began to justify their rule through wildlife conservation. Concerns had been raised about the conservation status of the okapi virtually as soon as they were discovered. The colonial government claimed that it had matters in hand, but these claims were exaggerated given the extent and remoteness of its territories in northeast Congo, and Westerners with an interest in conserving Africa's megafauna (after having decimated it in previous decades) worried about its ability to protect this famous species. There was concern over the vulnerability of okapi—to African hunters (the meat was valued, as were the skins), and to European collectors keen to "bag" an okapi, dead or alive (for a fee, and or an enhanced reputation) for museums and zoos. As early as 1902, Johnston had urged the British and Belgian authorities to take measures to prevent the extermination of the okapi, writing that

> the only human enemies of the okapi hitherto have been the Congo Dwarfs and a few Bantu negroes who dwell on the fringe of the Congo Forest. How much longer the okapi will survive now that the natives possess guns and collectors are on the search for this extraordinary animal, it is impossible to say. It is to be hoped,

however, earnestly, that both the British and Belgian Governments will combine to save the okapi from extinction.[10]

Colonial Wildlife Conservation

By the 1880s, the decline of African megafauna had begun to alarm British colonial authorities, and so legislation was enacted in the Cape Colony and then extended to other territories in the 1890s. Hermann von Wissmann's championing of protected areas in German East Africa in the same decade was vigorously debated.

In this period, Sir Clement Hill, a keen hunter and head of the African Protectorates Department in Britain's Foreign Office, used his networks to consult widely on the status of African wildlife. Ray Lankester at the British Museum responded with an alarming note on species extinctions, naming the blaubok and the quagga, possibly soon to be followed by white rhinos, bontebok, and white-tailed gnu. There was concern over the scale of elephant hunting for ivory in the Congo, and the status of white-thighed colobus monkeys on West Africa's Gold Coast.

These concerns chimed with wider fears over imminent extinctions worldwide, including bison in the United States, and the destruction of seals, beavers, and sea otters across North America for fur. Hill organized a conference at the Foreign Office in 1900, that was attended by all the European powers with African territories. The resulting Convention for the Preservation of Wild Animals, Birds, and Fish in Africa, urged colonial powers to introduce detailed game regulations, restrict ownership of firearms (particularly for Africans), control exports of wildlife and wildlife products, and create game reserves.

Many European powers did not ratify the convention, however, including the Independent State of the Congo, although the Belgians did enact their own resolutions on April, 29, 1901, and July 27, 1905, to protect wildlife. They also considered founding three nature reserves where hunting would be banned between October 15 and May 15 each year. Technically, okapi were covered by resolutions of the London Convention. However, their home ranges were unknown at this point, and the seasonal hunting bans intended to provide a protected breeding season for wildlife, were useless for okapi as their reproduction isn't tied to a specific time of year.

The Belgian authorities restricted hunting of okapi on September 17, 1902, including them with series B animals as listed in the decree of April 1901. Employees were requested to send back skins and skeletons to the museum in Tervuren. The colonial state developed a network of collectors that included doctors, hunters, missionaries, and soldiers. Rare fauna were bought from local Africans, who brought them in to colonial stations for rewards paid in salt and other commodities. The Belgians had permitted expeditions into the Congo like those of James Harrison and Cuthbert Christy, on the understanding that they would receive a share in any significant species collected.

As Auguste Lameere remarked in 1903, the Belgian strategy was a little haphazard, and, moreover, the Belgian authorities appeared not to be aware that "the British Museum and other museums, when they wish to procure valuable specimens, obtain them easily by the lure of an 'honest little reward'; for the okapi, the 'honest little reward' promised by certain museums or zoological gardens is currently, it seems, around a hundred thousand francs."[11]

Decree no. 272 of July 26, 1910 (which went into force in January 1911) limited hunting in the Belgian Congo to licensed persons, and this included okapi. The tax was only fifteen hundred Belgian francs, however, so hardly sufficient to discourage hunters from hunting okapi given how lucrative they were. In addition, the granting of licenses was arbitrary. A territorial governor could decide which species could be hunted and whether hunting should be banned altogether or just partially.[12]

Tensions between Collecting and Conserving Okapi
The Museum, the Zoo, and the Colonial Authorities

At the same time as new restrictions on killing or collecting protected wildlife were being enacted in the Belgian Congo, there was growing pressure from zoos to acquire the Congo basin region's rare and diverse wildlife. The zoo in Antwerp had been very active in the trade of wild animals until the First World War interrupted this, buying mostly from suppliers like Carl Hagenbeck, Auguste Charbonnier, and Charles Cordier. The zoo also sold or exchanged species and was not at this stage working mainly through official (colonial) channels. That said, while there were legal restrictions on the capture and export of wildlife from the Belgian Congo up until this time, the zoo was still able to procure wildlife owing to the derogations granted by the 1901 legislation, and the individual hunting permits allowed by the 1910 decree. Further, with regard to okapi licenses, there was room for maneuver, and liberties were taken in the interpretation of licenses and permissions.[13]

In the interwar period, under the leadership of Henri Schouteden (director from 1927 to 1946) the Museum of Belgian Congo became a significant actor in encouraging and facilitating systematic collecting in the Congo. Schouteden developed the museum as a scientific institution, aiming to make it an international hub for natural science expertise about central Africa, notably the fauna of the Belgian Congo. He founded the Cercle Zoologique Congolais to unite Belgian and other international scientists in creating a comprehensive inventory of central African fauna and in advocating for integrating scientific goals into the colonial enterprise.

In parallel, the Antwerp Zoo campaigned to extend its influence over collection and distribution of living animals from the Congo and to increase its collection by any means (scientific hunting permits became available starting in 1937). Its role in promoting the colony was linked with the museum, and animals that died in the zoo often ended up in the museum. After the First World War, the zoo received increasing numbers of wild animals from the Congo, facilitated by the fact that more Belgians were living there and sending back animals and by improved means of transportation, notably air links from 1935.

The colonial authorities assured Michel L'Hoëst, director of the zoo at the time, of their support. They issued notices in their administrative bulletin encouraging donations of animals for the zoo, complete with a list of desired animals. The Ministry for the Colonies and governor-general in Léopoldville supported the capture of okapi to satisfy overseas demand. The colonial authorities were, however, in reality torn between promoting the colony through exporting its wildlife for exhibition and protecting wildlife on the ground.

In the interwar years the okapi became established as an iconic species of the Congo: at the Universal Exhibition of 1935, the Palais du Congo included sculptures of two emblematic species, the okapi and the gorilla. Exporting and improving the visibility of the okapi was a means of promoting the Belgian Congo and Belgium's role there. On the other hand, the colonial authorities also created protected areas and an institution focused on wildlife conservation in this period, which were important developments for conservation in the Belgian Congo. Their mission was to protect the region's wildlife, particularly its endemic and endangered species. Between 1925 and 1953, four national reserves were created in Belgian Congo. The first, significant for okapi, was the Albert National Park (now Virunga), established in 1925 (arguably the first national park in Africa) to which significant additions of okapi habitat were made in 1935.

In November 1934, the Institut des Parcs Nationaux du Congo Belge was founded to oversee scientific research, park management, conservation of the fauna and flora, and tourism. Headed by Victor van Straelen, director of the Royal Belgian Institute of Natural Sciences, it became a force for integrating conservation efforts into the running of the colony. Wildlife was now regarded as a scientific, diplomatic, and economic resource.

PROTECTING OKAPI AND THE TROUBLESOME COMMANDER GATTI

One problem for okapi conservation was that because its discovery only became widely known in 1901, it didn't figure by name in older legislation or amendments thereof. It was only with the ratification by Belgium of the Convention for the Protection of the Fauna and Flora of Africa signed in London in November 1933 that in July 1935 okapi were included on the list of fully protected animals.

A new Belgian decree that came into force in 1937 replaced the 1910 degree that had prohibited anyone not possessing a license from hunting. Under the new law, the okapi was included with animals enjoying complete protection from capture or hunting (in line with 1933 London convention rules on class A species that comprised seventeen mammals, three birds, and a plant), unless permits were granted for scientific purposes. The fee for an okapi license was fifteen thousand Belgian francs (a tenfold increase on the "tax" that Attilio Gatti had complained of; only the licenses for giraffe (twenty thousand Belgian francs) and white rhino (twenty-five thousand Belgian francs) cost more.[14]

In a 1936 article in *The Times*, Gatti maintains that the Belgian colonial ministry mistakenly believed that few okapi were being killed, due to their scarcity and elusiveness. According to him, however, locals killed at least a thousand per year. In almost every forest village, he claims, you could see "dozens of large belts of okapi skin . . . besides several chairs . . . where entire okapi hides take the places of canvas." In Epulu, he had counted 129 belts and seven chairs in one village of 155 pygmy people. Many adult okapis were trapped and killed in pits, and young were also caught alive.[15]

Although the Belgian authorities explained to Africans that okapi were no longer to be hunted, Gatti points out that: "no man . . . can enforce such orders when he has to administer alone a region several times larger than the whole of Belgium and including immense expanses of forest for the most part still unexplored." Restating a familiar European colonial theme, Gatti argues that Africans were the greatest threat to African wildlife. The pygmy peoples, he claims, "under the impulse of their enormous appetite," forgot all warnings and killed and ate any okapi they encountered.[16]

Gatti also implied that the Belgian authorities were not adequately protecting okapi, though his own collecting efforts were frustrated by their licenses. According to the colonial ministry, from 1925 to 1936 only ten licenses were granted specifically for okapi, permitting fourteen to be captured or killed. A problem was the interpretation of licenses, which were requested for "a minimum of . . ." the species in question. If applications were made, as in Gatti's case, for the capture and export of live specimens, license holders tended to interpret any animals killed during capture efforts as collateral damage excluded from their allocated totals. Violette Pouillard reports that Gatti interfered with seven okapi, five of which were killed and two of which were released, in his efforts to secure his allocated two for export. Further, there were the okapi that fell into Gatti's pits after his license had expired, and Henri-Martin Hackars, the conservator of Albert National Park in the 1930s, feared that "a long time after the departure . . . of the hunter, many okapis will perish" because pygmy peoples would continue to use Gatti's pits after he had gone.[17]

Gatti was a source of both positive and negative press for the authorities. His expeditions were funded by the Belgian colonial government's Fonds Colonial de Propagande Économiques et Sociale as well as international figures like Walter Rothschild, and Gatti himself was a member of the influential Commission Permanente de la Chasse et de la Pêche, and a filmmaker and writer with international reach. The colonial authority's ambiguity about his activities, including their legitimacy as scientific collecting expeditions, is clear from the on-off nature of the permissions it granted him.

According to Ellen Gatti, in the early 1930s, an initial application by Attilio to capture gorillas had been refused, but he was offered two okapi instead, and London Zoo was delighted at the prospect of receiving these. The zoo, she reports, was "tremendously helpful, and everything was arranged, signed and sealed through the foreign office." Minutes of the Zoological Society of London council confirm that the zoo had agreed in August 1933 to contribute five hundred pounds sterling to Gatti's expedition, to be paid on taking delivery of one or two live okapi.

However, London Zoo was familiar with the risks of transporting wildlife overseas. It therefore required "a letter from some important official such as the Harbourmaster, a veterinary officer, a doctor or a British Government official stating that when the animals were put on board ship they appeared to him to be in good condition, taking their food well and packed in suitable crates." The council agreed that the secretary would apply "through suitable channels" to the Belgian colonial minister for a permit for Gatti to capture and export the okapi.

In August 1934, Gatti wrote to the Zoological Society of London explaining his difficulties in keeping captured okapi alive. He argued that to safely deliver live and healthy okapi to the zoo, he would have to undertake "much more elaborate, patient and long work" at the capture site, at his camp where they grew vegetables to feed them, and at another transit camp near the road out of the region. He proposed a new payment breakdown that would not cover his additional expenses, he claimed, but included an additional payment if the okapi survived in London beyond three months (giving "your Society . . . much bigger gate receipts than anticipated").

When the Belgian authorities discovered that the Duke of Brabant (Prince Leopold) had promised the Prince of Wales two okapi, however, they decided to rescind Gatti's permit. This was a disaster, as London Zoo had agreed to contribute to the expedition's expenses once they received the okapi, and Gatti's book and film deals depended on this too. Not one to take no for an answer, Gatti persisted, securing interim funds from Walter Rothschild, and got his permit restored, receiving this news in Usumbura in Uganda Protectorate, en route to the Belgian Congo.

In September 1934, the Zoological Society of London's council agreed to pay Gatti five hundred pounds on receipt of confirmation by the government veterinary department in Kenya that two okapi who had been shipped from Mombasa were in good health and feeding well. A further five hundred pounds would be paid to Gatti on their safe arrival at the zoo, expected between May 20 and June 20, 1935, providing they were still in good condition and feeding well, with a third payment of two hundred fifty pounds due should they still be alive thirty days after arrival. Gatti accepted these conditions in October 1934. However, Gatti failed to secure and preserve live okapi to export before being held to the terms of his permit to secure only two okapi, alive or dead.[18]

In light of locals' use of okapi and the collecting activities of foreign museums and zoos, Gatti recommended that in order to protect okapi sanctuaries (what he called "Whipsnades"[19]) should be established in areas that even pygmy peoples avoided (based on taboos, according to Gatti) and where okapi were particularly abundant. Law enforcement could be more effective in such delimited areas, and okapi would gravitate to these "absolute reserves," where they could be studied by scientists (saving them the trouble of finding them). Permits to hunt okapi could still be granted for areas outside these sanctuaries, Gatti suggested, though a fairer licensing system less skewed to Belgian interests and institutions was required.

Gatti asserted that should pygmy peoples particularly desire okapi, they could find them outside of the okapi sanctuaries. He maintained that the market for okapi skin belts among African villagers living around the forests required regulation and that indelible marks ought to be applied to all existing okapi skin belts and chairs. Rewards could be offered to Mbuti hunters who inadvertently caught okapi in traps if they were delivered alive to the territorial administrator, although it was objected that if the pygmy peoples understood the ban and found or caught live okapi calves, they might kill and eat them rather than bring them in and risk prosecution.

According to Nell Boeykens, what the colonial authorities took from Gatti's suggestions was that they should be concerned about the killing of okapi by Indigenous Africans. Van Straelen, president of the National Parks Institute of the Belgian Congo, argued that although the okapi might be more widely distributed than previously thought, it was still sufficiently rare to be regarded as threatened. Further, Gatti had written that it was in a "precarious state biologically" because its "own internal production of parasites was extremely developed." While hunting and trapping okapi had been forbidden in a recent extension to the area of the Albert National Park, which included the forest east of Beni, the major challenge, according to van Straelen, was that Africans regarded okapi as sacred and valued its skin for a variety of purposes.[20]

Africans' killing of okapi also undermined the Belgian authority's prestige as colonial rulers, since it implied it was unable to control locals, and raised questions about its rationale for limiting European hunting and collection, given that the locals were already killing okapi. The Belgians wanted to ban all African hunting of okapi, but there was internal debate over the negative cultural consequences for Africans of banning okapi skin adornments. The response ultimately was to try to take more direct control of hunting and collection.[21]

There was no move to establish sanctuaries specifically for the protection and study of okapi in areas with low human populations. The idea was criticized by Robert Leiper, a pioneering helminthologist and director of the Institute of Agricultural Parasitology in St. Albans, England, who was concerned about the proposed corrals for young motherless okapi. Leiper had been director of the Prosectorium at London Zoo, where he recorded the parasites of the okapi. He argued that keeping okapis in larger numbers in semicaptivity would result in an accumulation of infection with intestinal parasites. Intestinal parasites were already a problem even for wild okapi, which are solitary. Captive okapi would lose condition and die within a few years, he predicted (accurately).[22]

For the colonial authorities, there were in any case insufficient resources for such an undertaking. Decades later, around 1957, there was a discussion about establishing a national park specifically for okapi. However, it didn't happen owing to debates regarding the rights of pygmy peoples and the concerns of mining companies, until politically it was too late. By the time the country won independence in 1960, no colonial national park for okapi had been established. It wasn't until thirty years later, in 1992, that the Okapi Wildlife Reserve was created in the Ituri forest.[23]

Capture and Export: Frying Pans and Fires

A February 1936 note in *The Times* claims that "as all okapis which have hitherto been sent to Europe have died soon after their arrival there the export of the animals is now prohibited in the hope of saving this rare species from extinction." (According to stud book records, this ban was at best short lived.) To understand why so many exported okapi died, it is first necessary to know who was catching them and how they were caught and kept in the Congo. Many okapi that ended up in captivity between the end of World War I and the late 1930s had been caught by Africans and brought to Belgian colonial posts, mission stations, or to Patrick Putnam at Epulu.[24]

Buta, named after the Norbertine mission there (now the capital city of Bas-Uélé province), was delivered by local Africans seeking a reward in December 1918 (*The Times* story about her arrival in Europe claimed she had been caught three years before she arrived in Europe in 1919).

The wife (unnamed in reports) of André Jacques van Landeghem, a Belgian official, is credited as the first to successfully hand-rear an okapi calf using condensed milk mixed with fresh cow's milk that was suckled by the calf from a bottle. A former milkmaid, she had the skills to raise the animal and kept Buta alive and successfully weaned her before her husband was ordered to send the young okapi to Brussels.[25]

When Judge Charles Smets visited Buta in April 1919, he was astonished to see a nearly full-grown female okapi "walking freely and fearlessly about the Post." Supposedly the "shyest of forest animals," she gathered much of her own food, browsing on the rose mallows (*Hibiscus*) and croton decorating the post's gardens. Smets felt that "the courageous and resourceful" Mrs. van Landeghem deserved "hearty congratulations" for her feat, which indeed was notable compared with the attempts of Herbert Lang and Gatti.[26]

Buta was one of the rare animals brought back by district administrator (and naturalist) Georges Lebrun in 1919 to replenish the depleted collection of the zoo in Antwerp after World War I. It was an achievement getting the young okapi from Leopoldville to Boma and thence by sea free of charge on the Compagnie Belge Maritime du Congo's vessel *Albertville* to Antwerp. On receiving the okapi on August 9, 1919, the director of the zoo, Henri Schouteden, declared himself "full of emotion" to be "in the presence of this famous animal." Sadly, she only survived for fifty-one days at the zoo.

While the Zoological Society of London had tried to buy Buta for a large sum, *The Times* admitted that it was "proper that the owner, a Belgian lady, should have presented the first living example of the rarest of living mammals to the chief zoological collection of her native country," even if it was disappointing for the society "in whose scientific proceedings the okapi was first named, figured, and described."[27]

The Missionary Naturalist, Joseph Hutsebaut

Brother Franz Joseph Hutsebaut (Brother Hymelinus), stationed at the Catholic Norbertine mission in Buta, was an uneducated man who therefore could not be ordained. He had come out to help establish the mission in 1911, mainly to cultivate the vegetable garden, but he also had animal husbandry skills that he had picked up on his father's farm in Belgium. When Africans brought him an orphaned baby elephant, he successfully raised it. Soon, all animals found in need of care were brought to him, and so initially okapi capture in the Belgian Congo was centered on the mission at Buta.[28]

Hutsebaut went on to become an internationally renowned naturalist, collecting for the Museum of the Belgian Congo at Tervuren and the Royal Society of Zoology in Antwerp. Belgian wildlife filmmaker Armand Denis visited Hutsebaut at the mission before the Second World War and described him as a simple Flemish peasant whose head was not turned by his fame for his knowledge of Congolese fauna.[29]

A station for the capture and acclimatization of wild animals was set up at the Buta mission, for which it received annual compensation from the zoo in Antwerp (ten thousand Belgian francs by 1937, nowhere near the value of the animals supplied). Hutsebaut received young okapi from Africans, taming and keeping them at the mission (see fig. 9.2) before sending them on to Belgium. Queen Elisabeth of the Belgians was photographed with an okapi calf at Buta in 1928 during the royal tour of the colony. At the request of King Albert I of the Belgians, Hutsebaut accompanied this okapi, named Tele, safely to the zoo where she survived for sixteen years before starving to death in 1943 when food became unavailable during the Second World War.[30]

Hutsebaut later arranged the capture of okapi for the Duke of Brabant, including the animal named Congo gifted to the Prince of Wales, who donated him to London Zoo. For "his services in connection with the care and transport of the Okapi," the Zoological Society of London awarded Hutsebaut its bronze medal. However, the lay brother found himself in the absurd situation where he would one week hear about Africans being arrested for trying to catch okapi for him and the next be contacted by the authorities in Stanleyville urging him to capture another okapi for the king. According to Armand Denis, he complained

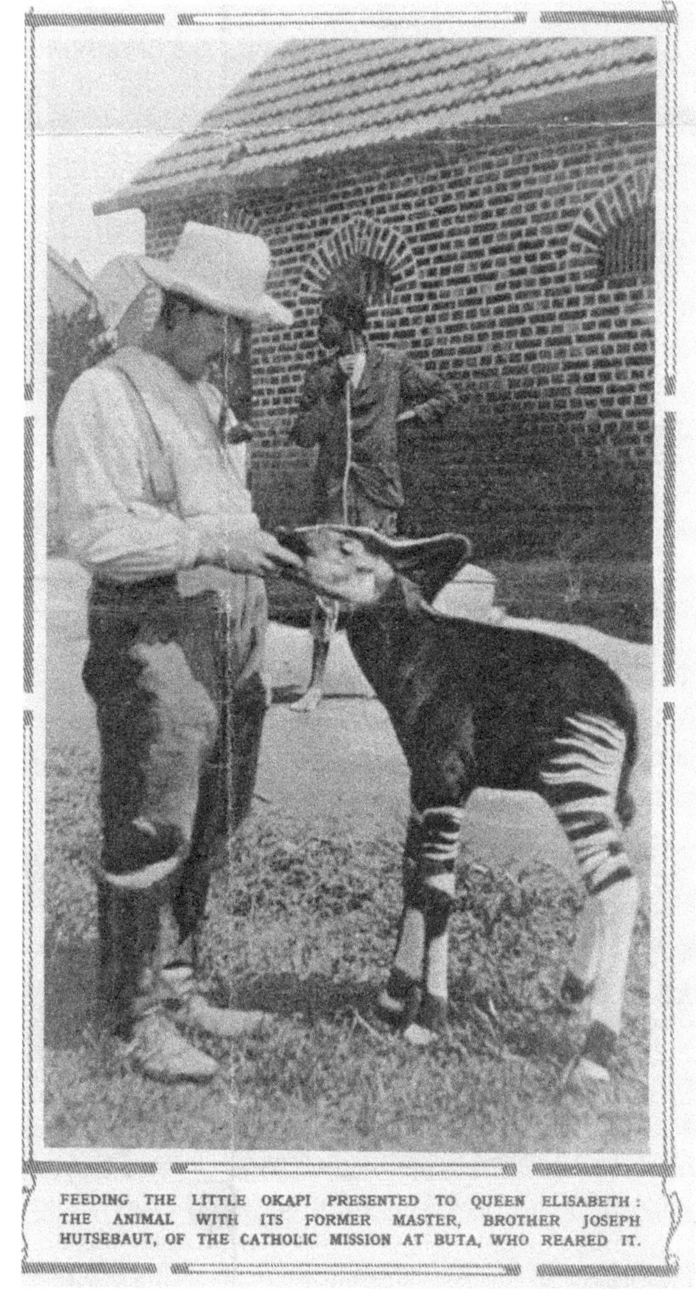

Figure 9.2. Brother Joseph Hutsebaut feeds an okapi calf he raised at Buta (reprinted in *Illustrated London News*, March 2, 1929, 349, Powell-Cotton Collection).

that "I never wanted any of this. I didn't come to the Congo to catch okapi. I came to worship God."[31]

The Eccentric Anthropologist Patrick Putnam

When in 1928 Hutsebaut had balked at taking on a formal role for the government in raising young okapi for export, the eccentric American anthropologist Patrick Putnam stepped in to augment Hutsebaut's efforts. Putnam was a tall, gaunt, long-haired and bearded Harvard-educated New Yorker who went to the Congo on an anthropology expedition in 1927 and settled there, ostensibly to produce a definitive work on the Mbuti people of the Ituri forest. In practice, he inspired others rather than producing anything substantial of an academic nature.

Putnam set up camp in the forest near where the road from Stanleyville to Irumu crosses rapids of the Epulu in 1928, with a concession for a hotel. A village grew up around his camp, known as Putnam's Camp, where he established a medical clinic for the people of the region. According to Pouillard, Putnam's teams of locals captured at least eighteen okapi between 1935 and 1940.[32]

Putnam kept the okapi until they became used to hand-feeding, at which point he had to hand them over to the Belgian authorities, who had their own operation by 1938. According to Armand Denis, "Putnam had a way with animals." When Denis commented that Putnam was "taking over where Brother Joseph left off," Putnam roared with laughter, "like some Old Testament prophet enjoying a good joke."

"Oh, you knew Brother Joseph, did you?" Putnam remarked. "It's a bit hard on the old boy's memory comparing him with someone like me. Besides, I could never handle an okapi as well as Brother Joseph." Both men agreed that Hutsebaut had become "by a strange quirk of fate . . . the greatest practical expert on the okapi that Africa had ever known."[33]

Putnam resumed okapi capture after World War II. According to his wife, Anne Eisner, who joined him in 1946, he did so for personal not commercial reasons (he kept a kind of zoo of local wildlife). Putnam bought wild animals the Mbuti and other local Africans brought in alive, but "most of all . . . he wanted an okapi," and he offered Mbuti people twenty dollars to bring in a live and healthy okapi.

Unfortunately, the Mbuti almost invariably injured the animals they brought in, which soon died despite Putnam's best efforts to treat their (usually spear-inflicted) wounds. According to Eisner, a local confidant named Faizi confessed that the Mbuti's instinct to kill the animals caught in nets for food was too strong to be deterred "by the lure of a reward." Local Africans including the Mbuti then began digging pit traps to catch an okapi for Putnam, and this was how the first was caught and brought in unharmed, in this postwar period.[34]

The Formalization of Okapi Capture

A hunting service was set up in the colony in 1936, following the British model in Uganda, at first comprising just two European officers, responsible for regulating hunting and destroying problem animals. One of these officers, G. L. E. Swaluë, was commissioned by Governor-General Pierre Ryckmans to capture and keep live okapi, which he did in the Kibali-Ituri district (later Orientale province) from August 1938, continuing until 1940.[35]

Swaluë used pit traps that were regularly monitored but developed a less labor-intensive and safer method of moving okapi to the trucks taking them to the capture center than had been used by others. Instead of keeping them corralled at the capture site while a long corridor was built to the road, a bottomless portable cage was built, and the okapi was led on foot inside the cage to the road. In his book of his 1938–40 expedition (the period in which Swaluë was active), Gatti claimed to have invented this approach.[36]

Swaluë caught thirty-two okapi, fourteen of which he managed to keep alive (two escaped). The survivors were later sent to the collecting station established in Stanleyville (a zoo would be set up there in 1951). Like Gatti, Swaluë believed that okapi were more abundant than previously supposed. And like Gatti, he was concerned about the killing of okapi by pygmy peoples as well as about the employees of mining companies who killed okapi for food. Worries about the effects of development on the region, including the introduction of better firearms and the opening up of the rainforest through roads and railways, were also expressed in the debates and journal of the Cercle Zoologique Congolais.[37]

Conclusion

European and American zoos had been keen to acquire okapis from the moment Harry Johnston's discovery was confirmed as a curious new species of giraffid in 1901, but this had proved more difficult than expected. The first live okapi to be exported arrived in Antwerp Zoo in 1919, but survived for less than two months. The mortality rates of okapi exported to Europe and later the US were grim, both en route, and after arrival. However, with extremely high prices being offered by Western institutions for okapi, reaching 150,000 BF in the interwar period, efforts continued. This was faciliated also in that the species was not fully protected by colonial conservation legislation until 1935. From 1937 a license was required to hunt one, with scientific justification, but the licence fee was considerably less than the money offered by Western zoos for safe delivery of a living okapi.

More significant was the establishment of a department for hunting and fishing in 1936, which set about centralizing and controlling okapi capture for export. Where colonial efforts to source okapi (living or dead, they were highly valued as diplomatic gifts) relied initially on company agents and licensed foreigners, and in the interwar period on the efforts of Joseph Hutsebaut at Buta and Patrick Putnam at Epulu, from 1938–40 they were taken on by operations of the colony's hunting officer Swaluë.

At this point, the Second World War interrupted capture operations. There are no records of okapi being exported between 1940 and 1948, and many captured okapi seem to have died young while waiting out the war at an institution in the Congo that the online Okapi Management Site refers to as Kisan Oka in its stud book (I can find no trace of it).

CHAPTER TEN

Zoo Conservation and the Deadly Journey to the West

[N]othing must be done that will hasten the extinction of any animal, . . . [and] the trapping of any species which does not thrive in captivity should also be discouraged . . . unless a really hopeful fresh approach is being made to methods of keeping in captivity healthily. This does not mean that no rare animals should be acquired. An obvious instance is that of a creature like the okapi, the wild stock of which is responsibly and effectively looked after in its own land.

—The Times, *March 1959*

After World War II, in a period of resurgent demand for rare species from the Congo basin region accompanied also by criticism of the high mortality rates of okapi captured in the prewar period, the government of Belgian Congo formalized the process of okapi capture by establishing an operation at Epulu under Jean de Medina. This operation was a turning point for the regulation of okapi capture and brought to an end the haphazard methods carried out by local Africans, colonial officials, and travelers, the shooting and collecting expeditions masquerading as scientific expeditions, and gradually also the operations of professional animal collectors servicing zoos. Still, many okapi continued to die en route to the West, and in zoos after they arrived there. This chapter explores why so many died in transit, and how Western zoos nevertheless forged ahead with ex situ conservation plans for okapi from the late 1950s into the postindependence period.[1]

Growing international environmental awareness along with an embrace of animal rights in this period was fueled by growing public concern over the fate of Africa's wildlife raised by wildlife documentaries shown in cinemas and on television and the books of Bernhard Grzimek and Joy Adamson and others. In Western Europe and the United States at least, activists advanced the case for the rights of domesticated and wild animals whether they were living in freedom or in human care. Zoos felt increasing pressure to justify this captivity.

Western zoos developed narratives about capturing okapi and other rare fauna to conserve them. The efforts of the Antwerp Zoo and the Royal Zoological Society of Antwerp were notable, including their efforts to improve capture and transport of okapi. The zoo, whom the colonial authorities consulted regarding capture and export, began to emphasize the philosophy of "capturer pour proteger" ("capturing to protect"). Walter Van den Bergh, director from 1946 to 1978, transformed the zoo into a scientific institution and set about restoring its depleted collection. He developed relationships with the colonial authorities in the Belgian Congo as well as international networks with other zoos and scientific institutions.[2]

As described in this chapter, one result of his work was the creation of a Depot of Zoological Specimens within the zoo in Léopoldville (capital of Belgian Congo since 1926, now Kinshasa) in November 1946. The aim was to centralize and enable the export of wildlife in the colony, and channel it to the zoo in Antwerp, an international hub for the wild animal trade owing to its convenient port and excellent colonial contacts.

It was only after Belgian colonial rule ended in 1960 and Western zoos established breeding programs able to sustain a (small) okapi population outside Africa that Western zoos' efforts to conserve okapi ex situ became independent of Belgian or Congolese influence. Civil war followed independence from colonial rule, and only a trickle of okapi were exported from the Democratic Republic of Congo thereafter, in 1965, 1967 to 1972, and then again from the late 1980s to 1992, after which exports ceased.[3]

Zoos as a Means of Ex Situ Conservation of Okapi

In the postwar period, conservationists began to argue for the establishment of a zoo population of okapi external to the Congo, to ensure their survival should they go extinct in the wild. In addition, veterinarians feared that the limited range of okapi along with possible fragmentation of this resulting from development in the region could result in outbreaks of infectious diseases (like Europe's 1953 myxomatosis outbreak) and decimate the wild population.

Recommendations to save the okapi from extinction included prohibiting capture and killing of okapi in their natural habitat and moving and acclimatizing okapi in other areas in the region (which never happened). Breeding them in captivity in the Congo and in scientifically managed zoos around the world ended up being the most widely agreed-on solution.

Bernhard Grzimek, charismatic director of the zoo in Frankfurt, West Germany, was a prominent advocate of such views. Tasked with rebuilding the ruined zoo in a bombed city in May 1945, he relocated it to the suburbs and began replacing its decimated collection. Okapi were high on his list of animals to acquire, even though he notes in his book *No Room for Wild Animals* that "one cannot help asking oneself whether, in view of the lack of success which has so far followed upon capture, it would not be far better to give up further attempts and to leave the animals in peace where they are." His justification was that development was destroying their rainforest home. Further, African workers were settling along the new roads being built through okapi habitat to provide access to mining operations, putting okapi at risk from hunting.[4]

In addition, Grzimek urged that more must be learned about okapi physiology and behavior before they could be safely translocated (even within Africa). The fact that okapi were reaching Europe and then dying in short order was helpful in a way (he maintained), because their corpses could be studied at "specialized European and American Institutes." The experts at these institutes had, according to him, discovered that okapi could succumb to trypanosomes not found in their home territories, which they picked up on their long journey across the Congo en route to Europe. Of course, if okapi were left in their home forests, this would be irrelevant, but ingeniously, Grzimek argued that this was vital knowledge in case okapi might in future have to be moved elsewhere in Africa where such fly-vectored trypanosomes existed, should their native rainforests be destroyed (he was prescient in anticipating mining as a threat).[5]

Grzimek visited the Congo in the early 1950s with his son Michael, collecting a small ark of rare species including a baby elephant, hogs, antelope, chimpanzees, and an okapi that they successfully transported back to Germany. In his book *No Room for Wild Animals*, Grzimek paints an apocalyptic picture of a planet being overrun by "the 'human locust.'" In Africa, he explains, in order to develop the land colonial authorities killed wild animals and decimated their habitats and helped Africans, whose labor they needed, to multiply by supplying medical treatments to help them survive the rigors of tropical diseases. Grzimek argues that zoologists like himself, who were unable to curb human population growth and its impacts, were "striving to save from destruction fellow-creatures whom we regard as just as fine and as worth preserving as ourselves."

One strategy for achieving this was to capture and remove wild African animals to safety in Western zoos.[6]

It is notable that most of these rare African species were going to European and US zoos. The pair Andre and Babesoa, sent to the national zoological gardens of South Africa in Pretoria from Epulu, appear to be the first okapi to be received in an African zoo outside of the Congo. Arriving in August 1956, neither survived their first year there. It seems puzzling to me that I have found no reports of similar operations across Africa since. Why are there no okapi in African zoos outside of the Democratic Republic of Congo (DRC)?

The Royal Zoological Society of Antwerp Steps In

As the first zoo to receive and display okapi in the West, the Antwerp Zoo was also the first to discover how difficult it was to keep them alive. Of seven received between the world wars, five died within months. Roubi survived nearly four years, but only Tele lived a respectable fifteen years. Attempts to breed okapi in captivity in the interwar period proved unsuccessful, and numbers in zoos were very low (in addition to Antwerp, two in London—the first short-lived, one in Paris, one in New York's Bronx Zoo). The first okapi to give birth in captivity were being held at Kisan Oka in the Belgian Congo, as a result of the interruption to export caused by the War. The calves, born in 1941, died very young.[7]

In 1947, the Belgian colonial authorities contacted the Royal Zoological Society of Antwerp and requested that one of its members be sent to Léopoldville to accompany the nine okapi it had in captivity on their journey to Europe. Lodewijk M. G. Geurden of Ghent University volunteered and was duly dispatched. However, a week after his departure, the zoo received a telegram from Eugène Jungers, the governor-general of the Belgian Congo, expressing surprise that the society had sent a representative to collect okapi when none were available.

Walter Van den Bergh raised this at a meeting of the International Union of Directors of Zoological Gardens (now World Association of Zoos and Aquariums). They concluded that the colonial administration was attempting to cover up the fact that it had caused the deaths of the nine okapi by transporting them under poor conditions. Van den Bergh recommended collecting and publishing all available data on the pathology of the okapi in three languages, Dutch, English, and French, thereby making information widely available that could help prevent such catastrophes in the future. (Although the resulting bibliography included titles in German, it was perhaps a little soon after the War to formally include these.) The attendees approved this suggestion and entrusted the job of compiling the information to the Antwerp Zoological Society.

Because Tele in Antwerp, Buta in London, and Congo in New York survived for fifteen, thirteen, and seventeen years respectively, it was judged possible to acclimatize, keep, and breed okapi in captivity. The Directors of Zoological Gardens (at the same meeting) therefore decided to develop a coordinated effort to discover how best to do so, for which purpose a stud book was established to keep track of efforts across different zoos. This work was entrusted to the Antwerp Zoo in 1947.

In 1948, Mr. G. Willems was sent to Stanleyville (now Kisangani) by the Antwerp Zoo to collect three okapi the colonial administration was offering to European zoos. However, once there he was forbidden to go to Stanleyville and forced instead to take charge of them only when they reached Léopoldville. One okapi died en route to Europe, while the other two made it to Antwerp and Copenhagen.

Van den Bergh traveled to the Congo in April 1948 to meet with Governor-General Jungers about improving this situation. The Royal Zoological Society of Antwerp, which was responsible for looking after okapi arriving in Europe on their way to their new homes in other zoos, had become so alarmed by the death rate that it decided it was necessary to set up an acclimatization station of its own in Stanleyville. Located in a region of the Congo that was the most biodiverse and rich in endemic species, including okapi, a station in Stanleyville would also care for animals owned by Belgians returning home on holiday to prevent them from being released into the wild after having become habituated to humans. The station would ensure that only healthy, parasite-free animals would be exported and only to zoos with experience in caring for okapis.

Unfortunately, Van den Bergh encountered only "systematic obstruction" when he met with Jungers. He had more luck in Stanleyville, where Ernest-Camille Bock, the governor of Orientale province, expressed his support. Bock agreed to the setting up of a holding station for zoological specimens that would be managed by an agent recommended and paid for by the Royal Zoological Society of Antwerp. This agent would however be subordinate to the governor. They would oversee the collection and dispatch of all animals captured by the game department to foreign scientific zoos and research centers. The government would provide housing, labor, materials, veterinary assistance, food, and transport for the animals. The holding station would also serve as a zoo for Stanleyville and supply the collections of the Léopoldville and Antwerp zoos.[8]

The Antwerp Zoo had to commit to making no material profit from the arrangement, and so it maintained it was investing funds with no expectation of a material return. The hope was that the expertise of the Society would contribute to better acclimatization and transport techniques and lead to increased survival of exported wildlife, which would in turn improve the reputation of the colonial government as well as that of the zoo. The colonial government resolved to consult the zoo when deciding where to send protected animals.

However, the colonial government did not in fact cooperate, and in 1949, it again forced the Zoological Society to collect okapi from Léopoldville rather than Stanleyville. Several animals including two gorillas and an okapi died en route. It was only in 1950 that the Zoological Society was finally able to collect its first okapi from Stanleyville, and when the minister of colonial affairs agreed to the proposed principles for the collecting station and received confirmation from the governor-general of the Belgian Congo that he was in agreement with this plan.

In June 1951, the association Les Amis de la Faune et de Flore Congolaises (The Friends of Congo's African Fauna and Flora) was established in Stanleyville, after which Van den Bergh submitted a proposal to the board of administrators of the Antwerp Zoo to construct a zoo there. This was immediately accepted by the zoo, but only approved and signed by the authorities in the Congo in early 1952.

The Mr. G. Willems who had previously traveled out to the Congo to bring back okapi took up this post in Stanleyville in January 1953. Unfortunately, Governor Bock had died suddenly the month before, and the new governor, Luc Breuls de Tiecken, created difficulties. He kept Willems busy expanding the size of the existing zoo, but plans for the holding station and related activities were ignored, and funds were not forthcoming, not even for food for the animals. In June, Willems learned that plans for expanding the city meant the government did not want the holding station to be built on the site originally proposed. No funds had been allocated to his project as of July 1953. The governor proposed sending several valuable animals to the Antwerp Zoo to compensate the Zoological Society for its losses, but the society refused this offer. In February 1954, the governor instead commissioned Willems to establish a new zoo four and a half miles (seven kilometers) outside of Stanleyville, near the airport.

Meanwhile, the Belgian Congo government ignored the agreement it had with the Antwerp Zoological Society and allowed other zoos to source okapi directly from the colony. In April 1946, London Zoo's superintendent Geoffrey Vevers initiated negotiations with a Mr. J. de Quidt, director of the Belgian

Ministry for Colonies, "to obtain a female Okapi as a mate to our fine male 'Buta' which H.M. the King of the Belgians presented to our King before the war." Cecil S. Webb, the zoo's curator-collector, who was at that time in British Kenya, had heard of six in captivity at a farm in the Ituri forest, and Vevers wondered whether "one of these could be spared to solace Buta in his loneliness." London Zoo took delivery of the female okapi Zendy in 1949 (though this particular delivery seems to have received the blessing of the Belgians in Antwerp).[9]

Van den Bergh grumbled that Ernst Lang, the director of the Basel Zoo, had bypassed him and procured an okapi directly from the colonial government in 1950 (according to the zoo, its first okapi, named Bambe, was a gift from the Belgian government on its seventy-fifth anniversary). In 1954, Van den Bergh learned that Grzimek had done the same when he fetched an okapi from the colony in May, jumping the queue established in consultation with the Antwerp Zoological Society (Frankfurt Zoo had ranked seventh in line).[10]

Willems reported that the game department set up by the service for agriculture and forestry of the Belgian Congo had established its own depot at its capture station at Epulu, whose director, de Medina, regarded the establishment of a holding station in Stanleyville as superfluous. The final straw for the Antwerp Zoological Society was the game department's decision in December 1954 to sell an okapi to "Barnum Circus"—not exactly a cultural and scientific institution. The society broke off the contract with the colonial authorities and terminated Willems's mandate at the close of 1954.[11]

The official book celebrating the first 150 years of the Royal Zoological Society of Antwerp gives an abbreviated account of these events. The "presence of foreigners, especially Swiss people" (probably Charles and Emy Cordier) in the colony is noted in the context of wildlife captures and exports. Regarding the termination of the zoo's efforts to establish a holding station and zoo in Stanleyville, Roland Baetens, the author of the book, writes only tersely that "in 1955 it became apparent that the enterprise was costing more than it brought in, and the Zoo withdrew." Perhaps it was also a business decision.[12]

Jean de Medina and the Groupe des Okapis

Despite the better legal protections for wildlife in place in the Belgian Congo from the 1930s, zoo managers had continued to buy wild animals from private individuals, violating numerous laws by exceeding quotas dictated by licenses, mislabeling rare species as more common species, and straight out smuggling. Even the Antwerp Zoo (among others) turned a blind eye to some of this (chimpanzees were much sought after and not always legally sourced). Refusals to grant or extend permits were regarded as administrative obstruction by zoos, and collectors justified ignoring these as "saving" rare species from Africans who would in any case kill them for food.[13]

To address these problems, the Belgian colonial administration established the Groupe de Capture et d'Elevage d'Okapis au Camp de l'Epulu on January 1, 1947. It was put under the direction of Medina, a former hunting officer, and located just up the road from its rival at Camp Putnam on the Epulu River. This group was conceived of as a professional, commercial operation, and by 1954 (by which time both Hutsebaut and Putnam had died), it had taken over the efforts of the amateurs at Buta and at Camp Putnam.[14]

The remit of de Medina's group was later broadened to include the capture of gorillas and bongo antelope and in 1951 it merged operations with the Station de Domestication des Éléphants run by cavalry officer Commandant Pierre Offerman at Gangala-na-Bodio north of Epulu (then a training center for African elephants that employed Indian mahouts and African capture experts). The new name for the combined groups was the Station de la Chasse.[15]

Where Hutsebaut and Putnam had bought okapi from locals or exchanged them for other rare animals, de Medina set out to control and improve the whole process from capture to habituation and transportation. He was an intriguing figure; he was born in the Congo and had ambitions for the colony and was in a way a go-between in the sense discussed in chapter 7. He had formerly worked as a guide and as a hunter. Armand Denis visited him in his home village of Banalia not long after he had been appointed official *conservateur des okapis* for the Congo and described him as "energetic and enthusiastic." De Medina "proceeded to talk non-stop about okapi" complaining good naturedly: "never try catching okapi for a living. . . . I don't mind leopards or lions or even elephants, but these okapi are quite impossible."[16]

At the time Denis visited de Medina, he had eight okapi in captivity to show for more than six months of effort. His teams were monitoring nearly two hundred pits, which trapped almost everything except okapi. Okapi were suspicious of the well-disguised pits; when water seeped into a pit, a bullfrog or two would get in, and the next night the okapi would hear the "the wretched frogs," and so "that's another okapi we don't catch." Two more okapi were caught while Denis was present, and he provides a vivid description of retrieving them.

Grzimek visited de Medina after the war. In *No Room for Wild Animals*, he notes that de Medina was the son of a Portuguese doctor and an African woman and had an African wife. At least in Grzimek's understanding, his African blood and appearance and the fact that he had an African wife meant de Medina enjoyed excellent relations with his African assistants. The Swiss zoo biologist Heini Hediger, who visited de Medina in 1948, believed his mixed origins meant de Medina had access to "the technical possibilities of Europe and the specificities of the Congolese forest" (Boyd Alexander had made similar assertions about José Lopes). There were fifteen okapi in captivity when Grzimek visited de Medina's capture station.[17]

Under de Medina, capture teams were formed consisting of five men, including the team leader. They dug a series of pits (about forty) over about 30 miles (50 kilometers), mostly 330–430 yards (300–400 meters) from a main road. Each team checked traps over 1.8–3 miles (3–5 kilometers) daily, and new pits were dug where fresh okapi spoor were sighted. A group might catch up to twenty-four okapi per year. Pits were approximately 7.2 feet (2.2 meters) long, 25.5 inches (65 centimeters) wide, and 6 feet 5 inches deep (2 meters), and were made less deep than Atillio Gatti's pits to minimize injury. Pits had a 20-inch- (0.5-meter) thick layer of leaves on the bottom to cushion the animal's fall (an innovation also attributed to G. L. E. Swaluë). They were covered with a lattice of sticks and layer of leaves to camouflage the trap.[18]

Once an okapi was sighted in a pit, another screen of vegetation was laid over the pit both to prevent it from escaping and to calm it. A large stockade of sturdy tall poles lashed together with thinner more flexible branches was built opposite the pit. The okapi would be released into this enclosure through a ramp that had been dug from the pit into the enclosure. Leafy branches were hung inside the enclosure, and a bucket of water was placed inside. Okapi were kept in this stockade for a few days to get accustomed to the presence of humans, while a fenced walkway was built to the main path. Okapi were then led along this to the truck and into a transport box, upholstered on the inside with plant material and jute.

At the capture station, the process was repeated in reverse, until the okapi was inside a solid pen of wooden stakes reinforced with a metal frame Throughout this process, no humans (in theory) would touch the okapi. Any wounds sustained during capture were dabbed with a cotton swab on a long pole, as no medication unsuitable for ingestion could be used because okapi can lick almost any parts of their bodies.[19]

Most of this operation was staged and filmed for the 1955 film *Bwana Kitoko*, directed by André Cauvin, which commemorated the visit of King Baudouin of the Belgians to the Belgian Congo. Until then, most Belgian films of the Congo (tightly controlled by the colonial office) had focused on the development of the country, while missionary films had focused on the supposed moral upliftment of Africans. Such films had previously neglected its wildlife.[20]

In the 11-minute sequence on okapi capture in *Bwana Kitoko*, there are a few departures from the actual process, for dramatic effect. De Medina, a portly figure in a khaki uniform and peaked cap, instructs a crowd of Mbuti people to go and catch an okapi, and they are filmed running along a road and then disappearing into the forest, where they dig a pit. Presumably, a captive okapi was chased along the forest path so that it would fall dramatically into the pit that was situated in a clearing where the camera had been set up to capture the moment. Its fall through the covering into the pit is shockingly sudden, even to the expectant viewer. This particular okapi, a large female, was in no hurry to clamber out of the pit into the enclosure, and so was helped along by a hefty kick from one of the Mbuti people hanging on the enclosure fence. The rest of the operation continues precisely as described above including the delivery of the okapi to the capture station, where several okapi are filmed in the pens being fed leaves by African staff and de Medina.

At the capture station, each okapi was looked after by a particular warden for about two months until it was sufficiently habituated to cohabit with a few other okapi, if female. Males were kept apart, or fights would result. Females about to give birth would be separated, as other females might try to take over the young, especially if another one was in season or had lost its own young.

Keepers brushed their okapis, promoting contact and keeping them healthy. Food was provided three to four times a day in the form of branches with foliage attached that were hung on the walls to prevent contact with dung and parasite eggs on the ground. Additionally, salt and water were provided. Dung was removed as soon as possible, as worms were a recurrent problem. Okapis are very cleanly, and would lick themselves, including their hooves, so constant cleaning of enclosures was essential.[21]

Agatha Gijzen claims that the catching and rearing station in Epulu had "perfected" a method developed from "the primitive method of the natives."[22] However, while the initial stages of the operation may have been a considerable improvement, in light of the fact that the death tolls remained dire it is hard to agree that the export of okapi had been perfected.

The Okapis' Route to the West

Eyewitness accounts of journeys made by captive okapis reveal why so many of them died in transit. The main challenge was simply the distance traveled and the use of a variety of unsuitable modes of transport, the transitions between which were hard on okapi. A Mr. W. Wendnagel, sent out to collect four animals destined for European zoos in 1949 (survivors of an original ten sent to Léopoldville), describes how first, the okapi were moved to Stanleyville, each in its own box with a keeper. Boxes were fitted with fly screens, but according to Wendnagel, these were mostly torn, and the wire fasteners were a risk to okapis' tongues. Floorboards of boxes were laid lengthwise rather than crosswise, making them slippery when wet. No straw was put in the boxes, so the animals often lay on the wet floor contaminated with excrement.[23]

Wendnagel recounts that from Stanleyville, these okapi were moved onto an open boat with a crane and from there pushed into a covered cattle barge, a process accompanied by considerable shouting. Concern over the effects of the shouting of African assistants and onlookers on okapi is a recurring theme in European accounts. The okapi were transported by river to the depot at Léopoldville, where they were housed in makeshift shelters. Léopoldville was in savanna country, more than 600 miles (nearly 1,000 kilometers) from the okapis' home territory. Conditions were poor; the facilities were old and cramped and the hygiene poor. De Medina himself expressed his distress at the resulting high mortality rate of the animals he sent there.[24]

About three weeks after the okapi he was traveling with arrived in Stanleyville, Wendnagel reports, they were loaded onto a railway wagon and transported overland from Léopoldville to Matadi. Here, the crates stood in the sun all afternoon before being loaded onto a ship destined for Boma. At Boma, Wendnagel and the okapi waited another five days for their ship to leave on the sea journey to Belgium, which departed on May 19, 1949.

On the sea journey, Wendnagel notes, the okapi got seasick and lost their appetites. The food was unsuitable, the leaves supplied at Boma having gone moldy, leaving only bananas, vegetables, and oat porridge. Four days after departure, one of the four okapi, which was destined for Amsterdam, died of what appeared to be a heart attack. Food taken on board at Santa Cruz on Tenerife arrived dried up and useless. A second okapi, this one bound for London, stopped eating and died on June 5. The ship finally docked in Antwerp on June 13. The surviving two okapi, Bambe and Zendy, were sent on to their respective zoos, Bambe to Basel (where he died two months later) and Zendy to London Zoo (where she survived for three years and three months).[25]

Wendnagel concluded that the journey was much too long, taking fifty-two days to get from Stanleyville to Antwerp. Along the way, there was considerable aggravation and stress for these sensitive, usually solitary rainforest animals. By the early 1950s, the solution was thought to be air transport. While it had been possible to fly from Belgian Congo to Belgium since 1935, and animals could be transported by airplane, the flight (with added local travel) still took four days. By 1950, the journey time had been reduced to two days.

In 1948, Esayo became the first okapi to be flown out of the Congo, en route to Copenhagen. Despite suffering from altitude sickness that resulted in alternating listlessness and great nervousness, Esayo survived the journey and lived on until November 1963. Biloto was delivered to New York in a Douglas DC-4 operated by Sabena in June 1949, and in his new home in the Bronx Zoo he went on to sire five male okapi, dying in February 1967. Epulu was the first okapi to travel to Germany; he was flown to Frankfurt in a DC-4 passenger plane chartered by Grzimek in 1954. Grzimek did his homework on cage designs, and measured up his to ensure it could fit onto the airplane, cleverly making it adjustable to give the okapi more room once aboard.[26]

Air transport wasn't always possible, however. In 1957, an okapi named Alafu was flown from the Congo to Belgium but then transferred to a ship bound for Rio de Janeiro after spending two days at the zoo. Accompanied by an experienced keeper and accommodated in a large stall that was 13' × 16' (4 × 5 meters) and featured heating lamps, Alafu struggled with the food. The tree branches rotted soon after departure, and Alafu refused the bread, bananas, and roots provided. His keeper sat beside him, petted him, and encouraged him to feed. Alafu temporarily recovered but then vomited, collapsed, and died. Addressing the high death rates in transit was hampered by the fact that it was impossible to perform autopsies en route.[27]

Once the okapi arrived at their destinations, the hardships were by no means over. The longer the journey inside a crate, the more likely the animal would be riddled with parasites, intestinal and liver worms proving most troublesome. Habituating okapi to new foods on arrival was also difficult. Many died within the first few months after arrival, and the attributed causes were bewilderingly diverse, including physical injuries like joint damage and injuries to neck vertebrae and a plethora of internal ailments and illnesses such as colic, enteritis (inflammation of the intestine), internal bleeding, cachexia (ongoing loss of skeletal muscle mass), cowpox, meningitis, peritonitis, pneumonia, sleeping sickness (trypanosoma), and tuberculosis. As Gatti reports, some okapi seem to have died of shock or stress.[28]

Okapi Survival Rates and Destinations in the West

Of the okapis born between 1918 and 1939 that survived up until arrival in Western zoos, only three made it to double figures. Half of the recorded okapi born up to 1950 did not survive long enough to make it out of Africa, dying in captivity in

Stanleyville or Léopoldville, with a couple dying at the official capture station at Epulu. Even by the 1970s, when more was known about okapi and reproduction in captivity was not uncommon, the success rate for rearing calves to more than one year stood at 41%.

Of the captured okapi born up to and including 1950 (all but two of whom were captured from the wild), 81% survived for less than six years, while 11% died before they reached a year old. The average age attained was a little over five years, but this average was heavily skewed by the 19% who survived for a decade or more. Of these, most notable were Copenhagen's and New York's Esayo and Biloto (respectively), who both survived to twenty-one years of age, Antwerp's Dasegela (twenty-three years), Frankfurt-am-Main's Epulu (twenty-four years), Chicago's and Busch Gardens' Aribi (twenty-eight years), and Paris's remarkable Dolo (thirty-two years), who sired eleven calves, six of whom survived to adulthood.[29]

Greater understanding of the importance of treating okapi prior to transport to reduce their parasitic load gradually improved survival rates, as did much shorter journeys thanks to air transport. A moratorium on export seems to have been in effect between mid-1950 and 1954 during which time attempts were made to build up a captive population and breed okapi in captivity in the Belgian Congo, in order to remove the need to capture wild animals for export. Building up a captive population would also mean that animals fully accustomed to captivity could be exported, rather than traumatized wild animals. Colonial capture operations for sales to zoos resumed from 1955 and continued as late as 1957.[30]

From the beginning of complete control of the capture and export of okapi in the early 1950s, the colonial administration had compiled a list of institutions eligible to receive okapis. Fourteen institutions were listed, including European zoos like those in Glasgow, London, and Rome, US zoos including those in San Diego, New York, and Washington, and the zoos in Rio de Janeiro in Brazil and Pretoria in South Africa. In total, twenty okapi were requested, including six couples.[31]

Okapis Born Abroad

New problems arose when okapi began to be born in overseas zoos, beginning at the Antwerp Zoo in September 1954. Little was understood about okapi reproduction and parenting, and most calves died. Some died as the result of maternal aggression. Dasegela, the first captive okapi in the West to give birth, did not accept her calf and injured him on the first day when he tried to suckle. Staff members removed the calf and tried to raise it themselves but failed. Gijzen speculates Dasegela rejected her calf because she herself was not reared by her mother. Dasegela went on to give birth to another five calves, none of whom survived to a year in age.[32]

Ebola, born to the wild-born okapi Irumu (and fathered by Dolo) in Paris in June 1957, was the first okapi to be conceived, born, and reared by its mother in captivity. She lived to twenty-two years of age. An okapi born at the Antwerp Zoo in April 1959 was removed from its mother and successfully reared on a diet of donkey's milk diluted with water, later sheep's milk, and solids starting at five weeks. Zoos were of course most interested in what to feed young okapi and also tried to learn from attempts to raise giraffe calves.

From the late 1950s on, okapi were giving birth and raising young in European and American zoos. Besides Mr. G at Brookfield Zoo born in September 1959, there was Nejoma born at the Bronx Zoo in October of the same year. Zamba was the first okapi born in England, at Bristol Zoo in 1963, but died of a fungal infection before reaching a month in age. Zamba II, born at the zoo in 1965, did no better. However, in September 1966, *The Times* featured a photo of an okapi calf at eight weeks of age at the zoo, "the first calf born in Britain to survive" (this was Katala, who lived nearly five years).

Lifespans of okapis in captivity gradually improved with better understanding of their physiology and behavior (at least in captivity). By 1974, there were fifty-three okapis in zoos around the world, forty-one of them born in captivity.[33]

Okapi in Postindependence Congo

Okapi were exported from the Congo again after the civil war that erupted one month after independence in 1960 and ended in 1965. The country was huge, and the political landscape was split. On the political left were radical nationalist parties, most notably Patrice Lumumba's Mouvement National Congolais, advocating pan-Africanist ideas and a unified state that would transform the social order to benefit Africans. On the right were more moderate, region- or ethnic-based parties, like Joseph Kasa-Vubu's Association des BaKongo (Abako), that advocated the creation of a federal state. The Belgians favored those on the political right, as they were less critical of the existing order and overseas influence. So did the Americans, who feared Lumumba would become another Fidel Castro.

When a mutiny of the army broke out and the mineral-rich province of Katanga seceded in July 1960, the Belgians supported the rebels, while the UN supported Lumumba's government. Lumumba (president) and Kasa-Vuba (prime minister) fell out, and the army chief of staff, Joseph-Désiré Mobutu, intervened, engineering the murder of Lumumba. A series of short-lived governments followed during a period of extreme instability, ending only in November 1965 when Mobutu once again deposed the government, taking control as president (he remained in power until 1997). Amid this turbulence, the eight remaining okapi at Epulu were moved to Kisangani and the station was apparently abandoned. Grzimek was told the buildings and equipment had been destroyed in the final phase of the civil war. Staff at the zoo in Antwerp had lost all contact with the facility, and the outside world was left in the dark about the status of okapi in the wild.[34]

Then in 1965, de Medina was asked to return to the Epulu station by the new Zaïre National Parks Institute and capture operations had resumed by 1967. De Medina was replaced by Jean Bosco, who remained director until the mid-1980s. Bosco had retired by the time the Swiss couple Karl and Rosmarie Ruf arrived in 1987 to revive the station, but he was still living in Epulu and helped them. Bosco collected no okapi between 1973 and 1984.

Karl had always wanted to work with animals, and as a young man, he had followed the advice of a professor at the Basel Zoo by taking up an apprenticeship as a butcher in order to learn about animals, inside and out. This got him a job as a keeper at the zoo.[35]

Through contacts he made at the zoo, including a Swiss man whose wife had nursed (and been befriended by) Mobutu's first wife, Marie-Antoinette, Karl was hired to travel to Gbadolite in northern Zaïre in 1979 to manage Mobutu's private zoo. Once a remote village near the border with Central African Republic, Mobutu transformed his ancestral home into his vision of a Versailles in the jungle, replete with three palaces, an international airport, and other trappings worthy of a national capital, including his zoo. (Although this zoo didn't have okapi, Mobutu did keep three roaming free at his residence in Kinshasa, with a private vet on standby.)

In 1984, the Rufs decided to leave Gbadolite, and drove east across Zaïre to see the gorillas in Virunga, passing via Epulu. Seeing the dilapidated state of the old okapi capture station, they conceived of a plan to raise funds to restore it to what it had been in its heyday in the 1950s under de Medina.

The Rufs visited several major zoos in the United States trying to raise funds, but as young unknowns they had little success until a contact at Miami Metro Zoo suggested they approach a wealthy private donor who supported the Bronx Zoo in New York. This turned out to the philanthropist Howard Gilman, who already kept okapi at his private White Oak Conservation Center. Described as "a very gentle, nature-loving person" by Rosmarie, Gilman supported the breeding and reintroduction of rare animals, including the okapi and the bongo. Having secured funds from Gilman and with the help of John Lukas (manager at Gilman's White Oak Conservation Center), the Rufs took on a five-year contract with the Zaïrean government's environment ministry to repair and run the center in Epulu. They reported directly to the director general of the country's national conservation organization, the Institut Congolais pour la Conservation de la Nature, in Kinshasa.

What became the Okapi Conservation Project started in 1987. Lukas, the Rufs, the Zaïrean conservation authorities including Jean N'lamba of the Institut Congolais pour la Conservation de la Nature (who became deputy director of the Okapi Conservation Project at Epulu), and local communities set about rebuilding the station. Since that time, project members have been working consistently to conserve okapi in their natural habitat, a complex and dangerous task given the political volatility and violence that have plagued the region. Karl and Jean N'lamba were tragically killed in a road accident in Uganda in 2003 on their return from negotiating with a rebel group active in their forest region.

In situ conservation in the Okapi Wildlife Reserve is discussed in chapter 15. The focus here is on ex situ conservation and the export of okapi from the Democratic Republic of Congo. Some sixty-six okapi were exported between 1947 and 1990, but the percentage that survived more than a year in captivity remained poor until the 1990s. Three from Epulu were sold to Cincinnati Zoo in 1985 to fund celebrations for the seventy-fifth anniversary of Virunga National Park (which has okapi). These okapi were sent to be quarantined at the Antwerp Zoo, first, where the two males died. Only the female survived and made it to Cincinnati.

After the Rufs took over the capture station, they decided not to export any more wild-caught okapi, favoring young animals raised at Epulu instead. The last three of these to be exported were sent in 1992 (two males, one female), one of whom was still alive in 2024 (therefore, more than thirty years old). Rosmarie was still at Epulu in 2024 (when I was completing this book), serving as in-country director of the Okapi Conservation Project. In February 2022, she had been awarded the Mambasa Patriotism Prize for exceptional contributions to the conservation of the Okapi Wildlife Reserve. She told me in September 2023 that it remains dangerous to travel in and out of Epulu by road and that access requires a series of regional flights. Okapi are no longer kept in captivity at the station in Epulu.[36]

Zoo Conservation

In an article on keeping animals in captivity published in *The Times* in connection with a planned remodeling of London Zoo in 1959 it was argued that scientific zoos must avoid anything that might hasten the extinction of an animal. This meant not encouraging the trapping of threatened species or of animals that didn't thrive in captivity. However, despite the high death rates, the article presents the okapi as an exception to such proscriptions on the grounds that wild okapi were being effectively and responsibly conserved in the Belgian Congo and that only surplus specimens were exported to selected zoological gardens. Nevertheless, for scientific zoos like London Zoo, the article concludes, the conservation of threatened species including okapi was regarded as a core mission.[37]

In the 1960s, the idea of the heroic white hunter familiar from Hemingway's stories and Hollywood feature films shot in East Africa began losing its sway and was superseded by new role models including conservationists, zookeepers, and natural scientists like Joy and George Adamson and Jane Goodall. Much more empathetic relationships between humans and wild animals began to be portrayed in books, and feature and documentary films.[38]

In addition, animal rights became a mainstream ethical concern in the 1970s. The movement accorded rights to individual animals, and so it became harder to justify the loss of individuals for the greater good of the species, an idea that is still debated in conservation circles today. Whatever the case, shooting okapi for sport or science was no longer acceptable.

Animal rights and conservation NGOs began to put pressure on zoos, accusing them of commodifying wildlife by their participation in the international trade in wild animals. Prominent authors including Jean Dorst of the National Museum of Natural History in Paris, Gerald Durrell who opened his own zoo on the island of Jersey in the English Channel in 1959, and Desmond Morris of London Zoo, criticized zoos for indulging in irresponsible practices to source rare species, and for how they kept wild animals in captivity. Zoo magazines phased out heroic stories about the capture of rare animals, and the *International Zoo Yearbook* dropped its list of animal dealers in 1962. Scientific zoos increasingly argued that captive breeding could replace their reliance on sourcing wild animals.[39]

The Convention on International Trade in Endangered Species (CITES) came into force in 1975. Zaïre signed up as a Party in October 1976, and overhauled its wildlife legislation in 1982, but Belgium only ratified the convention in 1984. After this, very few animals were sourced from the wild. Okapi were (and are) listed as Appendix 1 species, that is, species for whom no trade in wild animals is allowed and whose export is controlled by permits to be granted by the governments of countries of export and import.[40]

These limitations with regard to breeding stock have had consequences. Europe's captive population descends from only twenty-three okapi and North America's from twenty-five. There is a lack of genetic variation, with very few animals caught in the wild since the late 1980s. With these concerns in mind, the Royal Zoological Society of Antwerp had organized an international conference on captive breeding of okapi in 1977, starting a long process to better coordinate the breeding of a viable captive population of okapi. In the early 1980s it sought cooperation between European and American partners.[41]

In 1981, the American Zoo and Aquarium Association set up species survival plans to coordinate the collaborative breeding of endangered and threatened animals in captivity. The initial focus was on how to maintain a genetically diverse, healthy, and stable population where captive breeding was viable. Okapi were one of ten species chosen to have a plan developed for them. Keeping track of exchanges of okapi for breeding purposes was necessary: for example, in 1981 an okapi birth was reported at San Diego Zoo to a mother on loan from Cheyenne Mountain Zoo in Colorado and a father (Mokola) born at San Diego Zoo. In 1987, visitors saw a calf born in San Diego Zoo at 2 p.m. to a mother (Kenge) on loan from Brookfield Zoo sired by Mokola. Brookfield, San Diego, and Dallas zoos now had stable breeding pairs. Okapi were also held by eleven other US zoos.[42]

The "Noah's ark paradigm" which developed in the United States over the following decade led zoos to focus on conservation rather than simply recreation, but the downside of this paradigm shift was that essentially any species with a champion could be added to the list of those with plans. This was a problem— as William Conway, director of the New York Zoological Society, noted in 1986—because due to the limited space available, international zoos could only maintain a maximum of 330 species at a sufficient size (250–300 individuals) to ensure a viable population for two hundred years in captivity.

In 1990, the American Zoo and Aquarium Association taxon advisory group was formalized to address this proliferation of species survival plans (by 1995 there would be 77 such plans for 128 species) and return the focus to creating plans only for species for which complete, active programs could realistically be developed. The new selection process introduced additional criteria: there had to be sufficient spaces available for the animal as well as the resources to operate a plan successfully, including a credible plan coordinator, a stud book keeper, and a supporting management group. Education, professional training, and fund-raising for conservation projects in the field and for research were also incorporated into species survival plans.

In 1977, fourteen European zoos launched the European Breeding Preservation Program through which they cooperated with each other and the United States on the exchange of okapi. An international stud book was maintained for the entire captive population, which showed that this still fell well short of the minimum size judged necessary to ensure a viable long-term captive population.

By the mid-1990s, criticism of the paradigm of saving species by breeding them in zoos had become widespread. The considerable sums spent on attempts to breed giant pandas and Sumatran rhinos in captivity were denounced, respectively, by renowned conservationists George Schaller and Alan Rabinowitz, on the grounds these attempts had caused many deaths and yet only achieved limited reproductive success.[43]

One justification for captive breeding rested on arguments made in the 1980s by Michael Soulé, a founding figure in the establishment of conservation biology. He worried about avoiding inbreeding in small, isolated populations of wild animals. Captive animals could be used to offset these effects in beleaguered populations. However, the influential New Zealand population ecologist and conservationist Graeme Caughley countered that there was no evidence of wild populations going extinct because of such genetic problems. Rather what was in question was the genetic viability of captive populations.

A 2015 study finds that despite careful management of okapi in Europe, the genetic diversity of this captive population was not representative of the genetic diversity of the recently sampled wild populations from across known okapi range in the Democratic Republic of Congo. Mitochondrial variation within the captive okapi population is much reduced compared with that of sampled wild okapi. Further, genetic differentiation is much higher between individuals from the captive population and individuals from the wild population than it is between individuals from wild okapi populations from different regions.[44]

Others spoke out against captive breeding as a conservation strategy. Andrew Balmford and colleagues argued in 1995 that data suggested it was always cheaper to conserve mammals in the wild. Snyder et al. argued in 1996 that hardly any species continued to breed successfully in captivity for long periods of time and that most reintroduction programs failed to establish viable wild populations.

John Oates argued that many zoos were simply a drain on wild populations, which were best left alone in their natural habitats (where conservation efforts, he maintained, should be focused). The education and inspiration zoos offer (Oates had been inspired as a boy by visits to London Zoo) could just as easily be provided by nonthreatened species.

Most of these authors argued that zoos should focus on education and funding research and in situ conservation rather than captive breeding. The chances of releasing captive-bred individuals into the wild were and remain slender, despite the use of the term "insurance population" by the European Association of Zoos and Aquaria's ex situ program, and occasional publicity regarding captive births in zoos suggesting otherwise.[45]

Conclusion

While much was learned about captive care in the postwar years and the efforts were laudable, it is sobering that at the close of the twentieth century, there were only just over one hundred okapi in captivity at the Epulu site and in zoos worldwide. Ever since rebels killed the captive okapi in Epulu in 2012, none have been kept there. By 2014, there were 172 housed in fifty institutions worldwide, 66 of which were housed in Europe, and a decade later there were only 185 in zoos worldwide. This is well short of the target population for the international breeding programs—the 270 animals (or 220) judged necessary to guarantee a stable and genetically healthy population.[46]

More than a century after the first okapi arrived in a Western zoo, they are as yet unlikely to be saved for posterity by captive breeding should they suffer a terrible catastrophe such as a disease epidemic, or a combination of this together with escalating habitat loss and hunting. Zoos have made important contributions by publicizing the existence of okapi and explaining why it is important to conserve them and their rainforest habitat and by supporting research and conservation work in the Democratic Republic of Congo. Judging by how few people outside of the Congo region seem to have heard of okapi, though, it seems they still have much work to do.

CHAPTER ELEVEN

Nature of the Beast

The sensitive, delicate head, the gentle, intelligent eyes, the huge
ears pointed forward so alertly . . .

The eyes . . . glassy and cruel, the whole expression . . . [like] a
stupid, brutal force intent only upon its one obstinate determination to get away at any cost.
—*Attilio Gatti, in both cases describing the okapi "Beautiful," 1936*

This chapter and the next occupy a space, like the okapi in the rainforest, deep
within this book. Around it I have arranged chapters on human representations
of this enigmatic being, and their consequences. In the preceding chapters I have
suggested what little really had been learned about the okapi in the first half-
century after it had been discovered by Western science. This had been extrapo-
lated (mostly) from the durable remains of the animal, extracted from its rainforest home in the heart of Africa and sent for inspection to the metropolitan
centers of calculation in Europe.

The focus here—and in the next chapter, which covers the period after World
War II—is on what Westerners *have* learned about this elusive giraffid. This
knowledge goes beyond the skins and bones and teeth of previous chapters, to
include what is known about the okapi's physiology and behavior. This, too, is
limited and mostly based on brief encounters in the wild or on observations of
captive or deceased okapi. However, the outlines of how knowledge of aspects
of okapi physiology and behavior developed over the course of the century will
emerge. Much of what was gleaned and published is gathered here.

As the opening quotation suggests, okapi behavior could baffle and upset observers. Okapi confounded expectations and aroused contradictory reactions. This raises questions about the supposedly primitive, and simple, nature of this animal.[1]

First Intimations: 1890–1919

Stanley notes tersely in his book *In Darkest Africa* (1890): "the Wambutti knew a donkey and called it '*atti*'. They say that they sometimes catch them in pits. What they can find to eat is a wonder. They eat leaves." I assume Stanley was remarking on the dietary challenges for a presumed equid (grazer) living in the rainforest rather than on the Mbuti's trapping skills.

The literature in English until 1906 teaches little about okapi habits. The previous year, James Harrison had noted only that it was solitary and did not go far if startled. He claimed it rooted and ate mud like pigs and also consumed shoots of bushes and trees. George Gosling who accompanied Boyd Alexander on his expedition which included a stay in the Congo, but died there of tropical fevers, kept notes during his attempts to find an okapi in early 1906. He claims they were usually found singly or in pairs or occasionally in threes and occurred "in the forest contained by the rivers Welle, Libuati, and Rubi." They were "very quick of hearing," well camouflaged, and avoided human disturbances like rubber collecting. For habitat, Gosling writes, they preferred small streams surrounded by muddy ground, where the tall, large-leafed plant (identified as *Sarcophrynium arnoldianum* by Powell-Cotton, now called *Megaphrynium macrostachyum*) grew. Gosling claims that the young shoots of this plant, which he estimates at 10 feet (3 meters) tall, and Alexander estimates at 6–8 feet (1.8–2.4 meters) tall, were their favorite food and indeed "where the plant is not to be found the animal will not exist." Okapi sought the plant in its swampy habitat at night, feeding on a leaf here and there and passing on, until no later than 8 a.m., after which they retired to the forest until dusk.[2]

Powell-Cotton responded to Alexander and Gosling's observations published in *The Times* in 1906 while he was still in okapi habitat in the Belgian Congo, writing that okapi were typically solitary, though a male, female, and calf may frequent the same area. Calves, which he claims were usually "born about the month of May," were left hidden in undergrowth, with their mothers frequently returning to suckle them. He describes okapi as restless and ever alert animals that never lay down to rest in one place for long. In the Ituri Forest, okapi avoided muddy ground (contradicting Harrison and Gosling's observations), drinking at clear streams. They sheltered from rain and carefully removed mud from

FRANÇAIS
BAHR EL
GHAZAL
CONGO
UBANGI
UELE
BOMOKANDI
ITURI
UGANDA
FORÊT
MANYEMA
AFRIQUE ORIENTLE ALLEMANDE
Lado
Yei
Banzyville
Yakoma
Libenge
Businga
Monveda
Bomokandi
Suruango
Amadi
Bonrokandi
Anqu
Poko
Gombari
Nepoko
Adja.nu
Mahagi
Buta
Bumba
Ilmese
Mobeka
Nlle Anvers
Basankusu
Mobena
Basoko
Mawambi
Mundala
Beni
Irumu
Epulu
Makala
Karimi
Stanleyville
Mont Ruwenzori
Coquilhatville
Bolondo
Ponthierville
Lac Albert-Edouard
Lac Victoria
Walikale
Lowa
Lac Kivu
Sendwe
Nya Lukemba
Uvira
Lokandu
Lac Tanganika
Nyangwe
Kasongo
Lac Tumba
Ubangi
Mongala
Congo
Lopori
Lulonga
Ikelemba
Maringa
Busira
Tshuapa
Momboyo
Lomami
Aruwimi
Lindi
Tshopo
Ituri
Nepoko
Rubi
Itimbiri
Uele
Vere
Bili
Bomu
grande forêt
Dungu
Uele
Uel
Giri
Lua
Lilongo
Likati
Bima
équatoriale
Lac Albert
Ruchuru
Lowa
LÉGENDE :
Limite de l'Etat Indépendant du Congo
Limite de la Grande Forêt équatoriale
Limite Sud de l'aire géographique de l'Okapi, d'après Eriksson
Zones de captures d'Okapis
Poste de l'Etat
Cours d'eau
AIRE GÉOGRAPHIQUE DE L'OKAPI

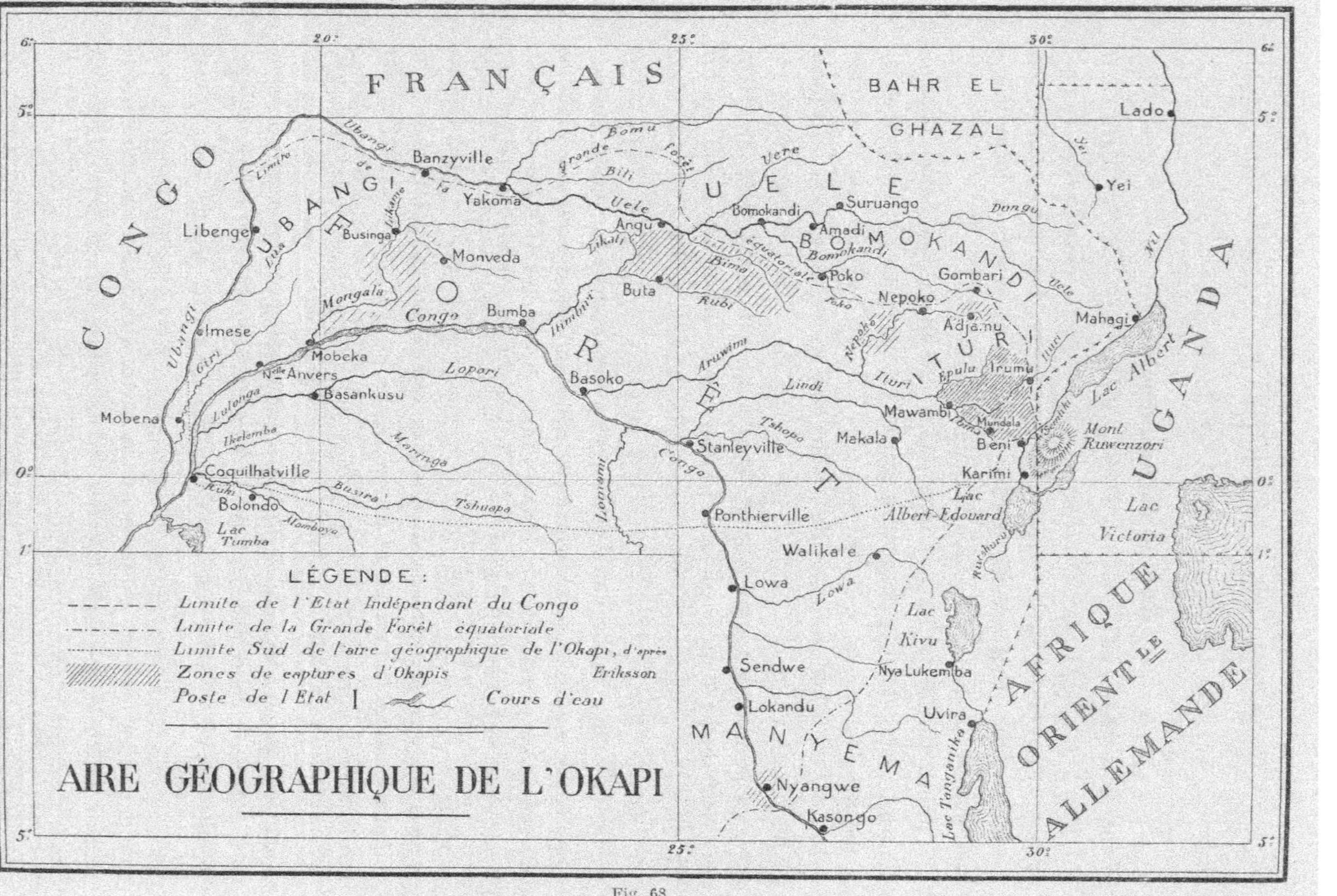

Fig. 68.

themselves by licking. Timidity and cleanliness were their most striking characteristics, and their senses of smell and hearing were acute. Powell-Cotton disputes Gosling's assertion that the *Sarcophrynium arnoldianum* was the Okapi's favorite food. In contrast, he reports that pygmies denied that okapi ate its leaves at all.[3]

Harrison replied to Powell-Cotton, disagreeing that okapi avoid muddy ground and claiming hunters always sought them in swamps. To bolster his argument, he notes that his skins had mud on their legs. Where Powell-Cotton claims okapi sometimes sheltered in Africans' dwellings when vacant, Harrison argues this was unlikely because the huts were too tiny to accommodate a ten-foot (three-meter) tall animal. (Actually, on average adult females stand 5 feet, 2 inches [1.59 meters] at the shoulder, and their necks are not *that* long.) Harrison defends his claim that okapi root for food in swamps and eat mud.[4]

Fraipont's 1907 monograph includes information reported by employees of the Independent State of the Congo, some of it contradictory. Eriksson says okapi grazed on grass in clearings in the forest with small streams running through them but avoided marshy ground. They also fed on leaves in the undergrowth. Anzélius reports they feed on the large, pointed leaves of an unidentified vine. Droppings were reportedly walnut-sized, greenish-gray, and shaped like goat droppings. There were claims okapi had been seen in herds around Adjamu and near Nepoko (by Africans and also by Second Lieutenant Léoni at Nepoko). Others including Anzélius report that okapi lived in pairs, sometimes with a calf and that the female always walked ahead of the male, who keeps watch. Some claimed okapi were—more rarely—found in herds, although Eriksson says that okapi never lived in herds. The okapi was said to be a nocturnal animal, feeding in clearings at the edge of swamps or streams. Anzélius claims they favored dark thickets. By nature, the okapi was regarded as a harmless, fearful, and cautious animal with very keen hearing that could be easily startled.[5]

Fraipont provides a map of the distribution of the okapi (see fig. 11.1), based on information from employees of the company, notably Commissioner Bertrand of Ubangi District in the far west of okapi range, and from the eastern extent of okapi range, Karl Eriksson, and Commander Albert Sillye. Sillye drew

Figure 11.1. (opposite) Okapi distribution map published in Julien Fraipont's monograph (*Okapia* [Brussels: Independent State of the Congo, 1907], 82). The two important early zones for procuring okapi were between Angu, Buta, and Poko (central) and between Mawambi, Irumu, and Beni (east). The presence of okapi in the southernmost patch near Nyangwe has never been confirmed.

on African knowledge, reporting that the okapi was "known by all the natives of Rubi-Uele, and as far as Nyangwe in Manyema" and was seen around Adjamu and in Nepoko.

Fraipont's map shows three main clusters of okapi. The westernmost area, located in what is now North Ubangi Province, extended up the Mongala River from Mobeka on the Congo River to Businga, and as far east as Monveda. Okapi skins were observed further north in the towns of Banzyville and Yakoma along the Ubangi River, but it was assumed these had been obtained through trade. The central cluster was in today's Bas-Uele Province, between the Uele and Rubi rivers (northern and southern borders), extending east from the Likati River and the settlement of Angu, past Buta with its mission, as far east as Poko. This was the region where the Alexander-Gosling expedition had hunted okapi. Okapi were also seen or captured in a few places further east along the Nepoko River and around Adjamu. The easternmost cluster was in the Ituri Forest bounded roughly by the Epulu River and Mawambi on the west, Irumu in the north, and the Semliki River and Beni in the south. This was where Johnston had found the first evidence for the existence of okapi. Only Sillye suggested that okapi could be found as far south as Nyangwe in Manyema District, and only Lomberg (drawing on information from Eriksson) claimed okapi range extended south of the Congo River, all the way west from Beni on the border with Uganda Protectorate, across the Congo River at Ponthierville (now Ubundu), to Coquilhatville (now Mbandaka) near the coast.[6]

Hermann Schubotz adds little to our knowledge of okapi behavior. He reports that the African hunter Etumba Mingi believed okapi wandered constantly through swamps and dense jungle, always alone except in breeding season, making them difficult to hunt. According to the Belgian Congo employee Andersson, okapi share the gait of the giraffe, putting their weight on the legs of the same side at the same time (e.g., right front and back leg pacing together). In contrast, equids walk differently, with a gait following a sequence: left back, left front, right back, right front. Lang confirmed the giraffid gait of the okapi.[7]

According to Herbert Lang, the okapi stood only five feet (1.5 meters) on the slightly raised withers, an estimate that is correct for males but two inches (five centimeters) short for females. Lang describes them as very wary, nocturnal animals that tended to live in ones or twos. Despite their reputation for shyness, they could live near villages if not hunted. He saw the tracks of two okapi which roamed near a village in the forest inhabited by a few old people charged with using a gong to communicate with a distant Azande chief. Lang claims that females always lead, and that okapi frequent the drier, more open, and hilly regions at night, moving to swampy regions with impenetrable thickets toward morning. During daylight hours, okapi would only move during raging storms, driven by the sounds of falling vegetation. They used clean brooks on firm sand as paths. Lang maintains that okapi never rested in the same place but also paradoxically notes that their habits were so regular that Africans said each pair of okapi "has its own village." He opines that its camouflage was of little use in the gloomy forest, arguing it relied on acute hearing for protection. Before Abawe captured a calf for him in 1915, Lang had not seen a live okapi, though he had once gotten close enough to smell its "peculiar odor."[8]

Lang disagrees with Gosling's assertion that the okapi's range coincided with the marsh plant "phrynium" (*Megaphrynium macrostachyum*), which others contended was a staple of the okapi's diet (and which was fed to officials' mules). Lang writes that okapi were browsers, feeding on a few common species of trees and bushes and that their range was not dependent on any one plant, but rather on physiographical conditions. They favored undulating drier areas and avoided extensive swampy areas. Okapi broke saplings and branches down to reach the foliage, he explained, tearing off the skin from their horns in so doing. In his telling, they had survived as a species by retreating into the disease-ridden rainforest, where they remained safe from Western hunters.

Lang includes a distribution map in his 1918 report on his Congo expedition. It shows a continuous area, its westernmost extent situated some 40 miles (64 kilometers) east of Libenge on the Ubangui River. From here, the northern extent of okapi distribution extends eastwards, staying south of the Ubangui until Angu, then dips southeast to Medje, stretches east to end roughly 30 miles (50 kilometers) west of Lake Albert, runs directly south to Beni, west to Makala, and then northwest to pass south of Buta and meet the Monkala River. In other words, Lang joins up the three clusters shown on Fraipont's map but omits Nyangwe in the Manyema Zone to the south.[9]

Abawe located a mother okapi with calf for Lang, tracking it to the calf's hiding place. Abawe seized the calf "firmly as if it were a struggling lion," calling for lianas to secure it. However, the calf was unafraid. According to Lang, he enjoyed affectionate attention. The calf became restless in the evenings, bleating "like a sheep" and searching for his mother, but gentle stroking calmed him. He adopted an African boy as a surrogate, with whom he slept at night. Although only a week old, the calf was active, running and jumping during play. After Lang's condensed milk gave out, he fed the calf a mixture of rice flour and water, but it died "ten days later" (it's unclear how long the calf survived with Lang; it was less than two weeks). Lang doesn't record many behavioral observations of the calf other than its mild nature, tameness, and pining for its mother—deduced from his bleating vocalizations and restlessness. Lang noted that while not annoyed by sunshine, the calf preferred shade. He had the gait of a giraffe, but he did not carry his head upright like one.[10]

By 1918, then, a little had been learned about the nature and habits of okapi. There were arguments over its diet, though most acknowledged it was a browser. Eriksson was wrong: the okapi does not graze. Gosling was also wrong about their diet: recent accounts have verified what pygmy people had told Powell-Cotton, which is that okapi do not feed on *Megaphrynium macrostachyum*. Some confusion arose from its alleged "rooting" behavior and eating of mud, which persisted until the 1930s when Attilio Gatti (discussed in chapter 12) worked out what this was really about. Lang was perhaps correct that the okapi range is not importantly influenced by the occurrence of particular plants but more by physiographical conditions. Gosling's description of okapi eating a leaf here and there and then moving on is accurate, and Lang also provided some accurate information on how okapi feed, though he failed to rear a calf for lack of appropriate food.

Western observers disagreed over whether okapi favored muddy or dry habitats and, by implication, over what some regarded as its characteristic cleanliness. They also differed in their assessments of how social the okapi was, though most reported that it was usually found singly or in pairs. (Reports of sightings of herds of okapi occurred but were rare.) Most assumed it was a nocturnal animal—unsurprising given how much difficulty they had in finding it, a circumstance most attributed to its acute hearing and timid, restless nature. The only accurate knowledge of okapi reproduction was that mothers secreted their calves in thick "nests" in their early months, returning occasionally to suckle them.

Some knowledge was gained from observations of Buta, the first okapi transported alive to Antwerp Zoo, though she survived for less than two months after arrival in August 1919. The problem with knowledge of okapi behavior gained from this young animal was that she had been hand-raised from an early age. It is unsurprising that this supposedly "shy, even wild" animal displayed a "familiarity" that Henri Schouteden found "truly disconcerting," approaching humans and seeking affection. Raised by humans, possibly lonely, and disconcerted by her long journey and new surroundings, she followed her keeper everywhere.[11]

The Okapi's Tangled Family Tree: 1920–40

By the interwar period, there were sufficient skeletal remains to make more detailed studies of the evolutionary history of the okapi. While it was agreed that it belonged in the family of Giraffidae, zoologists disagreed over whether it stood alone as a subfamily or was a genus of the subfamily of "primitive" medium-sized giraffids, the Palaeotraginae.

Degenerate or Merely Primitive?

At first described as a "degenerate giraffe," the okapi was soon classified as a "primitive giraffid," related, it was thought, to the genera *Palaeotragus* and *Samotherium*. While the Swedish paleontologist Anders Birger Bohlin, and French paleontologists Camille Arambourg and Jean Piveteau divided the family Giraffidae into subfamilies including the Palaeotraginae, Giraffinae, and Okapiinae (as a standalone subfamily), the American paleontologist Edwin H. Colbert classified *Okapia* as a genus of the Palaeotraginae.

Colbert's influential 1938 survey recognizes three main giraffid types amid the profusion of subfamilies and genera: primitive medium-sized giraffes with normal-length necks and limbs (genera like *Palaeotragus, Samotherium,* and possibly *Okapia*); large giraffes with elongated legs and neck (including modern *Giraffa*); and gigantic ox-like giraffes with short legs and neck and heavy broad skulls often with two pairs of highly developed horns (genera like *Sivatherium* and *Helladotherium*).

Colbert says the okapi has ruminant-like body proportions (as did *Helladotherium*), with slightly longer hind legs and a level back—unlike modern giraffes' greatly elongated front legs and sloping backs. The okapi skull generally resembles those of *Samotherium* and *Palaeotragus,* being long and low with horn cores above the eye sockets. One of the key differences between these two genera and the *Giraffa,* on the one hand, and the okapi, on the other, centers on the location of the orbit, or eye socket. The okapi's orbit is located centrally above the molars and premaxillaries (cheek teeth), whereas in the others, the eye socket is set farther back in the skull (see fig. 2.3 regarding these differences). Colbert argues that this difference identified the okapi as a primitive giraffid. The diastema (space between the front and cheek teeth) was comparable in *Giraffa* and *Okapia.* Both genera had developed enlarged frontal sinuses, forcing the horn cores towards the midline of the skull, but Colbert suggests this was parallel development, not a sign of genetic relatedness.

Okapi have very large auditory bullae (hollow bony structures enclosing parts of the middle and inner ear), distinguishing them from some fossil Palaeotraginae and the genus *Giraffa,* though not from all palaeotragines. In young giraffes, the bulla is large, but it gets proportionally smaller as the animal matures—and a version of "ontogeny recapitulates phylogeny" (the proposition that the development of an embryo goes through stages that echo the evolution of the adult form of its species) was suggested. Rothschild and Neuville propose that primitive giraffids, like okapi, lived in dense forests, in which acute hearing was more beneficial to survival than sharp eyesight. As different phylogenetic lines of Giraffidae evolved, a process characterized by a shift from forest to plains environments, the emphasis shifted from hearing to sight, with bullae of advanced plains-living giraffe becoming proportionally smaller.

Paleontologists disagreed on whether similarities in cheek teeth between *Giraffa* and *Okapia* were a sign of genetic relationship or simply retention of a common character from a shared ancestral form. Bohlin notes the fossil Palaeotraginae had a characteristic feature of the second upper deciduous molar, not shared by okapi, setting them apart from the Palaeotraginae and linking them to the Giraffinae. Colbert disagrees that this one difference outweighed other larger resemblances.

Okapia also differ from all other Giraffidae in having three instead of four tarsal (upper foot) bones, but Colbert argues (again) that this "single, isolated specialization" did not offset "the numerous correlated characters that mark the okapi as an essentially primitive giraffe." *Giraffa* show a major departure from *Okapia* in the cervical (neck) vertebrae, as Lankester observes. However, the cervical vertebra of *Palaeotragus* and *Samotherium*—apparently more related to *Okapia* in body proportions and skull structure—are structurally similar to those of *Giraffa*.

What to make of this goulash of giraffid resemblances and incongruities? It seems that none of the giraffids are separated by major differences. Indeed, they share many structural similarities across different subfamilies. Colbert attributes this overlap to the rapid development of this relatively young family of mammals from late Miocene ancestors. He insists that the okapi "is a truly primitive giraffid, in many ways more primitive than the earliest of the fossil giraffes." It was thus a "living fossil," surviving with "relatively few changes from the Upper Miocene period to the present day." Its "certain specializations" were "greatly overshadowed by the many primitive characters," and thus the genus *Okapia* belonged in the palaeotragine subfamily of giraffids.[12]

Anatomical Studies of Okapi

While some soft parts (female genitals and stomach) would be sent to Brussels in 1912, preserved in formalin, British scientists would have to wait until 1917 for the first account of some of the innards of an okapi, published after Cuthbert Christy sent the remarkable long blue tongue and parts of the digestive system to the Royal College of Surgeons for examination in the winter of 1915–16 (he also sold them an okapi skeleton). Comparative anatomist Richard Higgins Burne published detailed accounts of his examinations of the animal's tongue and larynx as well as general notes on its heart, lungs, and parts of its digestive system in 1917. It was only in the mid-1930s, however, that British scientists could get access to all of the parts of an okapi, after the animal Congo gifted to the Zoological Society of London in 1935 died and could be autopsied. A series of publications from the resulting work was interrupted by the outbreak of World War II (it was not resumed). Significant further work would begin in Europe in the postwar period, but this would have to wait until the 1950s.[13]

The Legacy of Congo, 1935–43

The death of Congo, London Zoo's first okapi, on November 4, 1935, led to a series of investigations resulting in further contributions to Western knowledge of okapi anatomy. The initial postmortem was performed by Professor Robert Leiper of the London School of Hygiene and Tropical Medicine, Lieutenant-Colonel Albert E. Hamerton (pathologist for the Zoological Society of London), and a few others. "Gerrard" (of the London taxidermist dynasty) had injected the body first with formalin and then later with a red medium for the arteries and a blue medium for the veins. A who's who of comparative anatomists were eager to work on this unusual specimen.

A committee for research on the okapi met in ZSL (Zoological Society of London) rooms on November 19, 1935, to formulate a plan for the dissection. The fourteen-person committee of physiologists and comparative anatomists (most also members of ZSL) was assembled and chaired by Martin Hinton, deputy keeper (soon-to-be keeper) of Zoology at the Natural History Museum (NHM). In advance of the meeting, Hinton circulated a list of all relevant papers on the giraffe and the okapi. A rough, unsorted list of 124 publications survives with the meeting minutes, which were compiled by Julian Huxley. Tellingly, nearly all the works on internal anatomy dealt exclusively with giraffes. Hinton supervised the organization of the work, the editing of resulting reports, and photography of all work completed. Of the few photographs in the NHM file, one is a haunting image of Congo's head and neck on the dissecting table, one ear crushed beneath him, his tongue protruding.

As revealed in letters written by or to Hinton days after the okapi died, Hinton chose committee members for their experience or in response to direct appeals from qualified experts. The team first sequenced the dissection plan so, in Hinton's words, "the efforts of one investigator will not prevent or injure the work to be done by others . . . a matter of considerable difficulty." Hinton planned to use the resulting publications to complete (or complement) Lankester's unfinished okapi monograph, as a second volume.[14] Work commenced in early December, continuing through into January. Manipulating the gradually dissembled corpse in and out of a tank and onto the operating table with pulleys proved difficult. When Leiper returned the liver in early December, he cautioned that it was "so pathological neither the size nor the contours should be regarded as normal." This highlights a problem with studying the anatomy of captive okapis who had died of infections and illnesses. Hinton replied dryly that he would relay the warning "to the vulture concerned."[15]

Only some of the dissection project's findings were published most likely because the work was interrupted by the outbreak of World War II. In 1943, NHM mammologist Reginald Pocock published his second paper based on the dissection of Congo (following a 1936 paper on okapi feet with their tubular scent gland). In the second paper, he explains that he had delayed publication of his full findings so that they could be included in the planned compilation on the general anatomy of the okapi. Hinton and his collaborators, including Pocock, had intended to publish this comprehensive work in the *Transactions* of the ZSL. However, nearly a decade later and in the midst of a protracted war, Pocock determined that "there seems to be no likelihood of this ambitious undertaking being accomplished." As a result, he "decided to wait no longer" and published all his observations on the external characteristics of okapi in the *Proceedings* of the ZSL.[16]

Leiper published a note on helminth parasites obtained from the okapi, and Burne published on the stomach, intestine, liver, and pancreas. Wilfrid Le Gros Clark, who had just taken over as Dr Lee's Professor of Anatomy at the University of Oxford in the year before Congo died, published on the brain. However, that seems to be all, and neither the hoped-for completion of Lankester's NHM okapi monograph, nor even a special issue in the *Transactions*, materialized. Solly Zuckerman, later a member of the ZSL council and its secretary from 1957–77, never published his examinations of the okapi—unsurprisingly given his high research output at the time and his deep involvement in the war effort from 1939.[17]

From the point of view of further okapi anatomical research in Britain, after Congo had been dissected, the problem was once again a dearth of materials. A significant body of information on the internal organs (viscera) and physiology of okapi began to accumulate after they were again sent to zoos in continental

Europe after World War II, often dying soon after arrival. The death of these captive zoo okapi would provide materials for post-mortems and dissections.

Okapis in the Field: The Interwar Years

What of observations of okapi in the field published in the interwar period? More substantial information on okapi behavior was recorded, notably by two men who spent extended periods in okapi habitat, Cuthbert Christy and Attilio Gatti. Gatti provided the most substantial information gathered in the first half of the twentieth century.

Christy published a memoir in 1924 based on two years spent exploring the natural history of the Congo a decade before. He clarifies that okapi were not the swamp-loving, marsh-feeding animals previously reported. Neither did they frequent dense, tangled jungle. The okapi's hooves were not adapted for marshy ground, and in tracking it, he observed it avoided mud and soft ground, crossing streams on firm substrates. During the day it kept to higher, well-drained ground, near a stream, under large trees where the undergrowth was thin. In such places sound traveled well, which made approaching okapi, with their acute hearing, very difficult. Christy also assumes they had a keen sense of smell, and he reports stories of pygmy peoples smearing themselves with okapi dung to approach the animals (although those hunting with him never did so). He describes okapi as shy and "ghost-like in [their] movements through the forest."[18]

Okapis were solitary, Christy observes, though an area of forest may be frequented by a pair of okapi for weeks, each going its own solitary way. When a female had an older calf, the two moved around together, and sometimes a male would accompany them. Christy asserts that okapi fed mostly in the late afternoon and early morning, continuing as late as 9 or 10 a.m., after which they lay down to rest in a spot that provided good visibility. Because of this preference, they avoided resting in thickets or other areas with dense vegetation.

Christy describes how the animal ate by stretching up to full height and using its long tongue to reach small branches and young leaves of undergrowth and saplings. It was not a grazer. Okapi sometimes frequented the large swampy areas in the Ituri Forest with characteristic patches of *Megaphrynium macrostachyum*—but not for feeding. Rather, like elephants, okapi seemed to rest in these spots because it was impossible for hunters to penetrate them noiselessly (it is unclear how this squares with this claim that they favored resting places with good visibility). Christy does not report hearing an okapi make a noise other than "a sharp 'blow' or snort on being surprised."[19]

Attilio Gatti on Okapi Behavior

Gatti lived in the Ituri forest for two years from early 1934. He caught several okapi and kept the calf "Toto" in his camp for fifteen days. In his two books, he supplies the most complete information on wild okapi recorded in the first half of the twentieth century. Most of this appears in his book *Great Mother Forest*, published in 1936.

In this work, Gatti is struck by the okapi's "peculiar fastidiousness," rating cleanliness as its "most striking characteristic." According to Gatti, the okapi "never tires of washing himself carefully," doing so with its long, blue prehensile tongue of "16 or 17 inches [40 or 43 centimeters] in length." Further, an okapi's head could reach its tail, by bending its long neck parallel to its body as if "joined at the base with a hinge." He observed these grooming behaviors at close quarters while looking after Toto. Gatti claims okapi began and ended their day by bathing in a river, although this was apparently an assumption rather than an observation. He reports that okapi bathed by dashing through a sandy-bottomed stretch of river, followed by drying off in a sunny clearing, using its tongue to dry its coat.[20]

Gatti reports that okapi were "confirmed solitaires and untiring wanderers," except for the short mating season and during the few months when a mother nurses young. A group of okapi might have randomly met in a sunny clearing to dry their coats after a bath in a river, but they would soon go their own ways. Diet did not determine okapi's movements particularly. Their free and solitary movements made them hard to locate. Gatti eventually concludes that another reason for okapi's solitary, peripatetic lifestyle was to minimize the risk of infection from parasites from their own dung or that of other okapis. He comes to this conclusion after a healthy adult okapi male he had captured died suddenly, and a post-mortem examination revealed numerous parasites. Gatti considers the okapi diurnal and surmises that every evening each found a sleeping spot wherever it had wandered. This resting place would be clean, dry, slightly elevated, and protected from rain. A "last careful toilette" preceded sleep, the animal's head laid daintily on a liana or high root, and its "hypersensitive ears thrust forward to safeguard his sleep."[21]

Okapi generally moved slowly and steadily, though they were very sure-footed and adept at avoiding mud and pools, Gatti writes. However, if necessary, they could gallop along at six or seven miles an hour (nine to eleven kilometers per hour), even through thick forest. He is impressed by their capacity to pass through thick vegetation by jumping over obstacles or by passing under them. In fact, he rates an okapi's hind legs as equal to those of "a good Irish jumper." Gatti admires the okapi's physical attributes, which served it well in its natural habitat. In particular, he cites their tough, "a quarter of an inch thick" skin, their "short, heavy mane" (actually, only juveniles have manes), and their "woodenish legs [seemingly] only of bone," all of which protected okapi from thorns, scratches, and injury. Okapi used their bony legs and the thick plate on their skull that extended from the first vertebra to just above the eyes to smash aside obstacles. He notes that the okapi used the latter like a battering ram and that anything impervious to a headbutt would be smashed by the okapi's "heavy artillery," a backward kick.[22]

Gatti lists several plants favored by the okapi, providing only Kinande (or Nande) African names: tall reeds called *matungulu, moodi,* which produce succulent red flowers and tasty young leaves, and the leaves of *sangatoto, anzararo, bahapopo, apopo-mongele,* and the shrub *memengano.* Gatti realizes that okapi ingested minerals through licking a reddish, saliferous clay called *bulongo* by pygmy people as well as charcoal from trees burned by lightning. Pygmy people told him okapi used these as medicines when unwell, and he reports seeing evidence of this in okapi dung.

Unlike Herbert Lang, Gatti is impressed by the camouflage afforded by an okapi's coat patterning. In addition to their excellent hearing, he claims their large eyes could move independently of each other like a chameleon, allowing "an extraordinary field of rotation." He also maintains that okapi had rudimentary vocal cords that limited them to uttering only a snort like a horse, though they could also grind their teeth loudly. Gatti does not mention Toto bleating, as Lang claims his captured calf had.[23] Notably, nobody else makes the same claims about okapi's eyes or their teeth grinding as Gatti does.

Gatti thinks okapi numerous, contrary to other accounts of their rarity. Although females bore only one calf at a time and—according to pygmy people only did so every four to five years—Gatti reports seeing many young. Soon before giving birth, a pregnant female would build a nest in a suitable place with tangled, abundant vegetation. She would force her way into the vegetation, using her body to create a narrow, twisting passage until she reached a central, elevated, and well protected spot. She would then enlarge this spot by kicking and headbutting the vegetation. Here, she would give birth to her calf. The calf would remain in this nest for three to four months until it was big enough and strong enough to follow its mother on her wanderings. A mother okapi would leave the calf all day, returning to suckle in the evening.

According to Gatti, not even leopard, the main predator of okapi, would enter these "caves" because, in his words, the "enormous, cruel leopard of the forest has a holy terror of the rage of an adult okapi." Instead, leopard would wait near the entrance to the nest to ambush the calf when it eventually emerged. Gatti believes okapi calves always followed their mothers, and therefore they could not be trapped in pits.[24]

Writing in the mid-1930s, Gatti claims to have no idea what kind or quantity of milk to feed the captive calf Toto, as no okapi mother had been captured alive with her calf. While this was probably true, Mrs. van Landeghem and Father Hutsebaut had previously successfully raised calves brought to them by Africans. Gatti had to improvise. Initially, he and his men force-fed Toto goat milk, though the calf soon fed by himself, three times a day. After a week, Toto refused goat milk, and Gatti brought in a cow with a calf from Irumu. Unfortunately, this cow tried to gore Toto and was removed. Gatti concluded that tinned, pasteurized cow milk would be best for unweaned okapi calves both because goat milk was often infected (which he deduced after other wild animals suckled on goat milk sickened and died) and because okapi seemed to dislike it. Further, a diet of tinned, pasteurized cow milk would allow handlers to feed the calf on the same food throughout its journey to Europe. Gatti taught Toto to eat salad leaves that his wife Ellen grew in camp, to drink water from a bucket (Toto splayed his front legs giraffe-like when drinking), and to lick salt off the ground. Toto was a quick study.

Nevertheless, fifteen days after his capture Toto fell ill and within two hours had died. An autopsy revealed he had succumbed to an intestinal disease, Gatti claims, though the infected goat milk was probably a contributing factor. Gatti notes that one of the goats who supplied milk to Toto lost two of her kids, and a buffalo calf fed on her milk also died. Gatti speculates that "medicines of the forest" may have helped save Toto. However, later research on rearing okapi demonstrated that okapi milk is significantly more nutritious than the milk of these other animals—and that Toto was too young to wean.

Gatti is sensitive enough to speculate that there may also have been psychological reasons for Toto's early death. He finds the affectionate young calf charming and delights in how Toto threw up his head and stamped spiritedly on the ground with a rigid front leg when something went contrary to his wishes. Gatti spent most of his day teaching Toto to eat and drink, to jump over obstacles, and to squeeze beneath lianas. He learned how Toto liked to be caressed, especially beneath his throat where his tongue could not reach.

Gatti claims Toto was too busy to feel homesick, except at sundown when, in the forest, his mother would have returned to feed, groom, and comfort him and to sleep by his side through the night. Then, all Gatti's attentions, including the little blanket tied around Toto at night, were "small consolation and quite a poor substitute." The goat that was left to keep Toto company at night irritated the calf rather than comforting him. Having someone sleep with Toto, as Lang had done with his captured calf, was not an option for Gatti because the Africans with him were afraid to touch Toto (and he doesn't seem to have considered this himself). The way Toto looked at Gatti with big, sad eyes at this time of the day "broke my heart." He justifies Toto's captivity to himself by arguing that Toto did not need to fear rain or leopards or falling trees in camp—but admits that Toto could not possibly realize all this.[25]

Gatti's observations on the nature of the okapi proved influential. He describes okapi at rest as sweet and gentle creatures who went through an "amazing metamorphosis" when alarmed or irritated. This assessment was based on his attempted capture of a large male he named "Beautiful."

When his men began to dig Beautiful out of the pit in which he had been trapped, the animal at first "grew violent," which Gatti interprets as a reaction to the unfamiliar odors, lights, and subsequent digging sounds. After a while Beautiful settled, taking an apparent interest in the digging and even allowed Gatti to stroke his head. However, once freed from the pit, Beautiful's whole posture changed from that of a sensitive, gentle, and intelligent animal to something more akin to that of a rhino, with his head stretched forward, his ears laid back, and his eyes "glassy and cruel." Now, Gatti claims, Beautiful's expression "altered to that of a stupid, brutal force intent only upon its one obstinate determination": to escape.

Gatti concludes that the supposedly timid, gentle okapi had a cold, courageous, and stubborn temperament. It was quite prepared to charge its human persecutors. In short, the apparent languor of the okapi, with its sluggish walk and air of "absentminded inoffensiveness," was deceptive. In fact, the okapi had exquisitely sharp senses and was an animal of swift decision when threatened. The least sudden sound could result in its legs flashing into motion, smashing its way through the most impenetrable vegetation. If cornered or confronted by man or other forest animals, it did not hesitate to charge. The only thing an adult okapi feared, Gatti thinks, was the sound of a falling tree, "against which no strength nor courage can prevail." The intense storms that blew through the rainforest regularly toppled huge trees, a fact that Gatti takes advantage of by faking the sound of a falling tree to flush out an okapi mother and reveal the hiding place of her calf, the ill-fated Toto.

According to Gatti, Toto was a sweet, endearing, and fastidious individual, who showed nonchalance toward other species, no matter how large. Toto displayed a "quiet and dignified seriousness" yet was spirited when things did not go his way. The calf enjoyed exploring camp, in particular the places where tent ropes were tangled and numerous, which Gatti surmises reminded him of tangled lianas in the forest. However, Gatti detects flashes of the okapi's dual nature in Toto's behavior, when his "sweetness" was "charmingly interrupted by a flash in miniature of the obstinacy so strong in the adult."[26]

Gatti maintains that okapi can fall into a "state of hopeless discouragement . . . which is . . . the greatest danger for an okwapi, in captivity."[27] He bases this assertion on his experiences with three of the okapi that he captured: Beautiful, Toto, and an unnamed female. Beautiful, once proud and aggressive, apparently slipped into depression, shown by his lying down and giving up his struggle on having been recaptured and manhandled for a second time, while Toto's "homesickness" may have contributed to his early death. The adult female Gatti had captured and had been trying to bring to camp when a ferocious thunderstorm broke out, interrupting the journey, died "of a nervous breakdown" in transit.

Gatti claims there were two races of okapi, naming a new race *Okapi kibalensis* found in the Kibali-Ituri forest regions east of the Epulu and Linde rivers, with *O. johnstoni* found west of this region. Gatti argued that nobody had noticed this distinct species before because, Harry Johnston's first okapi aside, all okapi remains to reach Europe had been collected west of this region. Gatti based his identification of a new race of okapi primarily on differences in the skulls (as had Ray Lankester in the early 1900s), which he illustrated with photographs in *Great Mother Forest*. According to Gatti, *Ocuapia kibalensis* had a longer, thinner, lighter head, with a straight line to the skull from horns to nose. The horns were small and not fused with the skull (it looked like the skull of a female). *Okapia johnstoni*, according to Gatti, was characterized by a massive, heavy skull with a smaller "ocular cavity." The horns were larger, fused with the skull, and inclined backward. The skull of *Okapia johnstoni* depicted in his photograph is certainly very different in appearance from the first one described, more closely resembling a giraffe skull (in 1910, Lankester had noted the high degree of variability in skull measurements between individual okapi).[28]

The smaller *O. johnstoni* was less nourished, less strong, and less pugnacious than *O. kibalensis*, Gatti maintains. The captive okapi at the Buta Mission, far to the west, were safely confined in a bamboo palisade which the okapi Gatti had captured would have effortlessly forced their way through, in his opinion. These Buta captives were furthermore mournful, subdued animals that obediently went wherever they were directed by shoves from a long stick pushed through their corrals. This quiescence offered a striking contrast to the strength, violence, and determination of the okapi Beautiful, who after forty hours without food in the bottom of a muddy pit, had smashed his way out of a palisade built of strong tree trunks.

Gatti also made claims about the differing temperaments of these "races," confirmed, according to him, by different attitudes to okapi displayed by Africans in the territories of the proposed races. In the Aruwimi, Itibimiri, and Uele regions, locals did not fear okapi, whereas in the Kibali-Ituri region, even experienced hunters considered okapi to be dangerous animals. *Kibalensis* was capable of violence and was much feared, Gatti maintained, whereas the animals he had seen at Buta (*johnstoni*) were placid and not feared. (They were, after all, captive and habituated.) Those who had helped to capture Toto would unashamedly flee at top speed should the (fearless) little calf lower his head and charge them.[29]

As his recommendation for a new scientific name suggests, Gatti was no scientist, so he hoped experts would confirm his proposal. They didn't: after he proposed his new race in a letter to *The Times*, Guy Dollman of the British Natural History Museum replied that given the limited geographical distribution of the okapi, it was "very unlikely that more than the typical form of okapi can be differentiated." Any description of a new race would require detailed inquiry and comparisons with type specimens in the Natural History Museum, especially as detailed knowledge of the many stages okapis pass through as they mature would be required. The metropolitan experts brushed aside Gatti's protestations that he had seen more numerous okapi specimens, alive and dead, than were available in Europe, invalidating his attempt to establish a new "race."[30]

Conclusion

By mid-century, some knowledge of the internal anatomy and physiology of the okapi had been gained. Almost all of this came from captive okapis (alive and dead). From 1919, the scant knowledge of okapi behavior in the wild was supplemented with observations of the first okapis in zoos, though in many cases this remained the tacit knowledge of zookeepers.

Only from the late 1930s was some research on okapi physiology published in the West. A difficulty experienced in the dissections of Congo in London and other specimens elsewhere is that most of the dissected okapis were diseased. Few lived to old age, and okapi were too rare to cull for research purposes.

Work on the ecology and behavior of okapi in the wild was difficult, and knowledge accrued slowly, much of it inspired guesswork or hearsay, with many disagreements. After Lang's field observations, the next notable studies were made in the interwar period when Gatti and Christy spent long periods in the field and observed live okapi at close quarters for short periods. All this knowledge remained fragmented, when World War II interrupted the passage of okapi to Europe and studies in the field.

CHAPTER TWELVE

Okapi Science, 1946–2015

We don't know much more about his life in freedom than
biological data.

—*Agatha Gijzen,* Das Okapi, *1959*

Scientific knowledge about okapi progressed significantly in the postwar period.
From the 1950s, much was learned about okapi behavior in captivity, notably on
reproduction, diseases, and other factors relating to survival. Antwerp Zoo's Agatha Gijzen summarizes this in her 1959 monograph.[1] Zoo director and conservationist Bernhard Grzimek synthesized developments up to the 1970s, as did
Richard Bodmer and George Rabb for the period to 1992. The center of gravity
of live okapi studies shifted from the Congo to zoos in Europe and the United
States, with a second monograph on okapi published by zoo-associated American authors in 1999.

I confess that in my own research, it took a concerted effort to interest myself in the pages of veterinary and other zoo-related accounts of keeping captive
okapi. However, their insights provide windows into the world of okapi, wherever they may be. That said, I realized that the okapi under observation were
mostly captive bred, habituated to humans and zoo routines. If to be wild is to
be free- or self-willed, what can you learn about wild animals in a captive
situation?

In his influential book *Wild Animals in Captivity*, Heini Hediger, director of
the Basel Zoo, argues that wild animals are much less "free" than people think.[2]
They adhere to behavioral routines and stay in the same territories, shaped by
their ecological entanglements. Nevertheless, he maintains that zoos should draw

on knowledge of species' normal lives in the wild to reproduce approximations of their habitats. Hediger visited the Congo in 1948 to inform himself on the wild lives of his captives, including a visit to Jean de Medina's okapi capture center (he saw only captive okapi).

Hediger was critical of the scientific ethology of his time, worrying that animal behavior researchers studied animals as inalterable machines, rather than as thinking and feeling beings. He also thought that they ignored animals' considerable capacity to understand and interact with their human observers. According to Hediger, these relationships influenced animals' behavior in captivity.

If it is the case that animals enjoy a less unconstrained life in the wild than we might suppose, it nevertheless seems to me that an important dimension shaping their identities is the totality of the environment they perceive and respond to. German biologist Jakob von Uexküll calls this their *Umwelt*. If zoos are to recreate this, they must base such efforts on knowledge of how the animals they keep behave in the wild and are shaped by their natural environments. For most of the twentieth century, little was known about the natural environment of the okapi. Although scientific knowledge has progressed the published evidence indicates that few can claim deep knowledge of the lives of wild okapi. To this day, in many zoos there is limited evidence of efforts to mimic an authentic okapi *Umwelt*.[3]

The first detailed scientific studies of wild okapi began late in the twentieth century. In the mid-1980s, John and Terese Hart started using tracking technology to study okapi in the Ituri Forest. Although tracking the development of scientific knowledge must always rely on incomplete evidence (much is not written or published), and work is ongoing, this chapter attempts an overview of what Western scientists learned in the period following the Second World War up until 2015, when political instability once again interrupted okapi studies in the Democratic Republic of Congo (DRC).

The Consolidation of Knowledge on Captive Okapis, 1946–74

In the first issue of the Zoological Society of London's (ZSL) magazine *Zoo Life*, published in 1946, Reginald Pocock summarizes current knowledge for the lay reader. He begins by asserting that "there are good grounds for the claim that the okapi is in several respects the most interesting animal now exhibited in the Zoological Gardens."[4]

Pocock narrates Johnston's discovery of evidence of okapi and explains how they were described, categorized, and named by Western scientists. He describes the key anatomical features of okapi, including bilobate (cleft) crown teeth designed for browsing, and their horns (ossicones). Pocock maintains the okapi, unlike the giraffe, had retained "a useless vestige of the primitive moist, naked 'nose' acquired very early in the evolution of the Mammalia," as well as scent glands in its four feet. He similarly claims that its stripes had "primitive" origins, suggesting (after Johnston, he says) they were "guide marks" enabling okapi to follow one another along dimly lit forest tracks, rather than camouflage to avoid detection (although it seems odd, then, that the front legs are striped, too).[5]

Pocock claims okapi (like giraffe) are habitually silent (after checking with Perry, Buta's keeper at London Zoo, regarding vocalization). Unlike giraffe, inhabiting open plains, okapi inhabit dense forest, which explain its "immense ears with their elaborate sound-channels." A newspaper story published during World War II claimed Buta could hear sounds inaudible to human ears, like mice squeaking. Zookeepers said Buta heard airplanes approaching long before they could, quipping that Buta "should be in the Observer Corps." Having said okapi are silent, Pocock also reports "dubious native sources" claiming they have a voice like a cow's. He speculates that this could be how okapi communicate in dense rainforest, given their keen hearing.

London Zoo's superintendent, Geoffrey Vevers, having been sent Pocock's proofs for comment, informed him that Ludwig Koch, a pioneering German animal sound recordist, had recorded okapi and giraffe vocalizations in Antwerp Zoo just before the war. However, Pocock chose to ignore this, replying tartly: "I should particularly like to know what stimulus induced those two reticent beasts to oblige a German particularly interested, probably financially, in collecting voice records." Pocock claimed that, by omitting any mention of this recording, he might provoke Koch (a Jew who had fled the Nazis to settle in England, where he became as renowned as David Attenborough now is) into contributing a note explaining how he had recorded these animals. I have no evidence of this plan working.

In his article, Pocock describes how, after the King of the Belgians gifted an okapi to King George VI, Vevers traveled to Antwerp Zoo in 1937 to choose an individual from three available animals. Vevers, familiar with the cause of death of the zoo's first okapi, Congo, and an expert on parasitic worms, examined the okapis' droppings with a microscope. One was mildly infected (this was the Congo later sent to the Bronx Zoo in New York, where he survived until 1952); one was heavily infected (and died a few months later); and the other was clear of infection. This was Buta, alive and flourishing more than eight years later at London Zoo, when Pocock wrote his article.

According to Pocock, Buta was fed on a varied diet, comprising mostly elm, oak, and poplar leaves in summer, supplemented with chopped root vegetables and cabbage or lettuce. The leaves were hung on a tree in his paddock for him to pick with his tongue, but he preferred to have his vegetables handed to him by his keeper. In winter, he was fed evergreen (or holm) oak foliage, supplied by Hannah Powell-Cotton from Quex Park (Percy had died in 1940) and by Thomas Coke, fourth Earl of Leicester.[6]

The *Evening Standard* reports that Buta's keepers were only anxious about him in January and February, when he had to subsist on "herbs" due to the acute shortage of leaves. Much preferring leaves, he lost weight over this period (when he was also confined, to prevent him from slipping on frosty ground and breaking his "slender leg-bones"). Well-wishers from Wales, the newspaper notes, would send Buta bamboo leaves for the coming winter. Buta never drank water, according to Pocock, getting his liquid from his food. He drank "evaporated milk" if offered (earning him visits from the catering staff's black cat).

Observations of Buta's temperament are anecdotal. *The Star* reports that on the advice of Attilio Gatti, who was at the zoo to see Buta arrive, "the walls of his den were painted black to resemble his den in the Congo forest." Buta was credited with "exercising his wits" one day after hearing his paddock gate being opened but not closed, while resting indoors. He strolled out into the zoo and located a privet hedge, which he cropped with enthusiasm, before being returned to quarters.

The Star, whose journalist insists on referring to Buta as "Johnny the okapi," reports that Keeper Perry had "carefully studied [Buta's] temperament," finding him "highly nervous, timid, frightened of sparrows, mice, and even of his own shadow. Though he knows me, he is still frightened of me." Keepers cleared out nesting sparrows who were attracted into Buta's warm quarters because their noisy quarrelling made him nervous. They were always worried he would be startled and charge off, colliding with a wall or door and injuring himself.

According to the *Evening Standard*, Buta was not evacuated to Whipsnade at the outbreak of the war because of his "temperamental make-up." "It was felt that it would be too risky to box so nervous an animal for such a journey, as Buta might panic and break a limb." The decision was regarded as justified, given that he appeared untroubled even when bombs fell nearby. So, the okapi seemed a paradoxical animal unperturbed by exploding bombs but frightened of squeaking mice![7]

Agatha Gijzen at Antwerp Zoo

In 1947, the Royal Zoological Society of Antwerp was tasked with collecting and publishing all available data on the keeping and pathology of okapi. Gijzen, a

Dutch biologist, historian, and archivist, began working as a zoologist at Antwerp Zoo in May 1947, following eight years at the Rotterdam Zoological Gardens in the Netherlands. She was part of zoo director Walter van den Bergh's drive to make the zoo more of a scientific institution. She became an expert on okapi, achieving some success with captive breeding and establishing the stud book.

In 1957, Gijzen was asked by the Board of the Royal Zoological Society of Antwerp to compile a survey of the okapi in captivity at the zoo, including reproduction, and to produce a comprehensive bibliography of published literature on the subject. Published in 1958 and widely circulated to zoos, this work informed her 1959 book *Das Okapi*. Published in German, Gijzen's was the first substantial monograph on the subject since Fraipont's 1907 volume and summarized knowledge to that point.

Systematic Western knowledge of okapi behavior comes from the postwar period onward. Most was based on observations of captive okapi. As Gijzen notes: "We don't know much more about his life in freedom than biological data . . . [T]he little that we have been able to gather about his psychology is largely about captured animals or was observed during capture and transport." One notable advance was the realization that okapis do not (as Powell-Cotton, Schubotz, and Gatti had believed) have a fixed breeding season. This observation arguably makes the notion of a "closed season" for hunting useless for okapi conservation.[8]

Gijzen draws deeply on Gatti's information, which would have pleased him, given that his attempts to establish two races of okapi had been dismissed as the suggestion of an unqualified amateur. Her description of okapi's cleanliness and how they bathe is derived from observations Gatti records in his book *Great Mother Forest*. She notes her observations of captive okapi confirmed that they do meticulously groom themselves, a behavior facilitated by their long tongues and flexible necks. Oddly, the okapi she observed in captivity always defecated and urinated inside including on the straw beds where they slept even though they had the option to move between indoor and outdoor areas. It's possible, Gijzen suggests, that this uncharacteristic behavior was induced by the unique conditions of captivity. Beyond this speculation, the reasons for this behavior are not well understood.

Gijzen quotes Gatti at length on the nature of the okapi, noting that many mistakenly believed it to be a serene, defenseless, gentle animal. Gijzen agreed with Gatti, observing that "anyone who has ever seen how an okapi changes from a peaceful and calm mother animal to a maddened devil in no time at all, destroying everything that comes within its reach, knows that one can experience the most unexpected things with this animal." She gives examples of okapi smashing apart transport crates. Gijzen warns that when okapi get bored or worried (for example, when being photographed) and begin to stomp their forefeet, it is time to leave.[9]

Gijzen also agrees with Gatti that okapi are prone to depression, which could be triggered by a shock during capture or transport and which "often leads to death." Her advice is to avoid "excitement" at all costs, including unnecessary noise or touching while handling captive animals. It was best if an okapi was always handled by the same people, in the catching station or zoo, so that the animal could get to know and trust them. Touch should be predictable, usually part of some routine, and mostly for the same reasons (grooming or checking hooves, for example). However, by the same token, keepers must avoid turning them into a "one-man-animal," frightened of anyone else.

Gijzen recommends speaking to the animal in a calm voice and using a "comfortable-sounding nickname" as well as giving them small treats. Regular touching of the whole body would prevent the animal getting a shock when it was necessary to touch it, and brushing was a good approach. A long-handled brush was recommended at first, to avoid getting kicked: okapi hooves "strike far and especially very hard and on top of that in all directions . . . and . . . rarely fail to achieve their goal."[10]

Gijzen argues that okapi are unintelligent because they run into trapping pits. She claims Jean de Medina said no other forest animals did so. (This was not so—indeed de Medina complained about how cautious okapi are and how attuned they are to clues like frogs in the concealed pits.) She bases her claim that they are unintelligent also on the assertion that, soon after captivity, they forget how to eat properly (including avoiding toxic plants) if their feed is changed.

Gijzen maintains that okapi's sensitivity to external shocks makes sense if we assume it is an animal of low intelligence. Her claim seems surprising, and I wonder whether a well-aimed kick had influenced her perceptions. That said, she was dedicated to the welfare of individual animals and courageous in handling potentially dangerous animals in need of medical care. Perhaps it was the many frustrations of trying to keep okapi alive in zoos that made her see them this way—much as Beautiful's determination to escape influenced Gatti's perceptions of okapi.

Anatomical Studies of Okapi

Where most earlier anatomical studies had focused on the skulls, skeletons, and horns of okapi, with a handful of studies following analyses of the okapi Congo who died at London Zoo in 1935, a sustained series of publications on okapi's internal anatomy and physiology emanated from zoos in Antwerp, Paris, and other European cities beginning in the 1950s. A particularly notable series of articles appeared in *Acta Tropica* in 1950.

Several of these were based on postmortem investigations of the okapi Bambe, whom the Belgian government gifted to the Basel Zoo in 1949 to commemorate the institution's 75th anniversary. Director Heini Hediger acknowledged this gift as "the biggest event in terms of animal population since the garden was founded." Unfortunately, Bambe died of a worm infection just over two months after arrival, at two years and two months of age. In line with Hediger's scientific approach, Bambe's body was investigated rather than simply discarded.[11]

Starting in 1958, Jacques Nouvel and his colleagues at the Zoo Vincennes (Paris Zoological Park), which had a good record of keeping okapi alive, published on okapi reproduction and mortality in the park. This work culminated in 1970 in a history of the family of okapi kept at the zoo. All these publications emanating from continental European zoos with okapi contributed new research on reproduction in captivity, parasites and diseases, organs of the okapi, and general challenges of keeping okapi alive in captivity.[12]

Postwar Reviews of Okapi Research

Significant work was done on okapi behavior in captivity in the postwar period. The bibliography Gijzen published in 1958 was expanded to 425 titles and republished in 1962 by Gijzen together with a veterinarian Dr. J. Mortelmans who had joined the zoo in 1958. Gijzen carried out a new survey of the 176 okapis in captivity (as of December 1970). She and coauthor Stefan Smet published their findings in 1974 in an article that also drew on their large 1972 review of information available on okapi since 1900. One new piece of knowledge gleaned since Gijzen's 1958 survey was the realization that captive okapi could breed at a much younger age than previously suspected. However, little new had been learned about wild okapi.

Several summaries of okapi biology and behavior drawing on studies of wild and captive okapi were published in animal encyclopedias in the postwar period, notably, *Grzimek's Animal Life Encyclopedia*, first published in German in 1968 and in English in 1972. In a chapter on giraffes and okapi (which he calls "the forest giraffe"), Grzimek discusses okapi taxonomy, the history of its discovery, how okapi were captured and transported (based partly his personal observations), okapi diet, diseases, and behavior in captivity. He admits that "very little is known so far about the life of the forest-type giraffe in the wild."[13]

Most of Grzimek's information is familiar from Gijzen's work. He adds that he thought de Medina had collected all the plants okapi ate, some 30 species from 13 plant families. He gives no plant species names, noting only (with surprise) the many *Euphorbia* and repeats the (incorrect) claim that okapi "eat a kind of forest grass." At Frankfurt Zoo, they could not supply this diversity of freshly cut branches except in summer. In other seasons they provided apples, bananas, cabbages, and onions along with clover and lucerne hay.

When Grzimek visited Epulu in the early 1950s, de Medina was in Portugal, and so his Greek deputy, a Mr. Marinos, was their host. There were 15 okapi in the capture station at the time, all named. Marinos ensured that all droppings were swept up and removed several times daily and that food was given as bundles of fodder suspended on trees six feet (1.8 meters) above the ground, to prevent okapi from getting infected by parasites. At Frankfurt Zoo, it was found that scrupulous cleaning of okapi enclosures did not eradicate the many worms, so a powerful floor heating system was installed in the indoor okapi pens and turned on after floors had been washed, drying the floors and killing any surviving larvae.

Grzimek confirms points on reproductive behavior based on studies conducted by Rothschild at Frankfurt Zoo. He describes the birth and early years of the first okapi born in Germany, the male Kiwu born at Frankfurt Zoo on September 9, 1960. He outlines key stages in calf development and discusses maternal behavior, notably their defense of their calves. He records the lack of specific imprinting of the calf on its mother, based on his observations at Frankfurt Zoo and in Epulu that they would nurse other calves.

While in Epulu, Grzimek observed surprising behavior in two baby okapi born to females who had been pregnant when captured. When he approached very close to them, after chasing the calves in their cage, they flopped down and pressed their heads to the ground, ears laid flat. His conclusion was that, like European deer, okapi calves were slower than their mothers, and so when they were chased, they would flop down while the parent attracted the attention of the pursuer, leading them away from the calf. Pygmy people sometimes found okapi calves in these circumstances.[14]

Grzimek observed that the female okapi which gave birth in September 1960 at Frankfurt Zoo "made constant vocal contact with the male in the pen next to hers" during the two hours it took her to give birth. He does not describe the sound. While at Epulu, he had seen de Medina imitate the lowing sound of a calf's distress call, to demonstrate how all the adult females in the pens immediately dashed over, threatening the men with aggressive stamping and kicking of their hind legs in the air.[15]

New Dimensions in Field Research: 1986–90

The most significant work on okapi ecology and behavior in the wild in the latter half of the twentieth century was undertaken by John and Terese Hart in the Ituri Forest between 1986 and 1990. Their main innovation was the use of radio telemetry in a study they undertook to track okapi movements and obtain more precise information on their habits.

John and Terese Hart's Radio-Telemetry Studies (1986–90)

Terese met John Hart in 1970 when they were both students at Carleton College, in the US state of Minnesota. They bonded over birdwatching but were swept up in the protests over Vietnam characteristic of their student generation. Like many other students taking anthropology courses across the country, they read Colin Turnbull's 1961 book *The Forest People* recounting his experiences with the Mbuti people of the Ituri Forest. Turnbull's romantic evocation of a nonmaterialist, peaceful society living contentedly in the rainforest was greatly appealing. John applied for and won a Watson Fellowship to spend a year in what Congolese president Mobutu Sese Seko had renamed Zaïre to learn more about the Mbuti people.

John contacted Turnbull, who was encouraging but who had just returned from (his final visit to) Zaïre and warned John that the growing political conflicts bordering the Ituri and in neighboring Uganda were likely the harbingers of the end of an era. Commerce would tear apart the Mbuti's way of life and curtail their freedom. Like Grzimek, Turnbull feared that the forces of modernization sweeping postcolonial Africa would destroy traditional African ways of life and damage human-wildlife relationships.[16]

John Hart arrived in Zaïre in 1973, moving to the Ituri Forest after a month in Kinshasa. Terese arrived about a year later to begin Peace Corps training in Bukavu in the northeast of the country. John managed to stretch his one year of funding to support himself for two-and-a-half years, and they returned to Minnesota in 1975, where they married. They then began training themselves in various subjects and skills required back in Zaïre.

The Harts returned to Zaïre in 1981 to begin their PhD projects, basing themselves in Epulu, where Putnam's Camp had been. The Okapi Capture Station, some distance away, had been taken over by the Zaïrian Parks Institute (the Institut Zairois pour la Conservation de la Nature or IZCN), which had granted the Harts permission to work in the country. They lived with their daughter Sarah in a rehabilitated colonial-era bungalow, developing deep relationships with Mbuti people over the next years and learning much about the ecology of the forest.

With independence in 1960, the Belgian botanists studying the region's flora retired. In addition, there was a shift in academic botany to cellular studies rather than field research. For these reasons, the Harts had to follow Putnam's example and collect the local fauna and flora themselves. They learned much from the Mbuti and sent specimens away for identification by American and European herbaria. John, who was studying the region's antelope, kept duikers at Epulu. It is clear from Terese Hart's wonderful memoir of this time that they saw themselves as the heirs of Patrick Putnam. Locals also saw them as continuing Putnam's legacy. In fact, the Harts found he had been absorbed into local oral history as no other Westerners had.[17]

During this time, the Harts learned much about Mbuti beliefs about the rainforest animals, notably large mammals, which received special respect when hunted. They heard stories about past hunts for elephant and okapi and learned okapi meat was said not to be sold or traded. (In the early decades of the century it certainly was, though perhaps only or mainly by non-pygmy African hunters.) Okapi were regarded by the Mbuti as at one with the forest. Adult okapi were seldom caught; it was usually the calf flushed from its hiding place that was caught in hunting nets (the opposite of the situation with hunting pits). After the birth of their second child, Rebekah, and the completion of field research for their doctorates in 1983, the Harts temporarily returned to the United States.[18]

The Harts' concentrated fieldwork on okapi began in 1986, when they initiated a radio-collar study of okapi for the New York Zoological Society (the name of which was changed to the Wildlife Conservation Society, or WCS, in 1993). Based at Epulu again, they established the Afarama Camp at the confluence of the Afarama and Edoro rivers, some fifteen miles (twenty-five kilometers) walk northeast of Epulu along forest paths, beyond the forest hunting zone. Operating in their Edoro Study Area, they caught and collared the first okapi, a young female they named Paskalina on Easter day in 1986.

Working with Mbuti people and Bantu villagers, the Harts trapped okapi in pits, collared them with Telonics radio collars, then tracked their movements through triangulation using hand-held antennae that they supplemented by direct observations. By 1988, they had captured and collared more than twenty okapi, following them over a study area of more than nineteen square miles (fifty square kilometers). The Harts established the home ranges of okapi: females ranged over 1.9–2.7 square miles (five to seven square kilometers), and males used larger ranges of 3.8–6.5 square miles (ten to seventeen square kilometers) but shared these with other okapi.[19]

The Harts identified more than one hundred plants okapi browsed on, many more than Grzimek had noted in 1968. They observed that okapi ate both young and mature leaves but not old leaves which are often covered by mosses and algae. They also did not eat the large-leaved herbs growing on the forest floor, making them the only hoofed animal in the Ituri Forest to feed solely on leaves in the shady understory. The Harts surmised that because these leafy plants have evolved toxins to protect themselves from browsers—each unique to that plant—okapi ate only a few leaves from each plant species each day. In this way they avoided ingesting a large amount of any particular toxin.

The Harts also learned that nursing mothers boost their leaf intake by up to 80% while suckling their young. The young hid out in "nests," as was well known. It was less well known that young okapi do not defecate in their first month, probably to avoid detection by predators. (According to Kurt Benirschke, American zoos had puzzled over this, many giving their young calves laxatives.)

John Hart thought that okapi avoided certain forest areas, including large areas between the Lomami and Lualaba rivers, because the poor soils yielded plants that were deficient in nutrients. More work remains to be done, but this observation may overturn the idea that okapi are not limited in their range by dietary considerations.[20]

In a 1988 paper, the Harts offer an okapi distribution map showing an area extending from the Ebola River basin near Businga in the west, eastward between the northern extent of the rainforest and the Congo River to the south, to the headwaters of the Ituri River in the east, and south as far as Maiko National Park (see figure 7.1 for reference points). They conjecture there may be okapi south of the Congo River below Kisangani, as far south along the river as the southern border of Maiko National Park (some distance away, east of the river). Due to deforestation and disruption, recent surveys found no okapi north of Virunga National Park (the population within which was judged small and possibly no longer viable) or around Beni, where Johnston had found the first evidence of the existence of okapi.

The Harts conclude that okapi are unevenly distributed across the region. They argue that more survey work would be required to identify hotspots, with the distribution west of the Congo River especially in need of investigation. They advocated for the creation of an okapi national park in the central Ituri Forest, where a thriving okapi population survived.

Efforts had been underway since 1986 to develop the necessary infrastructure and initiate legislative action required for this park, supported by WWF, WCS, IZCN, and the Zaïrean company TABAZAÏRE. Separate efforts were underway to improve facilities at the Station d'Epulu funded by the American NGO White Oak Conservation (Gilman International Conservation).

Alan Root filmed the Harts' lives during this okapi tracking project, and released the documentary *The Heart of Brightness* in 1991. The Harts also built a research and training center at Epulu, fulfilling Putnam's dream of attracting researchers from the United States, Europe, and other African countries to the area. Their project concluded in 1990, with over 700 hours of direct observations logged. In addition, they had initiated okapi surveys in other forests in the region. The Harts continue to work in the DRC, though did not return to live in the Ituri Forest after 1990 due to escalating armed conflicts. This is a pity, as efforts to establish a protected area specifically for okapi bore fruit in 1992, with the creation of the Réserve de Faune à Okapis, or RFO (translated as Okapi Wildlife Reserve, or Okapi Faunal Reserve).

Conflict Interrupts Okapi Studies and Conservation

The political situation in northeastern Zaïre (now Democratic Republic of Congo, or DRC) is complex, emerging from a long history of conflicts dating to colonial times. The period of civil war known as the Second Congo War (1998–2002) had its origins in the Rwandan genocide of 1994. Hutus fled into northeastern Zaïre to regroup, from where they continued to attack Rwanda. Rwanda and Uganda invaded DRC to tackle these disruptive Hutu elements in October 1996, and this escalated until Mobutu was forced to leave the country forever in 1997. Laurent Kabila was installed as leader in May 1997, restoring the country name to Democratic Republic of Congo. Regarded as a puppet by many Congolese, he expelled Rwandan and Ugandan troops in 1998, triggering a second Congolese War. This ended in a decidedly shaky peace with the signing of the Global and All-Inclusive Agreement in December 2002.

The research station at Epulu was sacked and looted during this long period of political instability and violence, first by Mobutu's retreating soldiers in 1996 and then by Kabila's pillaging and poaching forces. In 2002, during the final phases of the Second Congo War, Epulu and the RFO were on the frontier of the fighting between two rebel militia groups trying to take control of the eastern Ituri.

Miraculously, the records, botanical collection, some equipment, and fourteen okapi were preserved through these periods of fighting, as was the RFO itself. This was through the efforts of the Congolese staff who remained, notably at first Robert Mwinyihali at Epulu, and botanist and conservationist Corneille Ewango. Ewango never left the area, sometimes avoiding soldiers by hiding in the forest with Mbuti people. Ewango moved the entire herbarium collection to Uganda for safekeeping (by bicycle), buried data files and the project's new Land Rover engine to protect them, and negotiated with militia leaders to spare the okapi and reduce poaching of wildlife. He was awarded the 2005 Goldman Environmental Prize for his bravery and dedication to conservation and the 2011 Future for Nature Award for his continuing species protection work in the DRC.[21]

Problem Breeders and the Pooky: State of the Knowledge, 1992–2013

A wide-ranging summary of okapi science including the Harts' findings was published by Richard Bodmer and George Rabb in 1992. In 1999, Susan Lindsey and her coauthors published a book summarizing the state of knowledge of okapi at the close of the twentieth century. The most recent authoritative overview at the time of writing was published in *Mammals of Africa* (2013), by John Hart. While the following section reviews the results of studies of wild and captive okapi in this period, genetic studies are handled separately thereafter. The final section of this chapter focuses specifically on ideas about the behavior and nature or psychology of okapi, augmented with my own experiences of London Zoo's okapi. This informs a closing section considering the nature of wildness, and differences between free-living okapi and those in human care.

Bodmer and Rabb's Studies (1981–92)

Bodmer and Rabb's notable work on okapi behavior culminated in an authoritative coauthored summary of the state of the knowledge on okapi in 1992. Bodmer had conducted a small field study of the social structure of okapi in the Epulu area in 1982, where he was hosted by the Harts. This study was partly funded by the Chicago Zoological Society. He also worked with Rabb studying the development of two okapi calves born at the Chicago Zoological Park in 1981 and 1983. The goal of this research was to address the persistent high mortality rate of infant okapi in zoos, and it yielded important information on developmental stages characterized by distinctive activity levels and kinds of behavior, including play.

Rabb focused on the systematics and biogeography of reptiles and amphibians in his graduate work followed by a job as curator and coordinator of research at the Chicago Zoological Park (Brookfield Zoo) in 1956. On his first day at Brookfield, his arrival was overshadowed by the arrival of the zoo's first okapi. Later, as director of the park and president of the Chicago Zoological Society (1976–2003), Rabb was a leading figure in the movement to shift zoos from being places that displayed exotic animals for public entertainment to being centers of education and conservation. In the 1970s, he developed Tropic World at Brookfield Zoo, which attempted to show multiple species (including okapi) living together in "natural" habitats.

Rabb's vision included undertaking biological research in zoos, and he encouraged his staff to study animal behavior. His own research included the study of mother-infant relationships in okapi, which he summarized in a 1978 article, and he coauthored a paper on this subject with Bodmer in 1985. Rabb was active in international conservation efforts, chairing the IUCN's Species Survival Commission from 1989 to 1996 and contributing to the establishment of the RFO in 1992.[22]

Rabb developed an Ituri Forest exhibit at Brookfield Zoo where the goal was "to show the environment, the animals and the indigenous people who share forest with them" in order to motivate visitors to want to "protect both the animals and their wilderness." Putting the okapi and local Africans in the same (created) space—in a Western conservation framing—gave visitors the illusion of experiencing a more holistic view of the situation in the DRC. However, it absorbed locals into the natural landscape and ignored their quite different thoughts on living on the land and with wildlife.[23]

In their 1992 overview, Bodmer and Rabb summarize the state of Western knowledge on the physiology, ecology, behavior, taxonomy, and conservation status of okapi to around 1991. They report that okapi had been placed in their own subfamily Okapinae as they lack a dental feature characteristic of the Palaeotraginae to which Charles Forsyth Major and some later taxonomists had suggested they belonged. However, Geraads (1986) had tentatively suggested that *Okapia, Giraffa,* and *Palaeotragus* may all be members of the subfamily Giraffinae. Essentially, ninety years on, the taxonomy remained unresolved.[24]

Bodmer and Rabb confirmed that okapi have excellent hearing, a good sense of smell, and good eyesight attuned for low-light vision in the rainforest. The tongue was proportionately longer than that of the giraffe, extending for nearly 10 inches (25 centimeters) beyond the snout (which differs significantly from Gatti's assertion that the okapi's tongue extended 16 to 17 inches [40 to 43 centimeters] in length!). They note that, like the giraffe, okapi have only rudimentary vocal cords and a limited vocal repertoire.[25]

The state of knowledge about okapi internal organs was by now very detailed. The heart and digestive system were similar to those of other browsing ruminants. They reported that okapi milk had a high concentration of protein, a third greater than that of cow's milk, with a low fat content. This finding suggests that Gatti's recommendation to feed captured okapi calves pasteurized cow's milk would not have worked, although there were at least two occasions when okapi calves subsisted on doctored cow's milk. According to Grzimek in *No Room for Wild Animals*, Mrs. van Landegem reared a calf using cow's milk mixed with condensed milk, while Anne Eisner successfully raised a calf at Epulu using Klim powdered milk ("milk" spelled backward, developed for use in the tropics) mixed with sugar and water.[26]

Precise information on okapi diet and digestion was available. Captive okapi were known to require 9.48 to 11 pounds (4.3 to 5 kilograms) intake of dry matter per day. Variation in forest types and seasonality in plant phenology (leafing, flowering, etc.) were regarded as less important for okapi feeding habits than the occurrence of preferred food species. Clearings created by fallen trees were important feeding areas, as okapi favor sun-loving plant species. Bodmer and Rabb mention the Harts' findings that okapi browsed more than one hundred species and that forty-three species from thirteen plant families were fed to okapi at the Epulu station. They list five favored species mentioned by the Harts but Harrison's and other early okapi hunters' suggestion that okapi eat *Megaphrynium macrostachyum* was not supported.

Bodmer and Rabb also recount the Harts' description of the okapis' manner of browsing using their tongues. With their teeth, they clip off leafy upper portions of smaller understory vegetation. To the knowledge that okapi occasionally ingest clay and burnt charcoal as a mineral supplement, is added the Harts' observation that okapi also lick bat guano deposits in hollow trees.

Much was now known about okapi reproduction, mostly through observation of captive okapi. In the US, a cooperative program of research on breeding okapi had been undertaken by the Dallas, Brookfield, Oklahoma, and San Diego zoos. This included urine sampling and analysis over 24 months to monitor the breeding status of mature females and to inform attempts at artificial insemination and embryo transplant. By the early 1990s, it was known that estrus cycles occur around every fifteen days, year-round, in captivity, though they could be irregular with long periods of quiescence. There appears to be no seasonal periodicity in fertility of males.[27]

Courting and mating behavior in captivity were well known, with characteristic patterns of positioning, sniffing, licking, flehmen (males raise heads and curl their upper lip), and postures confirming intention and receptivity. This repertoire includes some aggression around non-receptivity of females (females kicking back at males; males striking females with their horns). The youngest female okapi to breed in captivity was one year and seven months old, and the oldest was twenty-six years old.

By the 1990s, as Bodmer and Rabb report, the longevity of zoo okapi surviving their first month was fifteen to twenty years. Causes of death in captivity included parasitism (notably by the nematode *Monodontella giraffae*), bacterial infections (including pneumonia and septicemia), fungal and viral diseases, and accidental trauma. While okapi breed readily in captivity, rearing calves had proven difficult, with around 50% dying in the first month of life until the 1950s. Although success rates improved, in the 1990s workshops on captive propagation still included okapi as "problem species." Gestation ranges from 414 to 493 days (well over a year).

Given these difficulties, zoo staff were very attentive and carefully recorded their observations. They found that newborns could stand within thirty minutes and most initiate nursing some seventy-two minutes postpartum. Infants were typical hiding ungulates, like giraffe: after one to two days of following their mother and exploring their environment, they settled in one place. Time spent in the hiding site and infant behaviors show clear developmental stages, with 80% of the infants' time spent at the nest for the first two months. If startled, calves dash away or freeze if on the nest.[28]

Nursing was infrequent but usually several minutes long. In the Ituri Forest, the Harts observed a lactating female increase her foraging time. In captivity, lactating females lost significant weight during the first two to three months of suckling. Infants were able to begin taking solid food by week three (Gatti tried to initiate this too early with Toto), with rumination (regurgitating and chewing already consumed feed) beginning by week six.

Infants did not defecate until one to two months after birth, doing so regularly from their third month, perhaps to reduce predation. Calves gained weight rapidly, doubling their birth weight of 30 to 66 pounds (14 to 30 kilograms) in their first month and tripling it by the end of month two. Newborns are around 28 inches (72 to 83 centimeters) tall (about the size of a large dog), growing 5.9 inches (15 centimeters) by the end of their second month, after which growth rates slowed as they moved onto solid food. Weaning commonly occurred at six months, though nursing could continue for over a year.

Breeding in captivity brought unique behavioral challenges. Benirschke, director of research at the San Diego Zoo, explained that zookeepers had not initially understood why captive okapi mothers licked their calves' rectal areas. Because captive mother okapi did not need to leave their calves to forage, they spent unusual amounts of time with their young. This unnaturally prolonged proximity combined with boredom meant that they over-performed behavior that was usually possible only once daily in the wild. Zookeepers eventually concluded mothers licked calves in this way to ingest harmful bacteria, against which calves were not yet equipped to defend themselves. Ingesting these bacteria caused the mothers' well developed immune systems to kick in, subsequently providing the necessary antibodies to the calf through their milk. Little jackets were made for calves to prevent over-licking. Benirschke explained that the San Diego Zoo stopped using the jackets once they realized the potential health benefits that the licking held for the calves (he doesn't explain how they reduced the amount of licking).

Young okapis' stiff black dorsal manes disappear at twelve to fourteen months, and horn development in males usually begins at one year of age. Okapis reach adult size in about three years, with males standing up to five feet (1.55 meters) and females five feet, two inches (1.59 meters) tall at the shoulder (at 15.2 hands tall, females are similar in size to a larger riding horse), with a head-body length of six feet, nine inches (2.1 meters). Males weigh up to 573 pounds (260 kilograms) and females up to 784 pounds (356 kilograms). For a rough comparison, okapi are generally at the small end of large adult horses in size, and lighter, but notably taller and heavier than red deer (elk).

Bodmer and Rabb note that although usually "tranquil and non-aggressive . . . [okapi] have several aggressive behaviors," including head slaps, head throwing, and kicking. However, aggressive kicking was often symbolic with no contact achieved. Dominant animals hold their heads and necks more erect than subordinates, and submissive behavior includes laying the head and neck prostrate on the ground.

The authors also highlight the fact that, when the Harts had quantified the size of okapi home ranges, they found that males (who ranged more widely than females) moved an estimated 2.5 miles (4 kilometers) in a day, contradicting earlier assumptions that all adults roamed over great distances daily. The Harts thought okapi distribution was related to ecological and physiographical conditions, particularly the need to avoid large swamps and open savanna, rather than to the occurrence of specific plant species. (However, the Harts did also suggest that okapi avoided areas with nutrient-poor soils.) Bodmer and Rabb note that okapi favored three main forest types in the northeastern DRC. Of these, two were upland forest types, and one was a lowland swamp forest. All three were characterized either by a specific dominant tree species or a combination of dominant species. Within these, okapi frequented stream beds, and secondary growth forest and gardens, as well as primary forest.

The distribution map in Bodmer and Rabb's paper shows an area similar to the Harts' map of 1988. A westernmost extension of the range beyond the Mongala River to Libangi on the Ubangui River is described as "former range," and a large bulge southwest of the Congo River between Kisangani and Ubundu is marked as "uncertain." Okapi were said to be restricted to altitudes between 1,640–4,757 feet (500–1,450 meters) above sea level. Their range was limited by swamp forests to the west, high montane forests to the east, savanna to the north, and open woodlands to the south. Fraipont's suggestion that okapi occur as far south as the Manyema district was not supported.

At the end of the 1980s, most information on the social behavior of okapi was still based on observations of captive animals. Information about wild okapi was mostly based on anecdotal reports and was inconclusive. For example, it still had not been determined if okapi were nocturnal, diurnal, or a mixture of the two. (The Harts thought they were diurnal and occasionally engaged in some nocturnal activity shortly after nightfall, especially on moonlit nights.)

Bodmer and Rabb state that males marked bushes with urine and that females sometimes used common defecation sites. Both genders rubbed their necks on trees, leaving a tarry dermal residue. Social grooming was common in captive okapi, focusing on hard-to-reach areas notably the earlobes and neck. However, in the wild, they were apparently solitary and avoided each other except when courting and mating or when mothers were raising calves. Three okapi together (mother, calf, and attendant male), as reported by Delia Akeley in the 1920s, were only sighted twice in five years during the Harts' research project. The Harts never saw more than three together.

Play behavior was observed in okapi of both sexes and in all age classes, but most frequently in infants. Play behavior included capering and gamboling, and also the delightfully named and unique okapi play behavior known as "the pooky." Here, the okapi adopted a head-low, forward posture accompanied by rapid tail wagging, which may be performed while standing, walking, running in circles, or spinning. Okapi also engage in "lie and rise" play, which entailed lying down or rolling onto their sides, then rising to their feet again.

Bodmer and Rabb assert that okapi communicated vocally more than giraffe do, using a "chuff" sound to contact other okapi. Infants bleated—usually when in distress—and males make a soft moaning sound during courtship. Rabb recorded hearing a whistle and a bellow from an okapi in acute distress.

Lindsey, Green, and Bennett's The Okapi

Dallas Zoo has a long history of keeping okapi (since 1960). In the mid-1990s, artist and docent Mary Neel Green and researcher Susan Lyndaker Lindsey were working with okapi and discussed publishing an illustrated book on the species. The zoo's head of research, Cynthia Bennett, was also drawn into the project. With the assistance of Dallas Zoo colleagues and okapi conservationists, the three produced a beautifully illustrated book, *The Okapi: Mysterious Animal of Congo-Zaïre*, with a foreword by Jane Goodall. The authors donated proceeds of sales to okapi conservation.

Published in 1999, this slim volume offers the most comprehensive overview available of the knowledge accumulated up to the close of the century in which the okapi was discovered. It summarizes the knowledge on okapi biology, ecology, and behavior, gleaned from research on (mostly) captive and also wild okapi. The authors explain that most knowledge was gained from (then) recent research on captive okapi, concluding that "in many ways, the okapi remains an animal of mystery."[29]

The longest section of the book focuses on okapi in captivity, with wonderfully illustrated information on okapi physiology and behavior, including courting, reproduction, and lifespans in captivity. They summarize the endocrinology and reproductive cycles of female okapi, gleaned from analysis of urine and feces. Knowledge of male reproductive physiology remained scanty, however. Sampling of urine and dung of other okapi by males using smell and taste had been observed. This possibly allows males to assess the reproductive status of females (which is also possible by sampling pedal gland deposits).

An interesting development that the authors report on was speculation about vocalizations within the infrasound range (inaudible to humans). This idea first arose following observations of synchronized behavior between okapi females and calves in separate enclosures at Dallas Zoo in 1990. Later, recordings of okapi at San Diego Wild Animal Park and White Oak Conservation Center in Florida confirmed that okapi produce low frequency sounds. Mother-calf pairs are believed to use infrasonic calls to initiate nursing. In dense rainforest, such sounds travel well, but are difficult to locate, reducing predation risk.

Lindsey, Green, and Bennett provided guidance on feeding and providing water to captive okapi. (It is interesting that Buta allegedly took no water at London Zoo in the 1940s—especially as film footage shows his predecessor Congo apparently drinking water.) Their book provides detailed information on pregnancy, birth, and the challenges of raising calves. The authors acknowledge the dilemma that many zoos face: on the one hand, first-time mothers may be aggressive to calves, and it is tempting to remove the calves from harm's way, while on the other hand, the success rate of hand-rearing calves has not been high. Successful attempts to raise calves using a mixture of water, powdered cow's milk, and canned evaporated milk were achieved at Brookfield and Dallas zoos and at Marwell in the United Kingdom, but other zoos had had difficulty duplicating these positive outcomes into the 1990s.

The authors reported that habituation of okapi to human contact was recommended in order to facilitate the treatment of hoof problems and the collection of physiological data. This contact was initiated with calves while they were on the nest and their mothers were elsewhere. Some calves were habituated even to rectal temperature taking, but individual temperaments and relationships with keepers were believed to be very important to the success of such endeavors.

John Hart on Okapi

Hart wrote the chapter on okapi in Jonathan Kingdon and Mike Hoffmann's definitive monograph series *The Mammals of Africa* (2013).[30] In a short preamble on the taxonomy, Peter Grubb reiterated that the genus *Okapia* includes only one species, *Okapia johnstoni*. On the wider taxonomy, he noted only that some authors refer to it as a living fossil, and some align it with the extinct Palaeotraginae.

Hart's overview is one of the two most recent comprehensive scientific overviews of the okapi available when I wrote this book. Differences to the living giraffes include its smaller proportions, lack of elongated cervical (neck) vertebrae, and five sacral vertebrae (giraffes have four). Only male okapi have ossicones, and they have only two, above and behind the eyes. Giraffe males and females have ossicones, and they can have up to five, with a median ossicone between and above the eyes. Okapi, unlike giraffes, have large pedal glands above all four hooves in both sexes. The large auditory bullae (hollow bony structures enclosing parts of the middle and inner ear) and large frontal and palatine (the rear of the hard palate) sinuses are also unique to the okapi. Adult female okapi are distinctly larger than males, standing 2.8 inches (7 centimeters) higher at the withers (the highest part of the back, between the shoulders at the base of the neck), on average, and weighing on average 7% more than males. In contrast, male giraffes are typically larger than females.

Hart also describes similarities between the okapis and giraffids. Okapi share bilobed lower canines with giraffes and other giraffids. Further, while the formations of these is different, they share with giraffes and their fossil ancestors hair-covered horn formations that develop from cartilaginous ossicones. These are found in no other ruminants. Okapi also share giraffes' prehensile tongue and their lateral pacing gait, where the body weight is shared by legs on the same side of the body as the animal moves forward.

Hart regards the history of discovery and naming of the okapi as emblematic of the power of preconceptions: Johnston ignored its two-toed track as he was looking for a three-toed hipparion or a new kind of "one-toed" forest zebra, resulting in Sclater at first assigning it to the genus *Equus*. Hart affirms that "there are no currently recognized subspecies."[31] However, he reports that okapi found west and south of the Congo River were said to be darker than those found to the east and north of the river. The genetic relationships between these populations remained to be investigated (further work has since been done).

Genetic work had been done on okapi since 1969, almost exclusively on captive animals. Where giraffe have 30 chromosomes, okapi can have 44, 45, or 46! The odd number results from Robertson fusion, where 46 are reduced to 45 through centric fusions (aneuploidy). While aneuploidy can impact on fertility and cause conditions like Down syndrome in humans, okapi with 45 chromosomes appear unaffected and can produce viable offspring.

Hart describes okapi's appearance, with some additions to previous work. For example, the pelage (fur) is lightly oiled from a reddish-brown secretion from the skin that accumulates abundantly in the ears and renders the pelage somewhat water-resistant. I can testify to the waxy feel and reddish color of this from having briefly stroked Edi the okapi at London Zoo. There is also a faint, not unpleasant odor.

Much of Hart's information on habitat repeats that mentioned by Bodmer and Rabb, also reporting that okapi forage in small forest clearings abandoned by shifting cultivators. Okapi also use seasonally flooded areas while still wet but avoid swamps and dense thickets. In hilly areas in the northern Ituri Forest, they move seasonally into rocky inselbergs to find plants which flower in the wet season. While okapi use elephants' clay and mineral excavations, they avoid their muddy wallows.

Hart provides detailed notes on radio-collared okapi in the Edoro study area in the RFO, northwest of Epulu. His data reveal that the okapi frequented an area with a closed but not heavy forest canopy that allowed some light to reach the understory, which favors the shade-dependent sub-canopy trees and shrubs on which okapi feed. Uniquely among large African ungulates, they depend on forest sub-canopy foliage for their diet, a challenging niche given its low productivity and the uncommon and dispersed nature of palatable understory foliage comprising mostly mature, chemically defended leaves.

The exceptionally fast early growth rate of okapi calves, despite these dietary limitations in terms of low nutrients, and the challenge of finding enough and sufficiently varied food, suggested to Hart that heavy predation has selected for rapid growth, facilitated by very nutritious milk. Leopards are the major predator, and over five years nearly 18% of the Harts' radio-collared okapi were killed by leopards. These were mostly male okapi, and also younger animals.

He suggests that okapi's morphological features evolved to facilitate their diet, with fine cusped, low-crowned teeth for cutting soft foliage and an elongated tubular muzzle with a long mobile and muscular tongue for selecting and stripping leaves from branches. Further, they produce copious saliva to counteract the tannins and toxins in their foods, with a very fast rate of food passage compared to other ruminants. Okapi favor the leaves of woody, dicotyledonous species and are highly selective, even within individual plants, touching and smelling foliage before selecting particular twigs or branches and stripping the leaves with their tongues. They like the foliage of fallen canopy trees that survive and coppice in treefall gaps, even those species they avoid while they are still saplings in the understory. When young, these plants invest in ungulate-deterring chemicals, which they abandon as they grow beyond reach up into the canopy.

Hart speculates that okapi evolved in dry though not strongly seasonal evergreen forest environments. As such environments are rare in the warmer, wetter late Quaternary climates of Africa today, okapi currently occupy relicts of the environments they evolved in, which are now sub-optimal for their species. Densities varied across the region, from 0.38 per square mile (0.45 animals per square kilometer) in the RFO (with an estimated population of 6,500) to lower densities in the Maiko National Park (estimated population 4,000), and even lower densities in the Rubi, Tele, and Ebola River forests. Factoring in habitat suitability and human settlement and hunting, Hart ventures a global estimate of 35,000–50,000 animals (equivalent to the number of human residents of a large town) over a range of 77,220 square miles (200,000 square kilometers).

Hart's information on social and reproductive behavior mostly repeats Bodmer and Rabb's. He adds that males' movements are highly irregular, in contrast to females which move steadily around established home ranges. He speculates that the stomping behavior of both sexes has infrasound properties in addition to expelling secretion from their pedal glands. Combat has still not been observed in the wild, though one captured male had gashes in his side, possibly from another male's horns.

Breeding is not seasonal—though the Harts suggested it may be most common during the two annual dry seasons. It takes place on females' home ranges.

Males approach receptive females carefully, uttering chuffing calls, and receptive females respond with submissive postures. In captivity where females cannot escape their attentions, males can become aggressive with females. Learning this from the okapi keepers at London Zoo was my first intimation that my preconceptions about the docile nature of these animals were incorrect. Hart says almost nothing about the nature or psychology of okapi, noting only that they are shy and secretive.

Genetic Studies and Conservation Status in the Twenty-First Century

Between 2010 and 2014, David Stanton and others undertook studies of captive and wild okapi and of museum samples. Stanton's research was completed in a partnership between ZSL's Institute of Zoology and Cardiff University as part of his doctoral work. In addition to showing that the genetic diversity of captive populations in Europe did not match that of sampled wild populations, Stanton identified six main genetic lineages. He found that (unusually) okapi are both highly genetically diverse and highly genetically distinct.

Stanton and his colleagues fix the most ancient mitochondrial DNA divergence in the okapi family tree at more than 1.7 MYA and hypothesize that the physical barrier of the Congo River might explain this split. Further lineages probably resulted from the fragmentation of the Congo basin into refugia across the Pleistocene Epoch (2.58 MYA–11,700 BP) because of climatic shifts. However, contradicting Gatti, the Ituri and Epulu rivers were shown not to have been barriers to okapi interbreeding. (As a Mbuti told Stuart Nixon, "of course" okapi are good swimmers).[32]

By using non-invasive genetic sampling of okapi dung, Stanton and his colleagues also confirmed the occurrence of okapi southwest of the Congo River, to the west of the Lomami River in what—as a result—was upgraded to become Lomami National Park in 2016. This work was important as previous surveys of this elusive species had relied on visual identification of dung, which was sometimes confused with the dung of the bongo (*Tragelaphus eurycerus*), a large reddish, white-striped antelope also found in the region.

Okapi Behavior and the Nature of the Okapi

Not all of this knowledge has been published in scientific journals or communicated in scientific circles. There is a parallel world of zookeepers, some with intimate knowledge of the animals (as related in newspaper stories about London Zoo's okapi Buta). While more recently, keepers tend to have academic qualifications, this was not always the case. Even now the intimate, personal relations they have with the animals they interact with, and tacit knowledge gained, tends not to be published in scientific journals.

Occasionally, veterinarians or animal behaviorists working in zoos publish memoirs or popular articles or contribute to popular film or television documentaries. A survey of these offerings could make a useful contribution to scientific knowledge of okapi behavior in captivity. Zoo journals circulate some of this knowledge, and recently, social media platforms like WhatsApp groups are also used.

Maryvonne Leclerc-Cassan's memoir about her experiences working as a veterinarian in the Paris Zoological Park between 1969 and 1978 provides an example of knowledge acquired through interacting directly with okapi. This book includes notes on the birth of an okapi and observations about how different okapi mothers are to giraffe mothers. For example, the former are aggressive to humans, even those known to them, in defense of their calves, while the latter are more docile. The words, gestures, and other measures keepers used for reassuring captive animals did not work on new okapi mothers. She also shared the zoo's convention of naming okapi by using the first syllable of their city of birth, hence the mother Paquerette and her calf Pastourelle.

Discussing okapi behavior in the first months, Leclerc-Cassan notes that, in her experience, they will only emerge from their heated galleries once the outside temperature reaches at least 64.4°F (18°C), as "these animals, strong and fragile at the same time, must not risk any cold snap." These observations suggest that analyzing non-scientific publications like the memoirs of veterinarians, zookeepers, and documentary filmmakers may be worthwhile. Although the okapi information is limited in Leclerc-Cassan's book, with many other animals discussed, the cover features a striking image of her with one hand cradling an adult okapi's cheek.[33]

What is omitted from many mainstream animal behavior studies for much of the twentieth century is a consideration of how the presence of particular human observers, and indeed the experience of captivity, shape the observed behavior. For example, the calf Buta delivered to Amsterdam Zoo in 1919 was familiar with humans and sought affection from them. She had been hand-raised, as was Tele, the second okapi to be delivered to Antwerp Zoo. The capture stations in the Congo habituated okapi to humans. Even the "wild okapi" observed by Lang and Gatti were habituated to humans, very early in their short lives.

Meeting Okapi at London Zoo

When I met okapi keepers Gemma Metcalf and Poppy Jewell at London Zoo, they explained that captive okapi are certainly influenced by how they are kept.

At London Zoo the keepers are very hands-on, grooming the okapi, accustoming them to handling. Their okapi are reasonably confident and enjoy being fussed over. In the keepers' view, if keepers don't do this, the okapi remain elusive and jumpy, as they have a very quick natural flight response. They also have moody days, which keepers learn to recognize.

The male okapi I met at London Zoo, Mbuti, definitely preferred female handlers. He took a long hard look at me before entering his enclosure for a nice scratch behind the horns from Metcalf but not before letting fly a sonorous fart of disapproval. (Mbuti was euthanized in January 2023 at eighteen years old, following an illness.)

Oni, the adult female who arrived from Antwerp in 2015, "can be loved by anyone," whereas her second calf with Mbuti, Edi (born in September 2020), is less trusting and much less inclined to engage with humans. She hovered outside until Metcalf led her into her enclosure by tempting her with some tasty leaves. I think she knew I was there. I circled around to a platform outside her enclosure, and she then took some tasty tidbits from me—I could see her tiny teeth and blue tongue in action from close up and appreciate her beautiful coloration. Mbuti and Oni's previous calf, named Ada, born in 2017, was presented as "Meghan" to the press to chime with then-current media interest in Prince Harry's impending marriage to Meghan Markle, but the keepers never used the name.

The keepers told me that the zoo's okapi are fussy about where they sleep, what they eat, and which other people and animals they interact with. They vary in individual temperament, and it takes time to build relationships with them. They prefer routine and consistency. Mbuti ate almost anything and liked to sleep on sawdust, while Oni prefers coir as bedding and is a fussy eater. Given that they are very selective feeders in the wild, feeding okapi in captivity is a real challenge: they need both quantity and variety. I saw mounds of food in the hoof-stock area, including bags of pellets, branches and foliage, and piles of herbs. They are fed sycamore, poplar, ash, and willow (which they love but can only eat in small amounts, as it is fairly toxic). They are fed pellets including vitamins and minerals and get alfalfa (lucerne) for fiber, with silage in winter. They also receive nasturtiums, basil, mint, and rosemary as well as vegetables like pumpkins.

The keepers noted that institutions differ in their approaches to raising okapi, which can complicate transfers between zoos. For example, when Mbuti arrived, he was unaccustomed to handling and would kick out, leaving keepers with many bruises. He eventually became used to their attentions. This also applies in reverse, and the keepers worry about transferring calves who are accustomed to handling to zoos which don't believe okapi should be handled. Ada's keepers reduced their handling of her before she was transferred to Chester Zoo, for this reason.

Keepers report that routine care—such as healing the hoof problems that okapi often experience, inoculating them, and treating them for other illnesses—is difficult with okapi unused to handling. They do very poorly under anesthetic, so rendering them unconscious for these procedures is risky. Because few zoos keep okapi, the community of keepers is relatively small. In more recent times, they keep in touch through workshops and WhatsApp groups, where they can share and learn about developments in nutrition, medicines, and care.

Guidance on okapi keeping is summarized on the Okapi Management website. Creating a suitable environment for this rainforest species remains challenging. Temperatures should remain within the range of 66–78°F (19–26°C), which requires heating during colder months in higher latitudes. As rainforest species, they are accustomed to high humidity, and at least indoors it is best to maintain 50%–60% humidity. Because weather conditions in most regions where they are kept are unsuitable for okapi, they spend up to 50% of their time indoors on an annual basis, requiring spacious indoor accommodations.

Okapi are not naturally social animals, and individual animals require around 330 square feet (28 square meters) of stall space each, with six-foot (1.8 meters) high walls as well as their own bedding and latrine area. Outdoor spaces require shade, with a minimum area for two animals estimated at 5,412 square feet (500 square meters). They should have good cover outdoors and an area to retreat from view. Noise is also an issue: as the Okapi Management website notes, "this shy, forest species fares better in quiet, low-profile settings."[34]

Interestingly, anecdotal evidence suggests that the *kind* of noise matters more to okapi than its volume or intensity. According to Oxford zoologist Jonathan Kingdon, when he visited London Zoo as a boy just after World War II, a keeper told him that when the zoo was bombed during the Blitz, many animals were terrified. The zoo suffered bomb damage on several occasions, but especially on the night of September 25/26, 1940. Several incendiary bombs fell on the zoo, damaging structures including several animal houses. The zebra and wild ass houses were hit (one zebra was later caught, heading for Camden Town), as were the flower beds opposite the antelope house, where the okapi Buta was kept. Buta, however—possibly used to the raging thunderstorms in the rainforests of the Congo that he experienced in his youth—was completely unperturbed by the explosions.

The conditions under which okapi are kept, variations in how different zoos handle them, and of temperament between different okapi, raises deeper questions about the psychological effects of keeping animals in captivity, and the nature of human-animal interactions. These issues were raised precociously early in the twentieth century by zoo director and animal behaviorist Heini Hediger in *Wild Animals in Captivity*. While the effects of captivity on animal behavior is not necessarily a problem for research on captive okapi, it most certainly is relevant for attempts to extrapolate from this to the nature and behavior of okapi in the wild (where they also sometimes encounter humans, in quite different circumstances).

Despite abundant literature on keeping wild animals and detailed published guidance on indoor and outdoor enclosure design, including its horticultural aspects, most of the okapi accommodations that I have seen in European zoos have outside areas that more closely resemble paddocks (some with a few trees) than rainforests. Further, European climates cannot approximate the heat and humidity of the Congo Basin. When I asked Metcalf and Jewell about the state of these kinds of okapi enclosures, they agreed that something like the elephant house at Zurich Zoo, with heating and lush vegetation, would be ideal (but expensive). In most European zoos, heated and humid interior enclosures are provided.

Keepers look out for stereotypical behaviors (repetitive, invariant, and functionless), such as tongue rolling and excessive licking, to gauge the animal's stress levels. I watched London Zoo's keepers work with Edi on target-training: she was given a tasty treat for touching and licking a mark on a wooden paddle. Her outside pen has apparatus for enrichment, and food receptacles secured to trees. This is of course no substitute for "learning to be an okapi" in a Congo rainforest environment.[35]

Wild and Captive Okapi

Visiting the okapi at London Zoo left me sure that they are in good hands. It also left me unsure about what kind of okapi these are. Like most other captive okapi alive today, they were born and raised in zoos and will die in captivity, too. They are not wild okapi, and while physically they are okapi and retain most of the genes and many intrinsic behaviors of okapi, they have learned to be a kind of okapi that lives around humans in captivity.

But what are "wild okapi"? Harriet Ritvo suggests that in an era of commonplace capture, handling, and relocation of wild animals, including icons of wildness like wolves, we should be as skeptical about ideas of "wildness" as we are of ideas about "wilderness."[36]

Irus Braverman argues that conservation has shifted from a focus on wilderness to a focus on wildness, where the wild is perceived as a matrix of possibilities within which species can occupy degrees of wildness on a spectrum from semi- to fully wild. However, she asserts that in reality this perpetuates a duality in which "more wild" is more valuable than "less wild," and in situ conservation always trumps ex situ.[37]

Braverman contends that, faced by the realities of a globalized world that is undergoing accelerating anthropogenic transformations, conservationists hanging on to earlier notions of wilderness and ecological integrity are being dragged, willy-nilly, into a brave new post-wild world. She maintains that the notion of a spectrum of wildness is being replaced by a messier practice of valuing life in contexts which are nowhere free of human influence and always evolving (which is, of course, the status quo in the Anthropocene Epoch). She concludes that classifying animals as more or less wild, more or less endangered, and therefore more or less worthy of conservation is to misrepresent the messy and entangled nature of the interrelationships of animals and humans in the twenty-first century.

Braverman's assertion fits with the broader narratives championed by Emma Marris and others about a post-natural world in which we humans and our fellow creatures inhabit novel or hybrid ecosystems.[38] These are bracing ideas advanced by smart and principled writers, but I struggle with them. I do agree that wild animals can live freely in transformed landscapes, interacting with humans. I work on human-wildlife coexistence and prefer to think about communities of humans and other beings sharing landscapes. Have I given up on "wild"?

Where no wild spaces (as in not primarily controlled and shaped by humans) survive for a species or where these spaces have become dangerously precarious, I accept we're dealing with a post-natural scenario. In the latter case, it seems sensible to create safe havens where species can survive in human care. Where there is still some space for wildness and opportunities to preserve it, however, I'm in favor of doing so.

While hesitant to value captive okapi less than okapi living free of human influence in the rainforests of the Congo, I maintain that they are significantly different. Okapi have evolved to inhabit African rainforest environments, which have shaped their coloration, their features, their physiology, their senses, and their behavior. They have long lived around people but keep their distance and remain elusive in the wild. Okapi see things, hear things, smell things, and know things which humans will never know.

This seems to me a very satisfactory state of affairs. Acknowledging it is to recognize a need for compassion for nonhuman beings, which implies a commitment to enabling their flourishing and autonomy. It requires humility in terms of the limits of what we can know as humans and a commitment to the diversity of life with no requirement for personal benefit. It also requires consideration of the wellbeing of those fellow humans sharing landscapes with okapi.

Conclusion

The place of okapi in the larger family of giraffids remains uncertain, which is unsurprising given that according to a recent review, "the origin, phylogeny, and evolution of modern giraffes (*Giraffa camelopardalis*) [remains] obscure." The authors propose that one of two subfamilies of early ancestors of giraffids, the Canthumerycidae, gave rise to the subfamily Palaeotraginae during the Middle Miocene (15.98–11.63 MYA), roughly when the giraffe and okapi parted evolutionary ways. The Palaeotraginae produced the genera *Okapia* and *Samotherium*. From the African Samotheres evolved *Bohlina* species regarded as the immediate ancestors of *Giraffa*, the genus of modern giraffes.[39] That is one plausible interpretation.

Observations of okapi in zoos fueled research after World War II, the first reviews emerging in the 1950s. From then on, a significant body of knowledge was developed on captive okapi behavior, covering reproduction, diseases, and other factors relating to survival. Since the 1980s, some progress has been made in understanding wild okapi behavior, using new technologies of tracking and surveillance and of genetic analyses. This knowledge was drawn together into a comprehensive okapi conservation plan in 2015, albeit hampered by ongoing conflict in northeastern DRC.

The tacit, experience-based knowledge of zookeepers complements this scientific knowledge but is seldom recorded in scientific journals. Instead, it circulates through keepers' journals, social media groups, and sometimes documentaries and popular memoirs. It is unclear how much cross-pollination there is between zoologists and ethologists and keepers.

While considerable progress was made in Western knowledge of okapi anatomy and physiology from the 1950s, and this continues to be an active area of research at the time of writing, so far as I can discover there is still no comprehensive single-volume account of okapi anatomy and physiology. There remains a lack of definitive information on their digestive physiology and metabolism, knowledge which is presumably important for keeping them in zoos.[40]

Keeping okapi in zoos raises issues over to what extent it is possible to replicate a suitable environment for them, certainly, but also raises more profound questions over what nature of beast the captive okapi is. How do captive and wild okapi differ? In addition to the obvious genetic differences between free-ranging wild populations and the small captive population, how does living in captivity in atypical environments in interaction with humans change the nature of an animal?

These complicated questions about what happens to a species over time can only be properly considered alongside the recognition that okapi are individuals with unique characters and life experiences. In addition, okapi are raised in different institutions with their own philosophies and regimes of okapi care, which are performed by individual keepers with their own ideas, approaches, and personalities.

The okapi I met at London Zoo are singular, fascinating animals. There is something uniquely "okapi" about them, despite sharing their carers and environment with other megafauna in the zoo's hoofstock section. I am not sure they would survive in the wild, and doubt any of their descendants will be returned to the wild. Perhaps there is a new "race" of okapi after all—the zoo okapi? It would be good if their existence could again be connected in some more effective way with their wild relatives, through raising awareness of the challenges facing okapi conservation.

CHAPTER THIRTEEN

Indigenous Africans as "Primitive Experts" on Okapi

I knew from experience that no one could attempt to go anywhere
in that infernal confusion of violent, luxuriant growth without first
securing the full co-operation of the pygmies, those amazing bits of
primitive humanity . . .

—Attilio Gatti, 1946

Local Africans played a central role in the discovery of the okapi by Westerners,
in its location and capture, and what was learned about its biology and behavior in the wild. In addition to general assistance in the forests, a few well-known
individuals were famed for their okapi tracking and capturing skills. The early
Europeans tracking the animal depended absolutely on them.[1]

Visiting scientific travelers and collectors like Cuthbert Christy and Herbert
Lang struggled to disentangle what they regarded as Africans' superstitions from
their natural history knowledge. Their attempts to do so were partly an effort to
assert their scientific authority, and partly a reflection of racist ideas they held
about primitive cultures and the superiority of Western knowledge. However,
these difficulties also raise more profound questions about how different, exactly,
scientific and Indigenous epistemologies really are when it comes to discovering
and communicating natural history knowledge.

Whatever their ideas about human races were, all of the Western writers surveyed in this book expressed their appreciation of the natural history knowledge and tracking and hunting expertise of pygmy peoples. They also worked
with professional African hunters from other communities in the region. While

I do not attempt to speak on behalf of any Africans from the Congo basin region, we can and should consider how Westerners portrayed these Africans and their conceptions of African identities.

I have explained in the preface to this book why I chose, after much deliberation, to use the term "pygmy." This is partly because the term is used in almost all of the literature I am analyzing, and my discussion engages critically with Western representations of these peoples. My focus is on what outsiders made of their encounters with the peoples of the Congo basin rainforest—those they referred to as pygmies along with the African farmers and hunters from Bantu-speaking or Sudanic- and Nilotic-speaking groups to the northwest and northeast, respectively. I am aware that some peoples referred to collectively as pygmies dislike the term, and I apologize for any offense caused to them. I am using the term for the reasons given here, but do not subscribe to the negative connotations that have become attached to the term in Western culture, nor to the oversimplification that they are somehow all very similar culturally or physically.[2]

Most writers on the region focus on the pygmies, particularly regarding natural history expertise. Even though some okapi collectors depended on hunters of Bantu or Sudanic origin, they don't suggest that their peoples had special knowledge about the rainforests. Such knowledge was usually presented as the preserve of the pygmies. My aim is thus not only to consider how African peoples were understood and represented but also to recover and acknowledge what was learned about okapi from local Africans, both pygmy peoples but also the other Africans in the region.

Historical Sources

This analysis focuses on the written knowledge that would-be collectors recorded about okapi to discover what they learned from local Africans. I also explore the skepticism of metropolitan experts toward information gleaned not only from African sources, but also from Western collectors lacking formal scientific training. The texts include those written by Western okapi seekers up until mid-century.

I describe the interest these Western collectors took in pygmy peoples, drawing on Christy and Attilio Gatti's accounts, both of which address the nature, origins, and history of pygmies. I also document the detailed account of African hunting techniques recorded by Lang in his descriptions of his relations with the African hunters Aposho and Abawe.

"The Native Account of the Animal"

Western knowledge of the okapi begins with Stanley's brief mention of an unusual donkey-like animal existing in the rainforest that he had heard about from the Mbuti of the Semliki forest (the exact circumstances are unclear), which Harry Johnston set out to further investigate when he had the opportunity to enter the Belgian Congo to return a group of kidnapped Mbuti pygmy people to their home territory. These Mbuti had been captured with the intention to exhibit them at the Paris Exhibition of 1900; in other words, were themselves regarded as a great curiosity, to be collected and exhibited in a similar fashion as exotic faunal rarities like the (soon-to-be-discovered) okapi.[3]

Upon receiving the bandoliers made and worn by Africans that were the first Western evidence of the existence of the okapi, Philip Sclater declared dismissively that "whether the native account of the animal . . . is precisely correct or not," the animal was not any known species of zebra and was thus undescribed and in need of a scientific name. That is, in his opinion, Africans' knowledge about the okapi was untrustworthy and so until the animal had been studied, named, and described by Western experts, it could not be said to be known. It is thus ironic that the Western scientific description of the okapi remains tied to the physical evidence of a pair of African bandoliers, still stored in the British Natural History Museum.[4]

Johnston's preconception that the okapi was a kind of primitive one-toed horse meant he mistakenly dismissed the evidence he was shown of two-toed okapi hoofprints by Mbuti people. In fact, he believed "the natives to be deceiving us," even though he had visited the rainforest because these pygmy men told him about the okapi, which they knew well. In Beni, the Belgian officers had ordered their native militia to show Johnston their bandoliers, waist belts, and "other parts of their equipment" (unspecified) made from okapi skin. The fragments he could find were those he bought from local Africans.[5]

Johnston's inspired illustrations of what an okapi might look like (the discoverer of the okapi never saw a living one) derived from the skin and skulls sent to him by Karl Eriksson which had been carefully removed from a recently killed okapi Eriksson had obtained from an African soldier (the smaller skull was sourced separately). Johnston had also questioned various "Congo natives" as to the shape and appearance of the animal.[6]

Johnston questioned local Africans about okapi behavior. He was told that the okapi was "a defenseless animal, the size of a large antelope or a mule, that it only inhabited the densest parts of the forest; the male and the female generally went together. The okapi was finally said to live mainly on leaves." This information, gained from Africans, was duly repeated by others, including Julien Fraipont in his 1907 monograph.[7]

When the clinching evidence for the existence of the okapi for Western scientists arrived in England in June 1901, *The Times* hailed Johnston's "new African mammal" (a phrase worthy of further reflection) and also more accurately described it as "entirely new to science." In another *Times* article entitled *"Ex Africa semper aliquid novi"* the writer made the extraordinary assertion—given local knowledge of the animal was acknowledged—that the "actual existence of the animal may . . . now be taken as established, although apparently it has never been seen by human eyes sufficiently curious to compass its capture alive."[8]

The okapi could now be "properly" described and named. Unusually, an African name—recorded and shared by Johnston—was used for the new genus created for it. But Western knowledge of the okapi was partial and mistaken in many respects, based as it was for many decades mainly on the remains of a few dead okapi.

Baron Münchhausen in the Rainforest

Metropolitan experts like Sclater often dismissed local knowledge about wild animals and also often disregarded the accounts of men like Paul Du Chaillu or Attilio Gatti on the grounds that they embroidered the truth or were unskilled amateurs. The dismissal of African sources is striking in the face of how clearly the first Western okapi seekers acknowledged their dependence on Africans for finding and capturing okapi. This included taking African "folktales" or "superstitions" seriously—though Westerners were also anxious to distance themselves from these, so as to appear to be objective scientific observers.

In *The Times* article *"Ex Africa semper aliquid novi,"* which shares the few known facts and notes the limitations of early scientific knowledge about okapi, the author compares the discovery with a report about a mythical beast, a horse-like amphibious animal asserted to be nine feet (nearly three meters) tall, which had emerged from the sea near Ostend and devoured twenty sheep before returning to the waves, that had run in the same newspaper in 1801. Of course, the Royal Society would have laughingly dismissed the idea of a carnivorous horse, the writer notes. But while not wishing to be "derogatory to the dignity of science," the writer can't resist comparing news of the new mammal (not yet the okapi, taxonomic status unknown, and which was "neither horse, ox, ass, zebra, nor giraffe") with this earlier report of a mythical beast.[9]

However, those Royal Society members who were anthropologists as well as zoologists would now find "some real and most attractive *pabulum* for the advanced scientific digestion" in such "grotesque" stories, the article continued. Because it was now possible to better "distinguish between folklore and fact," true stories could be uncovered by sifting through the many dismissed as tall tales "in pre-anthropological days." Indeed, the article quips that "the Münchhausen of our forefathers is as often as not, as Mr Andrew Lang and others have taught us, only an anthropologist misunderstood." By extension, presumably, the superstitious tales of "primitive Africans" might be parsed for true scientific knowledge.[10]

Western Okapi Hunters' Dependence on Africans

It is clear from the accounts of the first Western seekers of okapi that they depended almost entirely on local Africans. Because this has been filtered out of scientific knowledge about okapi, and was omitted from most media reports about the okapi at the time these okapi collectors and hunters were writing, it is worthwhile to restate them here. Following an exploration of the first facts about okapi to emerge from the rainforest, of how they were learned as well as debates over the accuracy of this knowledge in the period up until World War I, I explore in more depth the relationships between Lang and the trapper Aposho and the hunter Abawe.

Early Okapi Hunters, Agukki, and Metropolitan Experts

James Harrison was impressed by the hunting skills of the pygmy people he hunted with in the Ituri forest, though he was also frustrated by their use of dogs which he believed ruined the chances of capturing okapi (okapi were usually trapped rather than hunted). He hunted okapi, without success, on a series of expeditions between 1904 and 1910. Unfortunately, Harrison does not supply sources for the brief notes on the okapi he records in his diary.[11]

Boyd Alexander hunted for okapi from a base he set up in a "Mobatti" (Mbuti) village called Lobi, in early 1906. He praises local pygmy hunters for their hunting and trapping skills in his account of this in *From the Niger to the Nile*. He records they used nets and gins (traps) for smaller game and pits for larger animals. Hunters armed with spears and accompanied by a small dog wearing a wooden bell that rattled when it was hunting (or was stuffed with leaves to stifle the sound) might track animals for days. Their kills and captures, many of which Alexander bought, were usually delivered to the village chief and paraded around the village. His fellow expedition leader, George Gosling, recorded notes on okapi behavior learned from local African hunters.[12]

It is frequently difficult to tell what knowledge was derived from Africans and what came from Westerners' observations. That okapi are usually found singly or in pairs, and frequent small streams, is clearly attributed to African hunters. José Lopes was the only one of the Alexander-Gosling expedition party to see an okapi (whole and alive). It seems he deduced that this particular animal always followed a particular course after drinking or crossing a stream, inspiring him to set a trap along the path. Lopes used the locals' technique, catching it in a pit trap with their assistance. He was unable to stop them butchering it on the spot.

Percy and Hannah Powell-Cotton traveled to Makala Station in July 1905 to find the renowned African okapi hunter Agukki (see fig. 13.1). They hired him,

Figure 13.1. The only definite photograph of Agukki and his wife, taken by Percy Powell-Cotton. The famed okapi hunter is pictured with his rifle (© Powell-Cotton Trust).

securing a skin from an okapi he shot several days later. Percy (an experienced hunter) never managed to shoot an okapi himself but learned much from locals, acknowledging their natural history expertise. Percy related that he gained "a mass of information from the forest tribes [which would] add considerably to our knowledge of this curious animal." Through accompanying the Mbuti in the rainforest (see fig. 13.2), but also by "listening to [their] tales," he learned that okapi were solitary animals, left their calves in nests, and did not (contrary to other reports) feed on the large-leafed *Megaphrynium macrostachyum*. He believed that Westerners could learn much about okapi from the Mbuti.[13]

What did metropolitan experts make of such claims about Indigenous knowledge? After Powell-Cotton had disputed J. J. David's claims to have shot an okapi, citing information he had received from Agukki, Sclater retorted (at a

Figure 13.2. Percy Powell-Cotton with his pygmy guides Appiando and Appiuri (© Powell-Cotton Trust).

Zoological Society of London meeting) that David "certainly claims to have performed this feat." Sclater cited a letter of David's that was published in the *Basler Nachrischten* in which he reports that "also, Ende November schuss ich (als erster Weisser) eine Okapia," adding that he could not understand why "the statement of Capt. Powell-Cotton's native hunter (who alleges that *he* shot the specimens sent home by Dr David) should be preferred to that of a European scientific man."[14]

Despite Powell-Cotton's and others' statements regarding the extent to which they depended on local trackers and hunters and the extreme difficulty of even seeing an okapi, Sclater asserted that "the stories of native Africans on such subjects are not always reliable," disregarding the claims of Agukki on racial grounds and also because he was not a "scientific man" and insinuating that Percy Powell-Cotton was likewise not a credible scientific authority.[15]

Belgian Officials, Collectors, and African Knowledge

In his okapi monograph, Fraipont reports knowledge gleaned from Africans. The okapi skeleton he used as the "type" for his description of the species was an old male that an African had killed half an hour's walk from the Mawambi post. In his accounts of the distribution of okapi, Fraipont relates that according to Commander Sillye, okapi were known to "all the natives of Rubi-Uélé, and as far as Nyangwe in Manyema." He also noted that Africans had reported seeing okapi in herds around Adjamu and in Nepoko, while Belgian officers had recorded seeing okapi skins as ornaments in places like Banzyville and Yakoma, though it was thought likely they had been bought from inhabitants of the Rubi-Uélé region. Only chiefs were allowed to sit on okapi skins, which Sillye claimed had the equivalent exchange value to what slaves had once been worth.[16]

According to Africans, Fraipont reports, okapi are forest animals that hide in dense forest during the day. They come out at night to feed, which they do until the early hours of the day. They prefer clearings, and the edges of swamps and streams.

Fraipont reported that in 1902 Lieutenant Anzélius, the chief of post at Mawambi, was sent to hunt for okapi, accompanied by local Africans, learning interesting and correct information on their appearance and habits. Notably, he learned that only males have horns, a matter much debated by Western hunters and scientists at this time. Unfortunately, it is unclear what Anzélius learned from his own observations and what he learned from African hunters and trappers.

Not everything Anzélius learned from local Africans was accurate. Africans were said to believe okapi lived in pairs of a male and female, but that is not true.

Mbuti hunters also claimed that "sometimes three of them are encountered at a time," which radio telemetry tracking did eventually confirm, as males were discovered to occasionally accompany a receptive female with a calf. These errors are unsurprising: okapi are difficult to find, let alone study, for anyone however skilled, foreign or local. They simply have better senses than humans do, and are very cautious in their habits.[17]

The debate as to whether only male okapis have horns, or whether instead a larger species of okapi without horns and a smaller species with horns existed, was informed by African knowledge. In his 1910 article "Hornless Okapis," Lydekker reports that "although . . . no decisive evidence of this is afforded by the two English mounted specimens, hornless okapis are regarded as females. On the other hand, all the horned examples that have come under my notice are undoubtedly males; and since the Ituri natives affirm that the bulls are armed while the cows are defenseless, the existence of this secondary sexual difference may be at least provisionally accepted."[18]

The only successful okapi hunters, according to Lieutenant Karl Eriksson, were the Mbuti ("Wambatti"), who located hiding okapi and killed them with their spears. While this may have been be true of okapi calves, it was probably uncommon for adults. It is more likely the Mbuti caught adult okapi in pits, or possibly nets, and stabbed them there. (Lang believed some Mbuti had a taboo against stabbing okapi.) Gosling believed Africans usually killed okapi with spears or traps and occasionally with rifles. As Gosling did not witness a successful hunt, this is hearsay. Prior to Lang's capture of an okapi with the assistance of the hunter Abawe, it seems only Lopes had participated in all stages of the capture of an okapi with local Africans, and it wasn't until Attilio Gatti's expeditions in the 1930s that Westerners gleaned more specific information about their techniques.[19]

Hermann Schubotz went hunting okapi in the vicinity of Angu on the Uele River in 1911. Like Powell-Cotton, he reports that he turned to expert African hunter Etumba Mingi, after he tried to shoot an okapi himself but failed, as both Etumba Mingi and the chief of post predicted he would. It would be fitting if an acknowledgement of Etumba Mingi could be added to the displays of the taxidermized okapi Schubotz donated to the Senckenberg Museum in Frankfurt and the skeleton he sent to Hamburg (both shot by this African hunter).[20]

Expertise and Superstition: Lang, Aposho, and Abawe Hunt Okapi
Lang first knew he was in okapi country after he found African men drying okapi meat over a fire. After walking over a thousand miles in pursuit of okapi (he claimed), he concluded it was only the "cunning Pygmy hunter" and the "more

powerful Bantu negro" who could catch the okapi, the latter with snares and pit-falls. According to Lang, locals believed that wearing okapi skin would confer the qualities of the okapi to the wearer, notably its "cunning in eluding enemies" and its "power of retaliating upon those who slaughtered it without due ceremony." He collected several okapi-skin artifacts, demonstrating okapi skins were valued for decorative as well as magical uses.[21]

Local beliefs about the okapi seriously complicated Lang's attempts to collect okapi remains. According to him, locals in Medje believed that "whole villages [had] been wiped out by the spirit of the 'O-api' when a reckless hunter had failed to take proper precautions," one of which was ensuring the head of a dying okapi never touched the ground because if it did, the earth would receive a curse that would ever after bring death and destruction upon the hunter, his relatives, and anyone stepping in the tracks the okapi had left on the day it died. These locals therefore made sure that the head of a dying okapi caught in a trap was placed on a cushion of leaves or branches, which might sound like a recipe for the acquisition of a perfect specimen, but because it was believed that an okapi may not be speared or killed with knives or arrows, it was instead "disgracefully beaten to death with clubs" and so "the skull and even other bones are shattered."[22]

Lang described three methods of catching okapi used by Africans. In addition to the use of pit traps, strong nets were hung across the trails of okapi known to hunters. The okapi were driven toward these by small dogs with wooden clappers (bells) attached to them and loud shouting by hunters. When the okapi tried breaking through the nets, they became completely entangled. This technique was only employed by powerful chiefs, as numerous beaters were required. This use of nets was probably not specifically used to catch okapi, however, though they could be caught in this way (Harrison had thought that only smaller game was hunted in this way).[23]

The third method for catching okapi that Lang describes was relayed to him by Aposho, a famous Medje okapi trapper who boasted of having killed two or three okapi a year since his boyhood and as many as ten in more recent years, when meat had become scarcer. Aposho and his friend and fellow trapper Bope covered a certain territory that they divided between them, each setting thirty to forty traps that they visited every four days. They set the traps along narrow trails hemmed in by thick vegetation, often paths down to drinking places on clear streams. Before setting traps, Aposho consulted his instrument of augury. If the prospects of success were good, he and his helpers dug a circular hole a foot (30 centimeters) wide and 8 inches (20 centimeters) deep. They then cut a strong, flexible sapling which they stripped of leaves and branches, and fastened a strand of rattan measuring 30 to 40 feet (9 to 12 meters) long to the top end of the sapling, attaching a specially consecrated noose to its other end. Three men pulled the sapling down using the rattan rope and attached it to a release-actuating crossbar. Narrow bits of bark were arranged so that pressure on any part of them would trigger the release of the noose. An okapi stepping in the circle would spring the trap and be caught, usually by one ankle. Once the trap was set, Aposho hummed his "imprecations" and he and his helpers performed a ritual song and a dance. All traces of human presence were then "effaced for a distance of several hundred yards."[24]

Because trapped okapi were swiftly beaten to death, Lang's hopes of seeing a live okapi or of recovering remains in good condition from these traps were not realized (and he remarks "it would be dangerous for anyone staying in these regions to interfere with [local] methods of hunting"). He did see numerous signs of okapi, shown him by Aposho, as well as nineteen beaten-up carcasses of okapi caught in traps.

Lang's 1918 expedition report includes a photograph of Aposho (see fig. 13.3), showing the dignified sixty-year-old trapper wearing rope armbands to "propitiate the bad spirits attendant upon each successful killing." It also includes a photograph of a hunter holding up a battered carcass of a male okapi caught in a spring trap (the rattan wound around its leg is visible). Lang's account shows that while pygmy people might be knowledgeable and in tune with the rainforest and its creatures, this did not mean that they refrained from killing and mutilating prey animals, even those like okapi connected to spiritual beliefs.[25]

It was Abawe, the son of the Azande chief Akegne, who finally procured a live okapi calf for Lang. Abawe was "the most renowned hunter in the region, making many contributions to our collections of rare animals during the two months we spent with him." When Lang was informed one day that Abawe would catch an okapi the following day, it seemed to him "utter foolishness" (after so many months in fruitless pursuit), but he decided to go on the hunt anyway because "at all events there was a chance to see, if not an okapi, woodcraft unsurpassed."[26]

Lang was both deeply impressed with Abawe as a hunter who he believed possessed profound natural history knowledge, and dismissive of his "superstitions." In his expedition report, he describes Abawe's preparations for this hunt, which included avoiding food for fear of contamination by an item touched by a woman because that, he claimed, would ruin his luck. Abawe's "merciless rigor alone," Lang states, "brought success and all superstition with which he surrounded himself was mere chaff appealing to the mysticism of the negro

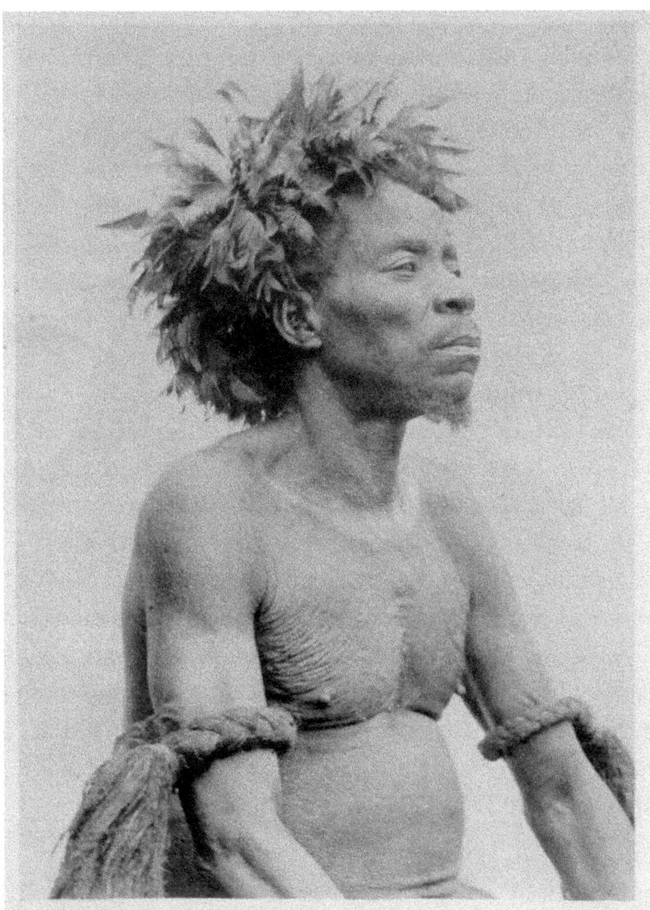

Figure 13.3. "Aposho, the skilful okapi trapper. Aposho was sixty years of age and had slain over a hundred okapi without coming to grief. The twisted ropes on his arms were supposed to propitiate the bad spirits attendant upon each successful killing" (Herbert Lang, "In Quest of the Rare Okapi," *Zoological Society Bulletin* 21, no. 3 [1918]: 1604).

mind, though sometimes he himself gave the impression of believing it." According to Lang, that is, Abawe pretended to believe in charms and indulged the superstitions of his people while knowing full well the real reasons for his success.[27]

Abawe had located the signs of an okapi mother with calf and found a regular crossing place at a stream near where he thought she secreted her calf while foraging for leaves. He posted two men at the crossing place on the appointed day for the capture, one of whom came back and informed Abawe and Lang that she had yet to pass that morning. The party therefore proceeded cautiously around the hill where he suspected the calf was secreted, until they heard a terrific crash. Stopping, Abawe awaited a signal from the other man at the crossing, which when it came revealed that "The calf is not with her!"[28]

The mother had bolted from just yards away from the party. Locating the signs of her flight, Abawe scrutinized the ground and found smaller tracks leading back along a well-worn trail. Very nearby, at the point where its mother had dashed off down the hill, they found her calf hidden in thick bushes. In support of his claims about Abawe's skills and his private skepticism about charms and good luck, Lang remarks that the capture "looked so like the rehearsal of a performance that evidently Abawe had spent the last few days in useful observation and had not trusted luck alone."[29]

From Lang's account of the seizing of the calf it seems these particular Africans had not caught a live okapi calf before. Having seized the calf, Abawe "called desperately for lianas with which to bind it," but in the event "everyone had to laugh" for "the terrible beast had no other desire than to lick the face of its captor." Abawe became concerned that the calf was so tame it must be bewitched and that he would secure no gifts for catching an animal as tame as a lamb (he was, however, duly rewarded). Lang thought his African hunting companions were disappointed to find that the calf of a creature supposed to be one of the "the shyest in the world" was unafraid of them and made no effort to flee. Indeed, it proved to be an endearing little creature, following a small African boy about camp. Despite this, "few dared to touch it."[30]

Christy on Mbuti Tracking Skills: Magical or Empirical?

Christy was one of the more successful pursuers of the okapi, both in shooting one and in acquiring more knowledge of the species. He began his researches by befriending forest-dwelling pygmy people in order to learn some of their bushcraft.

According to Christy these pygmies had formerly inhabited the depths of the rainforest, remaining untouched by outside influences, but encroachments on the forest and settlement by other Africans had changed this. This situation had been influenced by Henry Morton Stanley's coming, and before him the Arab slave trader Tippu Tip, with Wangwana (Moslem Africans, some with Arab ancestors) penetrating the forest with rice clearings.

According to this version of events, forest peoples learned to trade (or barter) with their new African neighbors, offering elephant ivory in return for tobacco, iron arrowheads and iron spearheads, and farmers' vegetables. They camped nearer the forest margins and struck up relations with particular villages. Christy found that African chiefs were jealous of such relationships with pygmy peoples, resenting interference by outsiders. Therefore, it had taken time for Christy to link up with pygmy people to learn their bushcraft.

Christy persisted because "the intimate knowledge possessed by the Pygmies of the private life of every animal great and small was a source of the deepest interest to me." As he began to learn their methods, however, Christy grew less impressed by their "simply marvelous" tracking skills. Like Lang, he felt he had seen through the apparent magic to the empirical nature of their tracking skills. Christy nevertheless gives an evocative description of forest people tracking: "wholly absorbed," moving with "cat-like stealth," communicating their minute observations "by almost imperceptible gestures," or if out of sight of one another, by imitating bird calls. Should someone (usually a "clumsy white man") crack a stick, the men instantly froze, standing on one leg if caught mid-step, "knowing that any animal within hearing is listening intently in the same manner for some explanation of the sound." Christy was impressed by their extraordinary skill with bows and arrows, and regretted being unable to discover the exact ingredients of the poison they applied to their arrows (a mixture of tree bark and crushed ants).

Christy learned that "without the assistance of the Pygmies the sportsman in these forests is helpless. He can neither find any animals nor get back to his starting-point." In 1913, Christy shot an okapi tracked down for him by two Mbuti hunters. He shot another in 1914, but the circumstances are unclear. His descriptions of shooting and recovering the remains of an okapi make it apparent that he could not have done so without the assistance of African hunters.[31]

Gatti's Noble Savages and Other Myths

The name of Gatti's book *Great Mother Forest* comes from an Mbuti song celebrating the forest as beautiful, a mother and giver of life and death, of food and drink, shelter, weapons, and medicines to all past generations of the Mbuti and those to come. His introduction depicts the pygmy people in a romantic way as habitually cheerful and carefree except at moments such as when singing this profound song in the evening during which an "unfathomable sadness" overtook them. "Perhaps in their childish primitiveness," Gatti speculates, "they are feeling how infinitesimally small and lost and crushed they are in the fearfully immense bosom of that forest that the ancient song calls their 'Great Mother.'"[32]

Gatti reiterates a common narrative about forest-dwelling pygmies: that they had inhabited the forest since "immemorable times." Bantu peoples arrived later, following the paths of pygmy peoples and learning to hunt, gather, and grow food in the forest. Finally, this story goes, Europeans, guided by pygmies and assisted by African villagers, established clearings and roads and eventually posts, towns, and mines. Despite this opening up of the great rainforest, according to Gatti, there were still thousands of square miles of unexplored forest. Here, even pygmies did not dare to cross certain "mental borders" created by "fear and superstition." Certain places were said to be inhabited by terrifying creatures, including a ferocious giant bird, wild hairy men, white pygmy elephants, and "an antelope somewhat resembling the okapi" but which also (in Gatti's fanciful description) sounds suspiciously like a unicorn, as he reports the pygmies described it as having a single spiral horn on its forehead.[33]

Gatti's thoughts on these "ridiculous fantasies of ignorant natives" resemble those of the anthropologically minded writer in *The Times* and Powell-Cotton. His long years of African exploration had taught him, he writes, that "these superstitions, these fears, are never born from thin air." On the contrary, "even if tremendously exaggerated by the minds of primitive men, there is always a basis, a reason which has caused them." The remarkable claims of such stories had, in Gatti's opinion, actual grounds in fact in the Congo rainforest region, which "belongs to the immemorial times of prehistory." In its depths even the European feels cut off from the rest of the world, "in time as well as in space." Of course, he maintains, it was the intrepid European explorer, with the "hard-headed persistency of the white man," who would reveal the secrets of the great forest.[34]

Africans may have described mythical forest beasts, but the forest actually did conceal wondrous animals, including pygmy antelope, buffalo, and elephant and giant aardvarks, hogs, and gorillas. Others were living fossils, according to Gatti. Of these, the "most striking of all [was] the okapi, almost indistinguishable from the Samotherium which lived in Greece something like fifteen million years ago." The stronghold of the okapi was in the "tabu" regions of the Ituri, Kibali, and Epulu forests, where maps failed, in spaces marked only as "limit of forest unknown," "very dense forest" and "unexplored," or "inaccessible country." With an operatic flourish, Gatti (sitting in London on a foggy day in 1933) maps "our base camp" onto one of these spaces, adding hopefully that "hic sunt Ocapiae."[35]

Gatti describes tracking okapi with Mbuti pygmies, notably a chief (Gatti's interpretation) whose name he renders as Makulu-kulu. He also worked with a chief (or *sultani*) named Kalumé, whom he claims was a M'bahio of the Mambuba people. Gatti provides much information on okapi behavior, but it is mostly unclear how he accumulated this knowledge. Some assertions, like his claim that an okapi used its head as a battering ram, are based on his own observations.[36]

The fact that Gatti provided Kinande names for the plants eaten by okapi suggests he gained this dietary knowledge from locals. Having paid attention to the "superstitions" of pygmy people, he confirms their belief that okapi lick the saliferous clay they called *bulongo* when they were ill. In this case, Gatti made observations based on Mbuti stories. It seems a little unclear why Mbuti peoples' observation and deduction should be regarded as a "superstition," whereas Gatti's description is presented as a qualitatively different factual observation and logical deduction.[37]

After a large male okapi was trapped in a pit dug for Gatti, there was general rejoicing among the Africans he had assembled, which Gatti believed was because they anticipated an end to their onerous and dangerous labors in the forest (most of them were not forest people). It was also, he speculated, that they celebrated an end to trying to catch okapi alive instead of eating them and using their skins for belts that could cure stomach ailments. Finally, the capture promised to deliver a big *matabisha* (generous tip) to reward their efforts.[38]

Gatti claims these men were frightened of the captured okapi, running away at every movement the animal made. They refused to dig the ramp to free it from the pit because this would bring them into close proximity to the okapi, leaving Gatti to do the shoveling. As Gatti worked, they clustered around the outside of the palisade they had constructed, suggesting he run away before the animal pounced on him. As soon as Beautiful emerged from the pit, the men took to the trees, and a great frenzy of shouting broke out, which according to Gatti provoked the okapi into its successful attack on the palisade.[39]

Gatti also describes Makulu-kulu's appeals to his god Muungu in order to ensure success in finding an okapi calf after Gatti gave up on the idea of capturing an adult. Gatti made fun of the offerings provided (claiming they were mostly eaten by Makulu-kulu) and implied the chief (or *sultani*) was colluding with Gatti to persuade his deity to fulfill their request. Makulu-kulu promised offerings including salt, francs, a pipe, and a blanket to his deity (Gatti took these to be hints to him to supply these items to Makulu-kulu as a reward for a successful capture). The deal done, so to speak, Makulu-kulu promised Gatti an okapi calf within a week, a promise he delivered on.[40]

Gatti was impressed that "among the thousand tracks of Okwapi that honeycomb the forest in every direction," Makulu-kulu found one that "he declared to be a female" and that "he judged to be a mother, as in the same day the tracks doubled back." The pygmies followed these prints "with indescribable patience and ability" all the way back to the okapi's well concealed nest. After a silent wait overnight, they captured the calf in a bag when it emerged with its mother the next day.[41]

The Africans there felt affectionate toward the calf, which they named Toto. However, they were also wary of him, fleeing whenever he initiated a mock charge. Gatti claims that they also knew only as much as he did about the daily habits of okapi (very little) and therefore could not advise him on how to teach the calf appropriate life skills. Gatti said the Africans in camp were amazed and amused by his efforts to educate Toto—instead of just eating him.[42]

On his subsequent expedition to the region (1938–40), again in pursuit of okapi and hoping to photograph and film them, as well as capture some, Gatti again resorted to working with pygmies. This time, he worked with a group led by a man named Kotu-Kotu (pygmy notions of leadership don't fit well with assumptions about hierarchical tribes and chiefs), with whom he had hunted before. As he wrote, "I knew from experience that no one could attempt to go anywhere in that infernal confusion of violent, luxuriant growth without first securing the full cooperation of the pygmies, those amazing bits of primitive humanity."[43]

Gatti also knew pygmies were deeply unpredictable: if they disliked you, you would never find them, and even if they did, their scouts would first report your presence to their chief. Following a discussion between the chief and all his hunters, Gatti maintained, a democratic decision would be taken as to whether to engage or not. Even then, they would wait for several days, watching the Westerner and all the members of his or her party, observing their disposition, behavior, and the goings on in their camp. If all was to their satisfaction, they would casually wander into camp, as if by chance, and negotiations would ensue. Patience was required at this point, and the stating of ambitious objectives too soon could be disastrous.[44]

Gatti was obliged to go through this process twice, with a chief Bwanasula whose village was on the road up through Beni to Irumu, to get access and porters to the hunting territory of Sultani Kotu-Kotu, "a true pygmy" living west of the territory of the "pygmoids" who inhabited a zone on either side of the road. Once in this "true pygmy zone" where okapi were to be found, Gatti had to wait nine days before Kotu-Kotu deigned to see him.[45]

Kotu-Kotu's hunters tracked down okapi for Gatti, but it was difficult to restrain them from killing them, and Gatti had to provide generous incentives not to do so in the form of salt, tobacco, and praise. Having failed to photograph a cornered okapi after it bolted fearlessly through their ranks, Gatti thought up an ingenious plan: when he next found an okapi standing in sunlight, he would, camera in hand, have his companions drape him in branches and leaves and advance on the animal as a "walking bush" until he was close enough for a good photograph.

Gatti set this plan in motion, and with an okapi spotted, he was having the final branches being fastened around him when Kotu-Kotu timidly whispered the word "bilulu" ("biting ants") in his ear. Impatiently dismissing this warning, Gatti crept up silently on the okapi, finally getting into position, at which moment he understood what Kotu-Kotu had meant: not look out for biting ants on the ground but rather that they were in the branches he told the pygmies to fasten about him![46]

After this amusing (in retrospect) fiasco during which Gatti abandoned all thoughts of stealthy photography in favor of tearing off the ant-infested branches, he and his helpers wracked their brains for another means of getting close to okapi. Gatti knew he could never approach as close as African hunters. Kotu-Kotu suggested exploiting the okapi's cleanliness and waiting for it to stop and urinate, which he believed it did seldom and so when it did, it would stay still for a long period for fear of soiling its immaculate white legs. This would be the perfect moment to surprise it (allegedly, this was a hunting technique for killing okapi).

The pygmies suggested several other ideas. One was to find a mother nursing and kill or trap her calf. Another was to pursue a mother and mobile calf with the pygmy peoples' small voiceless dogs until the calf was exhausted and they could capture it. Yet another was to use snares. Gatti dismissed all these suggestions as "repulsive," "cruel," and "stupidly savage to our way of thinking." Further, more pragmatically, it was unlikely to result in the capture of a perfect specimen. Worse, an accidental death would mean a deduction from the number of okapi he was allowed to capture according to his permit. Therefore, he resorted again to the *zemu*, or hunting pit.[47]

When Kotu-Kotu and his people would not move to dig pits closer to the main Stanleyville-Irumu road (moving west of their location required crossing into taboo territory), Gatti was forced to approach another chief, Abdullah, a descendant of Arab slave traders. Abdullah's village was west of Mambasa on the Stanleyville-Irumu road. He was a hunter, his men were healthy and strong from paddling on the river, and they often hunted in the forest accompanied by local pygmies. They did not hold the same taboos as the Mbuti and were persuaded to assist in reactivating old pits southeast of the Stanleyville-Irumu road. Eventually, a large male okapi fell into one.[48]

By this time both Gatti and his American electrician Charlie (employed to help operate the expedition's radio) were ill with fever and fully reliant on their

African companions. In the event, Kotu-Kotu guided Gatti northward and across a wide crocodile inhabited river using a bridge they built from lianas, to where the neighboring pygmies led by Chief Mitwa were delighted to receive them and take them to the pit where the okapi had been caught. It turned out a marriage was planned between a daughter of Mitwa and Makulu-kulu, who was the son of Kotu-Kotu, and it had been agreed that the abundant rewards promised for the capture of the okapi would help fund the festivities.[49]

Mitwa's people had built a palisade around the pit as instructed, and the okapi had charged it, opening a crack, which it was too weak to kick apart. Mitwa had then had his men place the cage built to transport the okapi against this fractured section, removing the broken trunks once it was in place. After they withdrew, the okapi sneaked through the hole, and the cage door swung shut, securing it. Abdullah's men, meanwhile, were cutting a passage through the forest and had built a strong stockade next to the road by the time Gatti arrived. The whole capture operation was a success, though sadly the male okapi would later die of an intestinal infection. This episode had been integrated into the life of the pygmies and repurposed to the mutual advantage of all concerned—except the okapi.[50]

Africans Assist Official Capture Operations

Officially sanctioned capture operations carried out by Franz Joseph Hutsebaut and Patrick Putnam depended mainly on Africans bringing in captured okapi. By the time Anne Eisner joined Patrick Putnam at Epulu, he was no longer pursing okapi for sale. Rather, he wanted one for his private menagerie. According to Anne, "I don't think [Pat] could have stood the work and monotony of caring for scores of natives at the hospital if he hadn't been able to relax, fooling around with the animals. He never seemed to tire of watching them and was always trying to bribe the pigmies [sic] into bringing in other rare specimens." The problem was, as other would-be collectors had found, each time the pygmies caught an okapi they couldn't help themselves and killed or seriously injured it despite the prospect of a reward.

Eventually, a calf was successfully captured without injury and put in an enclosure at Camp Putnam. Putnam ordered Abazinga "the animal man" to collect leaves from the twelve varieties of trees he knew okapi would eat. Abazinga duly gathered the leaves and dumped them in a corner of the stockade. The okapi sniffed them but wouldn't eat. By evening, Pat was frantic. "Pat, with his college education and long experience in the Congo," Anne remarks, "didn't know what to do." Neither did the Africans, including Abazinga, who, Eisner states, could "do more with an animal than anyone I ever saw."[51]

All were gathered around the okapi enclosure, wracking their brains, when up walked an Mbuti named Sale. After the problem was explained to him, he seemed amazed that they couldn't see the problem. "You do not put your bowl on the ground to eat, do you?" he asked. That said, he stalked off into the forest, returning with a length of vine, which he strung across the top of the stockade. Leaping over into the enclosure, he picked up several bunches of leaves and strung them along the vine, like washing on a line. The okapi stepped out of its corner, smelled the leaves overhead, and pulled off a bunch with its long blue tongue. Swallowing it, he moved along the vine, browsing on the remainder as if in the forest. Without Sale's knowledge of okapi behavior, the captive would have died.[52]

When hunting officer G. L. E. Swaluë began capturing okapis, he resorted to Africans' technique of using pit traps, as had Gatti and others before him. Jean de Medina worked with pygmies after setting up the Groupe de Capture et d'Elevage d'Okapis at Epulu in 1947, relying on pit traps for his capture operations. Pits were reportedly dug and checked by pygmies, who assisted in bringing the okapi back to the capture station and with sourcing the right plants for them to feed on. In addition, live okapi were still bought from locals in the 1950s, complicating relations between pygmies and those village chiefs who felt they had proprietary relations with them.[53]

Demand for okapi was high in the postwar period, partly as European zoos had lost okapi during the war. The Belgian authorities offered Africans living in okapi habitat rewards for live okapi. After the English anthropologist Colin Turnbull returned to the Ituri Forest in 1954, he came across an okapi calf near the Lelo River, concealed in a "heap of leaves." His Mbuti companion Ausu spotted it, and they took it to the Station de Chasse to claim the 500-franc reward. Turnbull used the story to illustrate interrelations between African farmers and pygmy people, complicated by these cash payments introduced by Europeans.

The complication was that Ausu had been injured some years previously hunting forest pigs, and an African chief André Somali with whom Ausu's family had a hereditary relationship, had nursed him back to health. Ausu felt he was cared for in the way a valuable hunting dog is cared for, and left as soon as possible. This André, according to Turnbull, despised pygmy people, but found them useful. He claimed the reward for the okapi calf because Ausu was "his pygmy." Following unsuccessful negotiations in which Ausu had offered André part of the fee, Ausu claimed the full reward with Turnbull present, then escaped

André's waiting henchmen by climbing over the back fence of the Station de Chasse and fleeing into the safety of the deep forest.[54]

While Westerners did continue to learn and benefit from local Africans' expertise related to okapi after the 1950s, the most intense period of dependence on African knowledge was from discovery to the formalization of capture operations in the Belgian Congo, which is why I focused on that period here. More could be written about the relationships between government capture operations and local Africans, that are depicted in the documentary films of the period. It would be interesting to know more, for example, about how local Africans incorporated opportunities for supplying food and other labor related to keeping okapi into their livelihoods in this period, and how they have done so since. While pygmy women did occasionally contribute physical labor in African farmers' fields, that pygmy men participated in the manual labor of digging pit traps represents a departure from traditional patterns. When it comes to hunting okapi, locals have continued to do so for various purposes, and they remember shifts in conservation management policies affecting hunting from the colonial and postcolonial periods.

Colin Turnbull on African Techniques for Hunting Okapi

In his 1965 ethnographic survey of the Mbuti, the anthropologist Colin Turnbull has little to say about okapi, unfortunately. He notes that the most important species hunted by the Mbuti were elephant, various antelope, pigs, and monkeys, adding that "the okapi is also hunted occasionally, though the Pygmies know that to do so is prohibited by game laws." The pygmies, Turnbull claims, regarded okapi as difficult to catch, and so they were pursued opportunistically when the pygmies stumbled across them by chance.[55]

Reviewing Paul Schebesta's accounts of Efe pygmy hunting, Turnbull notes that they never caught okapi on tracking hunts with dogs. The Mbuti did sometimes catch them in nets, though adults could break through (if there was time, nets would be lowered to prevent damage to them). Interestingly, Turnbull argues that "the pygmy has no use for traps or pits" because "these require too much work and are, by his standards, inefficient." The Western okapi seekers I describe in this book could be vague on this question of which Africans were doing the trapping. While Stanley, for instance, suggested pygmies used pits, those who provide more detail on hunting operations, notably Lang and Gatti, describe pit trapping as an activity practiced by African trappers and hunters like the Medje hunter Aposho, though Gatti thought that pygmies used them too.[56]

Turnbull claims that Mbuti men were not trappers but were rather either archers or net hunters. These two approaches required quite different levels of social cooperation. Net hunting required a minimum of seven nets, and in a camp of seventeen huts (considered average by Turnbull), most of the men and women and children joined the hunt. Men set up the nets, which were four to five feet (1.2 to 1.5 meters) high and 100–200 feet (thirty to sixty meters) long, in a wide semicircle. Meanwhile, the women dispersed in an opposing semicircle, and when the nets were ready, they started advancing toward them, beating sticks and branches and calling and shouting. They drove animals into the nets, and the men then seized the entangled animals. Each net's owner killed animals entangled in his net with his knife or spear. Larger animals were cut up immediately, and then the women carried the meat back to camp in their baskets. There were complicated rules for determining which body parts were given to whom. Once the meat had been shared out, the skins were stretched on frames or staked out on the ground to dry. It is worth remarking that okapi were not the target prey of net hunting, and no special knowledge of okapi was required to capture them in this way.[57]

In addition to these collaborative hunts, men did hunt individually or in small groups with bows and arrows, though they rarely killed large game with these weapons. Turnbull thought that if a man killed or found a large animal like an okapi by himself, the collective rules of division would still apply and that any meat surplus to the minimum required by his family would be shared. It seems unlikely okapi would be hunted in this way, though hunters that surprised an okapi might try to kill it, and young okapi would be taken.[58]

Of course, times have changed since Turnbull conducted his fieldwork. Although pygmy peoples still use traditional methods alongside firearms and wire snares, snares and guns are now the methods of choice for most non-pygmy hunters in the Congo basin region. Non-pygmy hunters tend to take less large prey but hunt a wider range of species, kill higher densities of animals, and sell two-thirds of what they kill, including okapi. Pygmies hunt larger animals but fewer kinds of species and sell about a third of what they take. Some pygmies do now use firearms, often provided by neighboring Africans in return for meat and skins, but it is not known whether they hunt okapi with them. The Mbuti are still net hunters.[59]

Turnbull has nothing to say about Mbuti taboos and related beliefs about okapi, and little has been published on this despite the alleged cultural importance of okapi to the Mbuti people. A few anecdotal statements were collected by Italian researchers Giuseppe Carpaneto and Francesco Germi in the 1980s, such as that okapi calls could be heard emanating from the depths of the forest, that the okapi was the king of the animals, that they were active in the day, solitary, kicked but did not bite, and roamed throughout the Ituri forest and slept among its rocks. They were said to be preyed on by leopards, and to eat fruits, leaves, and dead animals. Okapi destroyed nets but could be killed with arrows or spears. Their skins had to be given to the chief, and fetuses could only be eaten by elders (if a pregnant woman or her husband ate an okapi fetus, researchers were told, she would suffer a breach birth).[60]

African Involvement in Formal Okapi Research and Conservation circa 1960–99

Following independence, the Epulu capture station continued to be run by Congolese Africans notably Jean Bosco, until the Swiss couple Karl and Rosmarie Ruf arrived in 1987. Locals remained vital to the running of the station and as feeders and keepers of okapi, and Jean N'lamba was an important player in ongoing conservation efforts, becoming deputy director of the Okapi Conservation Program. The Harts learned much from pygmies about the plants and habits of animals of the Congo rainforest during their 1986–90 fieldwork and collaborated with pygmies and African villagers to trap okapi in pits and track them using radio telemetry. After civil war broke out in the mid-1990s, it was once again Congolese Africans who ensured the survival of the station.[61]

By this time, the Réserve de faune à Okapis had been established, and the Worldwide Fund for Nature, Wildlife Conservation Society, and White Oak Conservation were supporting okapi conservation in the region. Pygmy peoples were being recognized (and represented) as important role players for the future of conservation. They were positioned as rightful inhabitants of the rainforest, rather than the allegedly later arrivals loosely classified as Bantu or Sudanic farmers. The origins of this narrative is pursued in the next chapter.

Conclusion

Western metropolitan experts were skeptical of the veracity of both Africans' and Western explorers' and collectors' accounts of okapi biology and behavior. Nevertheless, much of their information was reported in the publications about okapi appearing in the first half of the twentieth century. It is possible in some cases to identify which knowledge was learned from Africans. Until the late 1930s, explorers and hunters did identify renowned okapi hunters and trappers in their reports, and so their names and stories about them can be recovered, but it is notable that since that time such stories about individual Africans and acknowledgment of their contributions to Western knowledge and capture and caring for okapi, have been few and far between.

Westerners struggled to reconcile their admiration for the natural history knowledge and tracking and hunting skills of Africans with their scorn for the spiritual dimensions of Africans' beliefs and practices. Even while learning about okapi from locals, they sought to define their own knowledge making as different and superior to that of Indigenous peoples. In sum, they weren't sure what to think about these "primitive experts" of the rainforest. They were also confused by the apparent contradiction between the skill that pygmy peoples displayed in finding and caring for okapi and their apparent callousness and willingness to kill them for food and skins.

While my survey of local knowledge of the okapi gleaned from Western writings on the species is inherently limited by their authors' omissions and distortions, it does facilitate the recovery of hidden histories. These include the lost histories of individual Africans central to the Western discovery and study of the okapi. Hopefully, new stories about okapi will begin to reincorporate these African voices.

CHAPTER FOURTEEN

Western Framings of the Peoples and Forests of the Congo

It seems safe . . . to assume that the Ituri Pygmies, whether primitive and indigenous, or introduced and degraded as regards stature, are the last remaining representatives of the inhabitants of forest-covered Africa. The fact that they inhabit only the Ituri forest is in itself significant . . . a partially isolated equatorial remnant of the vast primeval forests which at one time covered the continent. It is not surprising . . . that in it should be found the last remnants of the surviving descendants, not only of its ancient human inhabitants, but of its fauna. . . . [T]he little red buffalo, the okapi and the forest elephant, have remained like the Pygmies unmodified . . . to the present day.

—*Cuthbert Christy, 1924*

Okapi live in what for most Westerners are far flung, distant, and lonely places, mysterious and impenetrable to outsiders like themselves. However, they inhabit landscapes long lived in by Africans, and remain entangled in their lives and cultures.[1]

Westerners have long thought of pygmy peoples in particular—and the Congo basin rainforests they inhabit—as primeval throwbacks to an Edenic dawn of humanity, pure and unchanging until the onslaught of accelerating development in late colonial and postindependence Africa began to destroy their integrity, as the epigraph to this chapter suggests.

These ideas explain much about Western okapi hunters' attitudes regarding the capacities, character, knowledge, and expertise of their African guides. Further, misinterpretations of African ethnic groupings and identities continue to influence okapi conservation in the rainforests of the northeastern Democratic Republic of Congo (DRC).

The tendency to group the apparently endangered and rare okapi with similarly characterized pygmy peoples and to oppose them to local African farmers and hunters—has political and social implications. In struggling with the political challenges of wildlife conservation in the region, the foreign conservation NGOs that fund most okapi conservation efforts have drawn on ethnic characterizations of locals to shape their interventions in various ways.

In part, these characterizations fit with international ideas about Indigeneity and biocultural diversity that entered the mainstream in the 1990s. NGOs know they are under scrutiny to meet ethical standards in difficult local circumstances. This is complicated by legacies of colonial ethnology and racism. These characterizations fit less well with regional African states' interpretations of the rights of local peoples.

Exploring histories of how local peoples have been represented by Westerners in the past and comparing them to current thinking on the deep histories of these peoples, is therefore worthwhile. For similar reasons, it is worth examining myths about the rainforest in light of recent scientific thinking on the deep history of the region's physical environment. My intention is to disrupt comfortable mythologies of a primitive or primeval region that is best fenced off and left undisturbed to enable its native wildlife and hunter-gatherers to thrive in peace.

Races, Evolution, and Representations of Africans
Johnston's Types

Harry Johnston's writings on Africans provide a useful overview of ideas about race current when the okapi was discovered and during the first decades of Westerners' searches for it. He regarded himself as a student of the "natural history of the human race," as well as the "rest" of natural history. He had been exposed to cutting-edge European thinking on race science (or pseudoscience) in London in the 1870s.[2]

While a student making drawings in London's centers of comparative anatomy, he had encountered Thomas Huxley, William Flower, and others working out how Charles Darwin's theory of natural selection could be applied to explaining human evolution. Huxley defined five "races" of humans, and Flower developed calipers for recording human skull measurements, discussing the effects of environments and climate on racial traits including morals and intelligence.[3]

As Sven Lindqvist argues in his blistering attack on imperial justifications for racial hierarchies and genocide, this science was entangled in very dark narratives about human evolution. Seeing imperialism as a biological process (evolution) that would ultimately eliminate the "lower races" resulted in more than science fictions: the Tasmanians disappeared from the face of the earth.[4]

Natural history museums helped establish a pseudoscience of race, housing collections of skulls and skeletons of human racial types alongside their collections of other species. The Natural History Museum in London exhibited three skeletons in a row, that of an Akka pygmy woman, a European man, and a gorilla (Emin Pasha supplied the woman's skeleton, and Paul Du Chaillu donated the gorilla's). Many of these collections still exist, though are seldom exhibited: later, the Nazi's extermination of people based on so-called biological racial types rendered this kind of scientific essentialism unacceptable.[5]

Johnston wrote an overview of the African slave trade at the close of the nineteenth century—which he had campaigned against all his adult life. In it, in addition to voicing his moral outrage at the practice, he also represents slavery as in an obscure way justified—as a punishment of "the negro for his lazy backwardness." Those races that do not work or aim to improve every generation, he admonishes in a textbook passage of late Victorian moralizing, will be trampled on by those that do.[6]

Having been gifted freedom by the exertions of enlightened Europeans, Johnston insists, Africans must bestir themselves. Under European tutelage, they must develop the vast resources of tropical Africa or surely in time again face relegation to servitude. The African must not "squander away his existence with the heedlessness of those anthropoid apes to whom in a minute fractional proportion he is more nearly allied than we are, his present guardians." Africans' alleged backwardness was conceived of as being biological to a degree.[7]

Volume 2 of Johnston's monograph on the Uganda Protectorate focuses on the peoples of the region but also offers a wider history of African peoples. Johnston identifies three basic human types; the Mongolian, Caucasian, and Negro. These differed in characteristic skin tone and typical characteristics of the head and hair. Adjectival slips give the lie to the objectivity of these supposed "types": Mongolians have "coarse" hair, but Caucasians have "fine" hair; the Negro has a "poor chin," but Caucasians have "handsome features." The obvious problem for this neat system is that these main types have "interbred and produced hybrid peoples difficult to classify."[8]

Johnston believed that the "Negro type" originated in southern Asia, possibly India, where humans first emerged. According to him, these peoples migrated across the Arabian peninsula (then well-watered and vegetated) into East Africa, making their first permanent home within the borders of what was at the time the Uganda Protectorate (most modern thinking has the migration going in the opposite direction). The first radiation into Africa, he continues, was of a "dwarfish," less black subtype, who spread down along the Nile and into the Congo basin and was followed later by "black Negroes of ordinary stature" who spread west of the Nile into West Africa and down the east coast, avoiding the dense Congo forests. Eventually pressure from tribes from the north who farmed and manufactured iron tools forced them into the Congo forest. Johnston claims it was difficult to clearly distinguish among these types because they had interbred so much. One clear difference was that the "full-sized agricultural forest negroes" were agriculturalists, while the "Pygmy-Prognathous negroes" were not.[9]

Johnston summarizes what was known about pygmy peoples in his time. He reports that the average height of the men he personally met and measured was four foot nine inches (144 centimeters) and women four foot six inches (137 centimeters). Pygmy peoples had no language of their own, he thought, speaking rather a corrupted form of the language of the Negro tribe living closest to them with whom they associated. Nevertheless, he remarks that they showed a remarkable aptitude for learning new languages. He claims pygmies had virtually no formal religion but notes burial customs and remarks on their reluctance to discuss religious subjects. He asserts they had little culture, but he admits that he admired their singing and dancing. He was a keen artist and was impressed by their aptitude for drawing. In short, his discerning powers of observation and humaneness constantly undermined his intellectual prejudices. Despite making disparaging remarks on their culture and lack of government, in Johnston's opinion, pygmy peoples were "very moral."[10]

Even though he concedes that the intermixing of races and types over historical time complicated the project of describing enduring cultural and physiognomic human types, he stuck with it. This also required him to ignore the impacts of recent historical events, including those he had been a part of. Johnston also failed to acknowledge how imported ideas about race and ethnicity and their imposition on Africans by colonialists and missionaries may have shaped this history.

Western Representations of "the Pygmy"

In the late 1800s, Western explorers encountered peoples of strikingly different stature and appearance inhabiting the rainforests of Central Africa, whom they lumped together in the homogenous category of pygmy, a term derived from the mythical population of short-statured people described by Homer in the *Iliad* (various nationalities of explorer used other names too including dwarfs, *zwergen, nains, négrilles,* and so forth). It became a kind of competition to discover the shortest and "most primitive" pygmies, the archetype being hunter-gatherers living in beehive-shaped huts in the rainforest.

Explorers' quest to discover the remnants of a pure, prehistoric people meant that they largely ignored pygmies' cultural diversity. Among the first to recognize this diversity was Paul Schebesta, a Polish (Moravian) missionary, linguist, and anthropologist who lived and worked with Mbuti people in the Ituri rainforest in the 1930s. He published detailed accounts of their societies, languages, and lifestyles, noting several subgroups, each of whom spoke their own languages. Schebesta describes the "tiny pigmies and the timid okapi" as the two great mysteries of the Ituri; both were almost equally "weird and wonderful" and also elusive to the "white man." He notes that many Mbuti wore okapi-skin "aprons" and suggests that the animals were presumably not uncommon in the forest. Many early seekers of the okapi were equally interested in finding forest-dwelling pygmy peoples.[11]

James Harrison, for example, persuaded six Mbuti to accompany him to England in 1905. They stayed with him at Brandesburton Hall in between tours around east Yorkshire and the rest of England and Scotland (with an excursion to Berlin). They were exhibited in London at Olympia, the Crystal Palace, and the House of Commons and performed at the hippodrome. Allegedly, they recognized the okapi mount at the Natural History Museum and confirmed that the name "okapi" was authentic. In 1907, after they had spent thirty months in England, Harrison accompanied them back to the Congo.[12]

According to Percy Powell-Cotton, only the pygmy peoples could survive the rainforest without sickening. He describes them as "the most savage and primitive" on a notional scale of local African culture, at the other end of which were the "Mongwana, the followers and descendants of the Arab ivory and slave dealers," who had some Moslem civilization. Powell-Cotton's observations about pygmy peoples are similar to Johnston's; he notes, for example, their lack of government or central authority and their allegedly feckless existence living off the forest and despising agriculture and a settled life. For Herbert Lang, it was the Mangbetu who were "the most highly cultured natives of these regions," their culturedness being reflected, according to Lang and other Westerners, in their appearance and dress, sophisticated artworks (representational art), and centralized government.[13]

On his first expedition, Adolf Friedrich von Mecklenburg-Schwerin sought out "this exceedingly singular race," first the "Bugoie Batwa" and then the "true pygmies," the Mbuti. He was struck by their "big, intelligent eyes" and "good-humoured faces," claiming they "live by pillage, theft and hunting." He observed linguistic differences between pygmy peoples, notably, terms used for the okapi, which in Beni was "okapi" and "kwapi" but was called "kenge" by Mbuti in Mawambi. Like many others, he remarks on the "incredibly keen scent and sagacity of these pygmies" in hunting wild animals, noting their ability to "follow almost imperceptible indications which entirely escape the eyes of Europeans."[14]

It is unnecessary to recount all such descriptions—the point is that Johnston's views were widely held, if not always so clearly articulated. They were underpinned by often vague ideas about an evolutionary "ladder" of races and cultures, the lowest rungs of which were allegedly occupied by hunter gatherers like pygmies. Westerners used such representations of itinerant Indigenous peoples together with the concept of terra nullius (nobody's land, free for the taking) to dismiss such people's rights and claims to land. They dismissed the advanced natural history knowledge and field skills of pygmy peoples as unscientific, comparing it to the instinctive knowledge of the primate.

These ideas are not extinct. Maano Ramutsindela, Frank Matose, and Tafadzwa Mushonga argue that violence done to Baka (pygmy) communities in the name of wildlife conservation in the Congo basin in recent decades is facilitated by a colonial conception of Black people as being a "part of the natural landscape" and "closer to nature than their European counterparts." This alleged closeness (especially attributed to pygmies) does not represent a positive acknowledgment of their capacities as stewards of nature because it is informed by the idea that they are subhuman. They are accused of hunting "bushmeat," a term not used to describe the "venison" hunted in the West. Africans' subsistence hunting is redefined as poaching, and forest-dwelling pygmies are regarded as a threat to their environment, even though timber extraction, commercial agriculture, and mining clearly pose greater threats.[15]

There were however also positive Western representations of pygmy peoples mobilized by those searching for more just societies uncorrupted by European notions of civilization and their hierarchical societies. These representations are part of the long tradition of the noble savage associated with Jean-Jacques Rousseau and Michel de Montaigne. The consequences of these more positive conceptions was not necessarily more benign.

Turnbull and the Noble Savages

Many Westerners in postwar Europe who were sickened by the carnage wrought by their advanced mechanized society and by the environmental destruction it had caused looked elsewhere for better models of living together and in harmony with the natural world. Patrick Putnam and Colin Turnbull sought it among the Mbuti peoples of the Ituri forest. A later generation that was influenced by the emerging environmentalism of the 1960s instigated by the massive transformations of the global environment being driven by American capitalist enterprises abroad and rebelling against US involvement in Vietnam also looked to autochthonous (Indigenous or native) peoples of the Amazon and Congo rainforests for alternative ways of being in the world.[16]

Some were influenced by Colin Turnbull's hugely popular and somewhat romantic presentation of Mbuti life in the Ituri rainforest of the Congo in his widely read book *The Forest People* (1961). This is what first brought the pioneering okapi researchers John and Terese Hart to the Congo, and still later, anthropologists like Stephanie Rupp.[17]

Another consequence of a postwar revulsion against the idea of different human races arranged on a continuum of advancing civilization was a shift to the notion of a unified "family of humankind." The great ladder of being was kicked away and replaced by an evolutionary process perspective. The "hunter hypothesis" was proposed to explain what differentiated humans from their apelike ancestors, and physical anthropologists and others looked to so-called hunter-gatherers to understand this evolutionary transition. Scholars now emphasized the positive aspects of paradigms they had already developed about these peoples.[18]

There was a boom in hunter-gatherer studies in anthropology that focused on Africa's pygmies and San ("bushman") peoples, regarded as epitomes of this lifestyle. By the late 1980s, however, anthropologists began to challenge the idea that "hunter-gatherer" was a useful analytical category, arguing that it was a vague conception telling us more about Western romanticization of these peoples than about their lived realities or cultures.[19]

The idea of Indigenous peoples was adopted as a more acceptable conception and entered the mainstream in a period during which environmental concerns and attempts to address global poverty and underdevelopment came together, a moment epitomized by the 1992 UN Conference on Environment and Development held in Rio. This conception has guided the thinking of many transnational development and conservation agencies and conventions ever since.[20]

There were skeptics here, too, however. In 1988, anthropologist Adam Kuper caused a furor by arguing that the term "Indigenous" was really a euphemism for "primitive." He argued that it reflected a very long-held and illusory opposition between the values of mainstream Euro-American industrialized society and Indigenous values (the West and the rest).[21]

In 1991, conservationist Kent Redford also raised doubts about "the ecologically noble savage." He critiqued the rhetoric of both prominent conservationists celebrating Indigenous peoples who without exception lived in harmony with their surroundings, and Amerindian spokespersons claiming a Manichean cleavage between the corrupt and destructive Western way of life and the collective, communal, and environmentally sensitive Indian way of life. Redford points out that taking destructive aspects of the Western way of life to task didn't preclude Indigenous peoples from exploiting their natural resources: in fact, it was used to claim their right to do so (and why not?).[22]

What bothered Redford was partly that calls to learn from Indigenous knowledge were based on romantic misconceptions of precolonial Amerindian ways of life. Redford points to archaeological and palaeobiological evidence indicating that most tropical forests had been severely altered by humans before Europeans arrived, an argument also made subsequently by Jared Diamond. Were pre-European societies really good stewards of the land, or did their supposedly sustainable practices only work "under conditions of low population density, abundant land, and limited involvement with a market economy?"[23]

Arguments about the "ecologically noble savage" were fueled by developments in the Neotropics, but they also had consequences for the Congo and its peoples. The concepts of hunter-gatherers and Indigenous people became virtually synonymous with pygmies. While most people who referred to Indigenous people did not define what they meant by the term, a guiding definition was proposed by the International Labor Organization in 1989. It defined indigenous people as those differentiated from other communities in their region owing to their prior occupation of the territory, cultural distinctiveness, subjugation by political structures external to their own, and their self-definition as Indigenous. This definition has since guided the policies of major international organizations like the United Nations Environmental Program, the World Bank, and the International Union for the Conservation of Nature.[24]

Unfortunately, at least in the Congo basin region this definition can result in forest-dwelling peoples who are not defined as pygmies or hunter-gatherers being sidelined. Policymakers may simplify the complex economic, ecological, social, and political relationships that exist between diverse communities in focusing on their chosen clearly defined community: the pygmies. For conservationists, pygmies are guardians of the forest, and their subjugation at the hands of villagers and farmers (which does occur) is believed to mirror the subjugation of the forest and its resources by these other African communities.

Africans themselves have contested these framings, notably the idea that while all Africans are Indigenous, some are more Indigenous than others. Some Africans have, inevitably, appropriated the concept of Indigenous or "autochthonous" peoples to channel and benefit from government and NGO resources allocated to peoples so defined.[25]

Unravelling the Essentialized (Mythical) Pygmy

Paul Schebesta noted linguistic diversity among pygmy groups in the 1940s, and in his 1961 book Turnbull recognized two Mbuti groups distinguished by their hunting practices (one group being net hunters and the other being bow-hunters). However, in a 2017 review, Serge Bahuchet notes extensive variations "in virtually every facet of Pygmy life and culture."[26]

The singular or "essential pygmy" is, like the "bushman," a figment of the Western imagination. In part, this idea originated in a search for an other, an opposite to Westerners' imagined selves. The concept of ethnicity itself requires opposition to "an other." As Kuper notes, "there could be no barbarians before there were Greeks, or later, Romans, or Christians, or Europeans." In a 1578 essay titled "On the Cannibals," the philosopher and statesman Michel de Montaigne used European accounts of Amerindian peoples in Brazil to reflect on how Europeans saw themselves and others. He suggests that although Brazilians were obedient to different conventions (like ritual cannibalism), they were not so different to Europeans as the latter might suppose (neither is truly rational). Montaigne thought Brazilians were probably closer to nature and less corrupted by civilization than Europeans.[27]

The idea of Cannibalism brings these ideas into focus. It was certainly much on the mind of Western explorers of the Congo. Henry Morgan Stanley describes traveling past African communities who tried to net his men from their boats in order to eat them. Lang dismissed his armed escort to improve relations with locals, but he feared that he might be eaten as a result. Lang claimed locals would stroke his arms and remark "Nyama mingi! Nyama nzuri!" ("Lots of meat! Very tender meat!"), even as they reassured him killing him wouldn't be worthwhile because his rare meat would go to their chiefs. Lang even speculated that okapi were at increased risk because once the Belgians had banned cannibalism, local Africans turned to the larger animals of the rainforest for meat. He described how the Mangbetu chief Zebandra had used eight hundred drivers on a hunt which caught eleven okapi in a single week.[28]

As Harriet Ritvo reflects in her discussion of taxonomies of animals which were based on which should and which should never be eaten, cannibalism was associated with "savagery," defined as the opposite of "civilization." By extension, "if cannibals were savages, then all savages were cannibals." British explorers often assumed the worst of peoples they encountered, at least initially, assuming for example that Fijians, Māori, Tahitians, the Congolese, and others were all cannibals, which, of course, also rendered them in urgent need of civilization.[29]

Joseph Conrad portrays these imagined Congolese cannibals in his novella *Heart of Darkness.* Conrad contrasted the degeneracy of the employees of the colonial Company for Trade, once free of the constraints of their society, with the remarkable restraint shown by the hungry cannibals his protagonist Marlow hires to work on the steamer he is sailing upriver to meet that apostle of European civilization and depravity, Kurtz.

While Europeans posited that there were many "tribes" of African farmers, among whom they identified cannibals, they often assumed there was a singular, "true" pygmy type. Attilio Gatti describes a scale of African types isometric with distance from outside influence. "Degraded" Bantu peoples lived along the roads through the Ituri forest. Adjacent to these roads was a zone inhabited by clans of pygmoids ("products of years of intermarriage between real pygmies and usual natives") who hunted for themselves and on behalf of the Bantu farmers and guided white men wishing to get "a taste of the jungle." Beyond this pygmoid zone were "the immense hunting territories of the true pygmies." Most officials and okapi hunters never traveled beyond the pygmoid zone, he explains, and hence the misconception that okapi were rare, when outsiders had not entered the territories of the true pygmies.[30]

The so-called true pygmies were generally not identified as cannibals by those who met them, though the aesthetic practice of tooth-sharpening frightened Westerners. The best-known example for Americans was Ota Benga, photographed as a teenager with a cheeky grin revealing sharpened teeth while on display in the United States in 1904. Described as an African pygmy, he was shamefully exhibited in the monkey enclosure at the Bronx Zoo in 1906.

Recent Thinking on the African Peoples
of the Congo Basin

Pygmy peoples are the focus here because they are consistently identified as the Africans possessing the most profound indigenous knowledge of the rainforests and explicitly linked with okapi conservation. However, their histories are inextricably entwined with those of villagers of Bantu-, Sudanese-, and Nilotic-speaking origins, who also inhabit the region. The history and nature of the interrelations of pygmies and their (mostly) farmer neighbors, has become important since the work of Turnbull and has been much debated by anthropologists (see Online Materials, appendix 5).

Beginning, then, with the pygmy peoples, recent scholarship recognizes around twenty ethnic groups that make up those collectively referred to as "pygmies" and at least three ethnic groups within the BaMbuti (Efe, Asua, and Basua). Some of the groups identified as pygmies live out on the savannas, others in mountains or swamps. Some short-statured groups are not hunter-gatherers, while some taller peoples are. However, while the diversity of pygmy peoples is acknowledged, the literature on the peoples of the region has nevertheless tended to focus on people of short stature who have a seminomadic hunter-gatherer lifestyle.[31]

Population geneticists began studying the diversity and origins of pygmies in the late 1960s. By the 2010s, extensive sampling and statistical analysis had established the outlines of a distinct pygmy genetic origin: ancient pygmy and non-pygmy populations diverge around seventy thousand years ago (confidence intervals are high). It is unclear where this divergence occurred; that is, it might not have been in Central Africa. Eastern and Western ancestral pygmy populations diverged around twenty thousand years ago, which falls within the known period of human occupation of the Congo basin. This may be linked to forest fragmentation associated with the Last Glacial Maximum. A further divergence within the Western and Eastern pygmy populations occurred around three thousand years ago, possibly linked to a climate crisis four to five thousand years ago, during which forest areas contracted and became isolated from one another. This climate crisis has been linked with the expansion of agriculture and Bantu languages into the region—and, it is assumed, consequent intermixing of pygmy and non-pygmy populations.[32]

In West and Central Africa, the Late Stone Age extended from circa 40,000 to 3,500 BP, at which point hunter-gatherers working stone gave way to Neolithic farmers entering the region from the Sahel. From circa 2,800 to 2,500 BP, Iron Age farmers, whose slash-and-burn farming was more effective than that of their Neolithic farmer predecessors, spread down into Central Africa. Their iron tools and fire use along with the great quantities of charcoal they required for their smelting helped spread savannas into forest regions. In this period, only a few typically cultivated Indigenous plants, including oil palms, cowpeas, yams, and bambarra nuts, could survive rainforest conditions.[33]

While violent clashes may have characterized the first encounters between Bantu farmers moving into the Congo basin region and its pygmy peoples, it seems that peaceful relations of exchange and alliance were subsequently established. It made sense for the decentralized Bantu groups moving into new, unknown territories to seek cooperation, given that pygmies were intimately acquainted with their local environments and had much knowledge to share. Bantu peoples shaped by Niger-Congo beliefs systems may have seen pygmies as autochthons or "first comers" who had a connection with the mystical forces of the land that they did not. Precolonial Bantu political authority depended on such mystical foundations, and so Bantu leaders would have integrated pygmies into a social formation where they could perform important ritual functions that would give them access to these spirits. Pygmies appear to have been open to such alliances.

Particular pygmies achieved renown among other peoples as master hunters or healers. In addition to being directed to pygmies as natural masters of the hunt, okapi hunters were often directed to particular skilled individuals or associations of individuals ("segments"). Expert hunters or trappers might be supported by their segment for auxiliary purposes like checking traps. There are histories of individuals and their actions to be recovered. Pygmies are peoples with particular historical traditions, cultures, and individual life stories, they are not unchanging primitive collectives.

Historical or Prehistorical?

Much research on pygmy peoples treats them as survivors of the dawn of humanity rather than as historical peoples. And yet it seems clear that while retaining some distinctive elements of social organization shaped by enduring cultural principles, they have also undergone significant changes through a long history of interactions with other African peoples (see appendix 6).[34]

Linguistic and genetic studies suggest there was considerable interchange between pygmy and non-pygmy peoples in the precolonial period. Common terms for marital relations are notable, and there is genetic evidence of admixture in all pygmy groups. However, it has been hypothesized that once farmers had settled and began cultivating crops—particularly bananas, which produced much higher yields than yams, and starchy tubers introduced from Southeast Asia by Arab traders, which allowed the production of surpluses for trade—hunter-gatherers were repositioned into relations of economic dependency and social inferiority. Together with iron production, agriculture led to the development of larger political aggregations among Bantu peoples, starting around two thousand years ago. This resulted in regional trade, economic specialization, increased competition between groups, and war. Pygmy peoples were sidelined by these developments. They lacked interest in material wealth and resisted developing larger economic or military groups. This loss of status occurred between 1,500 and 500 BP, ending a period remembered in their oral traditions as a golden age.

It is unclear what impact the early Atlantic slave trade had on this scenario. The introduction of a suite of crops from the Americas by the Portuguese—including maize, cassava, sweet potatoes, peanuts, beans, and squash—certainly boosted horticultural possibilities in the Congo basin. This possibly reinforced regional trade, empowering settled farmer societies and reducing farmers' dependence on forest produce. Pygmies responded by disengaging from the larger hierarchical social order and withdrawing into the forests or other marginal spaces. They developed stratagems for maintaining their autonomy within their new subordinate client relationships that were well documented by the first Europeans to encounter them.

Pygmy peoples cultivated a restrained, subservient way of behaving when they visited the villages of local Africans, which they immediately abandoned when back in the forest. They stayed in camps outside the villages they were in exchange relations with and restricted their contact with such villages to a month or two at certain times of the year, initiating it only when life was hard in the forest or when they were making specific exchanges or participating in rituals and events. They broke off relations if they felt mistreated, readily forming new alliances. Pygmies sought ways of improving their access to resources while maintaining their autonomy. Following the formation of Bantu kingdoms, individuals managed to establish advantageous relations with royalty, securing prestigious court positions and therefore better standing in society. Pygmies became suppliers of important forest products to the regional trade, using their skills to hunt elephants and okapi and to gather honey.

By the later 1800s, the violence and instability associated with the slave trade and elephant hunting following accelerating European demand for ivory finally reached the northeastern Congo forests. African hunters began to usurp pygmy peoples' role as expert elephant hunters. Bantu peoples no longer venerated pygmy peoples as "first people" with privileged access to the spirits of the place as much as they had in the past. Pygmy peoples subsequently used their strategies of avoidance and selective engagement to take advantage of opportunities arising with the arrival of colonial explorers, traders, administrators, company agents, and missionaries. To the present day, they continue to provide their skills, on their own terms, to conservation and development projects in the region. It is perhaps this determined adherence to their autonomy, combined with boundary-crossing entrepreneurialism that makes it difficult for outsiders to pin down just who they are.

The Territorialization of African Identities

Colonial borders such as those between Belgian Congo and French Congo and the Uganda Protectorate (Great Britain) and German East Africa (now Rwanda and Burundi) established in principle (and subsequently fiercely disputed) during the 1884–85 Conference of Berlin effected arbitrary boundaries between previously integrated and fluid regions of exchange.

Traditional systems of land allocation were replaced by a formal statutory system of land tenure that allowed for land to be bought and sold. Customary systems of social regulation and land allocation were further complicated by new colonial administrative regions and the officials and soldiers that accompanied them. Chiefs were recognized based on treaty negotiations and other political considerations for the colonial overlords. These changes resulted in what has been called the territorialization of ethnicity. The dispersed villages of the rainforest farmers could not satisfy Belgian colonial aspirations to create an agriculture-based market economy, so authorities forcibly reorganized them into centralized permanent villages along roads. Regional chiefdoms were created or formalized starting in the 1930s to help coordinate the development of infrastructure, the process of rubber extraction, and to facilitate access to laborers who would plant cash crops like cotton and rice. Traditional land allocations systems were replaced with ones that served colonial commercial farming interests. Enforced collection of forest products was passed on by African farmers to local pygmy peoples.[35]

Following the First World War, a vast medical operation unfolded across the colony. Belgian doctors and Congolese nurses set about testing and treating people for tropical diseases like sleeping sickness and malaria. Whole villages could be moved out of tsetse fly zones, but mostly, this meant people were restricted to their own villages. Africans now required special permits for travel, which instilled a new sense of ethnic identity and fixity.[36]

After the colony was transferred to the Belgian government, it was decided real progress could be made only if the authorities learned more about its native peoples. The Bureau International d'Ethnographie was formed in 1908, and it commissioned a massive ethnographic study intended to produce an encyclopedia of the "black races." Conceptually, the peoples of the region were sorted into distinct blocs, delineated by their customs, ethnic characteristics, and identities. They were studied in isolation from one another and from multilingual communities known to have existed in the past. The frequent interactions between various groups were forgotten or ignored in the interests of achieving a clear taxonomy of tribes.[37]

The enforced immobility of the colony's peoples under the new postwar medical regimes reinforced ethnologists' idea of a static map of discrete ethnic groups tied to and shaped by local environments. Each tribe was ascribed certain characteristics (more warlike, industrious, and so forth) partly to facilitate colonial control and targeted labor recruitment.

Just as Johnston had ignored the confounding influence of historical movements and intermarriages of Africans, notably those with Europeans like himself, so these ethnologists were trying to nail down defined identities linked to specific areas, even though the colonial state was also moving people around within the colony to work on plantations of commercial crops, and later to service mines and infrastructure projects. They similarly largely ignored both pre- and colonial interventions that interfered with their cataloguing schemas.

In the Ituri forest, home to the okapi, these developments changed farming practices and labor needs. A case study of Lese farmers and Efe pygmies reveals the impacts. Efe (pygmy) women were forced to do more work in agricultural fields, with the result that the Efe became more dependent on crops. Efe hunting became concentrated around settlements, their foraging constrained by roads and permanent settlements. When a colonial police force was created, it recruited farmers, which, along with the advent of a cash economy, put farmers in a dominant position over the Efe. Whereas crop farmers earned cash, the Efe worked only for the barter of things. Farmers paid the imposed hut taxes on behalf of the Efe (who earned no cash), making the Efe in effect indentured laborers.[38]

In the social turmoil following independence in 1960, the market economy and plantation agriculture largely collapsed. Farmers fled into the forests, abandoning crops. They began living with pygmies, and social relations adjusted. While pygmies have not regained social parity, they are once again less subservient to farmers. Some pygmies have adopted limited farming themselves, and occasionally they lease land from villagers, which allows them to be less dependent on farmers for food. Some have entered the cash economy by selling crops commercially, undermining essentialized Western notions of pygmies as nonmaterialist hunter-gatherers.

Endangered Peoples?

In his book *The Mbuti Pygmies*, Turnbull acknowledges the Mbuti's resilience to historical change, notably the ravages of the early European colonists and Arab slave traders. However, he ascribes this to their ability to retreat into the rainforest. Over his thirty years of visits to the Ituri forest, the first of which he made in the late colonial period in the 1950s, he witnessed significant changes. He also saw how postindependence government attempts to support the Mbuti in becoming settled agriculturalists (a failure) had alienated them from the villagers they used to interact with.[39]

Despite this, according to Turnbull, the Mbuti retained their fundamental identity and focus on the quality of social life over economic life, on the spiritual rather than the material. However, he feared they had met their nemesis in the conversion of the rainforest into farmlands and mines and the penetration of its sanctuary by roads and railways and renegade soldiers. He did not see how the pygmies would be able to retain their identity if the forest disappeared and they were prevented from inhabiting its remnants.[40]

Reading Turnbull's account of this vital connection with the rainforest pushed me to finally confront my ignorance about the rainforests themselves, which amounted mostly to a sense they were vast and home to an entangled fecundity. Taking on the rainforests' deep history to determine whether they were actually unchanging primeval forests, outside of history, or if not, to understand how imminent their much-predicted demise in the face of accelerating human development was, seemed like a good challenge for me.

I was still struggling to shake off my postviral fatigue, my return to clarity amid amorphous stretches frustratingly slow. I remembered how the lack of a clear path to recovery had been made worse by my inability to remember my progress. And so, the disorienting small steps toward the light followed by lapses back into tangled shadows seemed obscurely analogous to the challenge of fogging out rainforest deep time. It was some sort of target to aim for, once I could function well enough and recall what I had learned.

The Rainforests

Grappling with the nature and history of the Congo basin rainforests requires acknowledging the powerful hold these forests exerted over the imaginations of the Westerners who described their journeys through them. Seasoned explorers were overcome by the scale of the huge trees, the silence, the profusion of vegetable life, the darkness and apparent impenetrability of the deep forest.

In *Heart of Darkness*, Marlow describes traveling up the Congo River as "like travelling back to the earliest beginnings of the world, when vegetation rioted on the earth and the big trees were kings." Harry Johnston describes its heat, humidity, and decay as reminiscent of "Miocene times, . . . scarcely suitable for the modern type of real humanity." Stanley writes that if you entered its depths "an awe of the forest rushed upon the soul and filled the mind" and that "one became conscious of its eerie strangeness, the absence of sunshine, its subdued light, and marveled at the queer feeling of loneliness." It was, he concludes "as if one stood amid the inhabitants of another world." This awe and wonder, however, soon gave way in the experience of that sturdy soldier Boyd Alexander to a sense of a "haunting fear that rests on the forest like a spell." Lang proposed that the okapi remained unknown to Westerners for so long because of the vastness and monotony of the forest, which was cursed with appalling heat and humidity and violent storms. Delia Akeley wrote that "there is something weird and frightening about the Congo forest, with its abnormal and distorted growth of vegetation, strange animals, and stranger human beings."[41]

The difficult terrain, obscuring vegetation, and resulting disorientation, the physical discomfort caused by the heat, humidity, and frequent frighteningly intense thunderstorms, the myriad biting insects, and the fear of sudden encounters with hostile animals or humans (possibly cannibals) overwhelmed many seasoned travelers. Worse, the forests' most dangerous denizens entered travelers' bodies and attacked them from within: few escaped fevers and other tropical maladies.

The forest seemed to strip away the advantages of Western civilization and technology, rendering binoculars and rifles cumbersome and ineffective. Tall, large-framed Europeans were clumsy and noisy in the quiet of the dense forest; they were put at the mercy of the diminutive pygmy peoples, who moved through the forest's trackless entanglements with ease and tracked its creatures expertly. As a result, few accounts of the rainforests provide nuanced descriptions of the particular places explored. At least until the 1950s, it is hard to find any historical analyses of what most travelers presented as a timeless, boundless, obscure, and primeval space.

The Deep History of Rainforests in Central Africa

The existence of rainforests depends on climate variables, and long-term fluctuations in these have shaped the historical extent and nature of rainforests in Central Africa. During the most recent Ice Age, much of the region's present-day rainforest was covered in savanna-grasslands, the forest retreating to refugia from which slower-spreading species are still reemerging. The rainfall regime of humid Africa is near the lower threshold of viability for rainforests, so small changes in rainfall can have large effects, with drier seasons resulting in expanding savanna and shrinking rainforests.[42]

After the Ice Age, the region rapidly warmed and became wetter, introducing a period known as the African humid period (circa 11,000–4,000 BP). Solar forcing caused intense heating of the Sahara during the northern summer, which strengthened the West African monsoon and sucked higher rainfall deeper into the Congo basin. Africa's forested zones shifted more than three hundred miles (five hundred kilometers) northward during this period, but from 4,000 to 2,000 BP they retreated again in the face of a drier regime resulting from a shift in solar orbiting. Scholars debate the causes of rainforest retreat and the expansion of savannas in this period, acknowledging that human activity became a significant factor, too, even as they underline that the timing suggests that climate change was the underlying driver.

This history of climate-linked vegetation disturbance may account for lower plant diversity in African rainforests compared with that of rainforests in Borneo and Amazonia, which dried but never ceased to be forest. On the other hand, Central African rainforests have a significantly higher mean above-ground biomass (429 metric tons per hectare) compared with that of Amazonian forests (289 metric tons per hectare). The above-ground biomass of Central African rainforests is comparable to that of the famously high biomass of Bornean rainforests (445 metric tons per hectare).

African rainforests also have a lower mean stem density than Amazonian or Bornean forests do—that is, they have proportionately more larger-sized trees. It is not clear why, though one intriguing argument is that whereas Amazonia lost its megafauna around 10,000 years BP, Central African rainforests have retained their megafauna into recent times. These megafauna cull smaller trees and disperse the large seeds of high-wood-density species.

Historical Human Impacts

Only in the past twenty-five years have mainstream ecologists accepted that past human impacts have significantly shaped the earth's ecosystems. In the 1950s, Paul Richards warned ecologists working in African rainforests not to ignore human modifications of their study areas. James Fairhead and Melissa Leach's work on West Africa in the 1990s sparked debate over human influence on the region's forests, bringing anthropogenic influences into the frame of studies of long-term vegetation change. Research on African tropical environments has since become increasingly interdisciplinary.[43]

Migrations of peoples, farming and iron smelting, burning practices, the slave trade, internal conflicts, colonization, disease epidemics, and biological introductions all shaped long-term shifts in forest composition and dynamics. Natural history and climate change are just two aspects of the environmental history of the region's rainforests.[44]

Archaeologist Richard Oslisly and coauthors describe two major waves of incursions into African rainforests by metalworking farmers. The drier climate of these rainforests compared to most of the tropics made them more suitable for agricultural crops than other rainforest regions. The first of these incursions occurred circa 2,800 BP, peaking in around 1,900 BP, before fading out in a human population crash from circa 1,600 to 1,000 BP possibly caused by disease outbreaks. The forest recovered during this period, only to give way again when a new wave of metalworkers began pouring in circa 1,000 BP.

A subsequent human population crash around 500 BP was possibly triggered by the Atlantic slave trade. Later, colonial policies of forced resettlement and pressure on farming as labor was diverted by the rigors of rubber collection meant that some forest areas were spared from direct land use. Traces of disturbed, low-diversity forest remain in an area of sandy soils and lower rainfall between the western and eastern regions of the Congo basin. Satellite-based analysis reveals evidence of past forest retreat and cultivation in this zone.[45]

Julie Morin-Rivat and colleagues have studied sunlight-loving large canopy tree genera, research triggered by their concern over the poor regeneration that has been typical of central African rainforests since the mid-1800s. They identify a wet period (1300–1850 CE) characterized by high anthropogenic disturbances including clearing and burning that favored tree regeneration. Then came a period of reduced disturbance from 1850 to the present that is correlated with a lack of regeneration of important canopy tree species.[46]

Their study attributes this reduced forest disturbance to European colonization. Europeans discouraged nomadic lifestyles and coerced African peoples to settle along rivers and roads, which led to increased mortality from diseases and was a drain on their time in the forest as they were forced to work on plantations and other projects. Starting in the 1930s, access to education contributed to urbanization among Africans in search of wage labor and resulted in further abandonment of the forests.

This study along with others counters the claim that human beings' actions have had only negative effects on biodiversity in the rainforests, suggesting that limited slash and burn agriculture followed by long fallow periods actually favors canopy tree regeneration in the Congo basin forests. Closure of the canopy following depopulation of the forests under conservation-minded interventions appears to inhibit their regeneration. In short, the Congo basin forests have fluctuated significantly throughout the Holocene, shaped by climatic shifts and varying levels of human impacts. But have Turnbull and others' grim predictions about the African rainforests come true?

Recent Threats and Perturbations

A high-resolution survey of deforestation across African rainforests from 1990 to 2010 found deforestation rates were a quarter of those observed in Latin America and slowing. At regional scale, this was attributed to the absence of the agro-industrial-scale clearing driving deforestation elsewhere in the tropics. While logging concessions have been granted to extensive regions of Congo basin rainforest, the logging that has been undertaken is not the clear-felling of stands of low-diversity forest more typical of temperate-region logging but rather removal of one or two valuable timber trees per hectare. Forests rebound quickly from this kind of logging, recovering biomass after twenty-four years, though the highly valued timber trees take much longer to recover. The real damage is more collateral, affecting other trees and resulting in increased drying and vulnerability to fires introduced by logging tracks.[47]

Hotspots of deforestation have been identified at the fringes of the Congo basin, notably in the northeast. Overall, deforestation is clustered along transport networks (roads and rivers) within five-hour travel times to major markets and within twelve-hour travel times to cities where charcoal and wood fuel are heavily used. Mineral and oil extraction have resulted in economic booms, attracting labor away from rural areas and driving urbanization and deforestation in surrounding peri-urban regions but slowing deforestation in remoter forest regions.

The biggest impact of logging on biodiversity has been indirect. The Democratic Republic of Congo's road network deteriorated in postcolonial times, but the opening of logging roads opened up rainforest areas with surviving wildlife to commercial bushmeat hunting. The fast-growing urban centers that had booming populations and cash economies created significant markets for bushmeat. Smoked wild meat, which provides an inexpensive source of protein (it is cheaper than other animal protein, except pork and caterpillars) was sold to poorer families, while fresh meat and sought-after species were sold to upmarket restaurants.

An estimated 178 Central African species are hunted for bushmeat, more than half of which are threatened by this hunting. Tens of millions of individual animals (more than two million metric tons) are extracted from the Congo basin annually. The low risk of penalties, certain demand, and ease of market entry have driven the marked increase in commercial hunting over the past fifty years.

Recent studies estimate more than 90% of villager hunters now sell some of their catch. Only smaller, fast-breeding species like duiker and large rodents are resilient to these commercial hunting regimes. Overall, this consumption is not sustainable.[48]

The impacts of hunting on forests, while invisible to remote sensing, may be considerable. The negative effects of overhunting disrupt food webs and the resulting trophic cascades disrupt ecosystems functioning. Most of the hunted animals are larger, seed-dispersing, fruit-eating mammals, many of which consume large seeds. Most tree species are dispersed by these animals. The decline in larger-bodied "forest architects" and the mesofauna means top-level predators like leopard leave. The small-bodied seedeaters, preferring smaller seeds of lower-density plants, proliferate. The animal-dispersed trees are mostly the slower growing, longer-lived, high-density trees. They are the significant species for carbon-storage in tropical forests.

Research on defaunated or "empty" forests shows that it is the lower-density, faster-growing, and abiotically dispersed (by wind or other means than animals) plants that are favored. This means less food for large fruit-eating mammals, which in turn further disadvantages long-lived hardwood trees, inducing a positive feedback loop that reduces both biodiversity and global resilience to climate change. The forests lose resilience to drought and possibly fire over the long term.

Climate Uncertainties

Although it is the second most important convective engine driving global atmospheric circulation, the Congo basin's climate is poorly understood. The political chaos that was unleashed in the 1960s, has resulted in a long-term breakdown in weather reporting. From 1990 to 2010, less than ten rain gauges were reporting data for this immense region. In 2013, only three meteorological stations reported to the Global Telecommunication System.[49]

The models of the region's climate that draw on available satellite data show little agreement on rainfall. While drastic change such as forest dieback resulting from climate change is not predicted for the Congo basin region, it is unknown how these forests will respond to rising temperatures and possible drying, or to more extreme variability. Historical data suggests forests can shift rapidly to savanna when minimum thresholds for rainfall are crossed. Retreating rainforest can cause feedbacks whereby water is not recirculated to the atmosphere, further reducing rainfall and driving forest retreat. Given human incursions, the ecological effects of commercial hunting, and increased vulnerability to fires, policies and management based on interdisciplinary research into the ecological and social challenges of conserving the region's rainforests are clearly needed. The okapi would not survive their loss.

Conclusion

Pygmies are not peoples without a history, as the cliché of the noble savage—typically ascribed to hunter-gatherer peoples—would have it. If they were the first peoples of the region, that was a very long time ago, and the history of their exact geographical origins and patterns of dispersal remains obscure. They have very long histories of interaction with the other African peoples, who have been migrating into the region for at least four thousand years. The assumption that pygmies have sheltered in unchanging primeval rainforests from the demographic and ecological changes sweeping the landscapes and cultures on the plains beyond their fringes is mistaken. Climatic shifts and human incursions and retreats have impacted these forests across space and time.

Colonial settlement schemes for African farmers and for pygmy peoples, the introduction of cash crops and enforced wage labor, medical interventions, and anthropological surveys all had an effect on regional ideas about identity. Unfortunately, enforced immobility and education in mission and colonial schools persuaded many Africans that they had distinctive characteristics and differences along the lines of those described by the anthropologists. It is hard to estimate the influence of this colonial legacy on current conflicts in the region. Certainly, ethnicity is a powerful factor in shaping alliances and determining who is the enemy.

Agencies working in the region have appropriated aspects of these historical narratives about identity that suit their goals. The claim that if the rainforest belongs to anyone by right of first settlement it is the pygmies risks disenfranchising other later arrivals. It ignores centuries-long interactions and shifting relationships impacted by climatic and environmental conditions, trade, and the arrival of Europeans. And yet this story strongly persists, thousands of years after other "immigrant" communities of Africans arrived.

The Amazon rainforests and its Indigenous peoples have captured the popular imagination in the West. The Congo basin rainforests, meanwhile, occupy a much less prominent and more ambiguous space—they truly are the forgotten green lung of the planet. This relegation to obscurity for the Congo rainforest is influenced by the terrible history of Western interference in the region. Western conceptions of the rainforests, its peoples, and wildlife remain entangled in ideas of the primitive, of violence and death. Conrad's *Heart of Darkness* famously shines a light on the first of many violent extractions of the region's resources that have included ivory, rubber, the uranium used in the atomic bombs dropped on Japan, and the copper and cobalt used in munitions in both world wars. The film *Apocalypse Now* transposed Conrad's book to Cambodia and Vietnam, where Congolese copper was used to deadly effect. As the film *Virunga* and Siddharth Kara's book *Cobalt Red* (2023) show, this legacy of external meddling to access the DRC's astonishing wealth in natural resources continues to date.[50]

Apparently endemic violence and instability in northeastern Democratic Republic of Congo means few Westerners now travel to the region and get to test their prejudices on the ground. Scientific research is hampered by lack of resources, data, and armed militia. Conservation efforts must contend with poor infrastructure and rebel armies and compete with far more lucrative activities like mining for the copper, cobalt, and coltan that goes to China and the West.[51]

Conservation organizations therefore struggle to raise attention and funds for a region it seems the West would rather forget. Wildlife documentaries and travel agents favor the open savannas of east Africa, with its easier climate, fewer tropical diseases, and the charismatic big cats and other megafauna of the plains. While the Congo is blessed with the great apes and forest elephants, many of its fascinating forest species, including the forest buffalo, the bongo, and the okapi, remain little known outside of the region.

It is unsurprising that fundraising efforts focus on charismatic species and appeal to persisting romantic ideas about hunter-gatherer pygmy peoples, often linking the two, one effect of which is to consign other local peoples to the background or portray them as a threat to biodiversity. This framing privileges certain areas, species, and peoples and marginalizes others. It usually associates Indigenousness and ecological integrity with an unchanging "pure" state that can trap locals into adhering to externally imposed ideas about who they are and how they should behave. It causes social friction between human communities afforded more or less recognition and resources based on others' ideas about their cultural identities.[52]

Recognizing the existence of biocultural diversity has not often translated into an understanding of exactly how cultural and biological diversities help each other flourish. Developing such an understanding requires learning to value and utilize nature through participative community processes and a commitment to allowing these processes to inform conservation policy and actions. It requires participating agencies from outside of the region to acknowledge their historical biases and ongoing impacts.[53]

It will take a global village to preserve the okapi and its rainforest home.

CHAPTER FIFTEEN

Clashing Worldviews in a Crucible for Wildlife Conservation

Today, the okapi is a precious national treasure, featuring on the logo of my organization, ICCN, in popular culture and on our banknotes, but it retains its enigma. As this status review highlights, we still have much to learn about the species, including exactly how it is faring in the face of multiple threats.

—*Cosima Wilungula, Director General, Institut Congolais pour la Conservation de la Nature, 2015*

The okapi is the national animal of the Democratic Republic of Congo, where it is recognized as a conservation priority. However, political instability, armed militias, and a lack of resources have hamstrung conservation efforts in its home range in the northeast of the country. In this context, external conservation organizations and funding have been important in assisting okapi conservation efforts on the ground.[1]

Several overseas NGOs have been instrumental in the creation of protected areas for okapi. Their operations in the region are politically complicated, and they are aware that their human rights records are being scrutinized both by international conservation and human rights groups, and organizations working in the country. Their conceptualizations of the rights of local Africans in the context of wildlife conservation have been influenced by international ideas about Indigenous peoples, notably in regarding pygmies as first peoples and "true" forest people, influencing both their external-facing fundraising messaging and how they interact with locals. These ideas don't necessarily align with regional ideas about the status of pygmy peoples, or whether distinctions should be made between indigenous and local African peoples.

In this chapter, I consider how this impacts on okapi conservation, focusing on the Okapi Wildlife Reserve as a case study which is emblematic of these challenges for conservation efforts in the region.

The Establishment of a Reserve for Okapi and the Mbuti and Efe Peoples

In 1992, the Institut Congolais pour la Conservation de la Nature (ICCN) together with an American paper magnate's Gilman Investment Company, Worldwide Fund for Nature, and the Wildlife Conservation Society succeeded in getting 5,290 square miles (13,700 square kilometers) of the Ituri rainforest designated as the Réserve de Faune à Okapis (Okapi Wildlife Reserve). It was listed as a World Heritage Site in 1996 and added to the "List of World Heritage in Danger" in 1997. Once it had been legally recognized, conservationists then had to work out what its designation as a multiuse, inhabited conservation area meant in practice.[2]

As of 2024, the Okapi Wildlife Reserve is co-managed by the Wildlife Conservation Society (WCS) and the ICCN. WCS reached an agreement with the government of the Democratic Republic of Congo (DRC) in 2019 that enables it to work with government partners on management, governance, and logistics for conservation in the region. The director of the Okapi Wildlife Reserve is appointed by WCS and runs the reserve with a staff of ICCN ecoguards who enforce regulations and monitor biodiversity and agricultural expansion. WCS works on conservation capacity, best practices, strategy, and goals, which include maintaining good relations with government, supporting the livelihoods of the people who live in the area, and reducing conflict. The reserve is seen as an anchor for stability, a location for community development, and an opportunity to improve social cohesion.[3]

Wildlife Conservation Global runs the Okapi Conservation Project, which was set up with field headquarters at Epulu in the south of the reserve in 1987. It provides the ecoguards with food and health services and promotes good relations between them and communities. The Okapi Conservation Project also operates a community program focused on agroforestry and environmental education. It collaborates with WCS and the ICCN. Its primary aims are to conserve okapi and their habitat in the reserve and to protect "the lifestyle and culture of the indigenous Mbuti and Efe people," whom it describes as "among the last true 'forest people' left on Earth" and the "true hunter-gatherers and deep forest-dwellers living traditional lifestyles as they have for millennia." Both WCS and the Okapi Conservation Project stress the importance of the role of Mbuti pygmies in okapi conservation.

According to Okapi Conservation Project president John Lukas, who has been active in okapi conservation since 1987, "the Mbuti pygmies and the okapi have a relationship that extends back over 40,000 years, sharing the Ituri Forest as their home, living in harmony with each other. In Mbuti culture, it is taboo to harm okapi and chimpanzees as they embody important spirits of the forest." Lukas reminds us of the Mbuti's historical role in helping Harry Johnston discover the okapi for Western science and is determined to involve them in okapi research and management.[4]

While the more than five thousand pygmies living in the reserve should have a say in reserve management, as the Okapi Conservation Project argues, there are complications around ideas about "true" pygmies and their first rights to the forest. Local non-pygmy farming people should not be sidelined as a result of stereotypes of pygmies as the only legitimate "first people."

Rights, Claims, and Identities

The idea that pygmies have an intrinsic right to access their ancestral lands (usually understood as pristine rainforest), based on their unique cultural identity and behavior as hunter-gatherers, seems important for their survival as distinctive cultural groups in the Democratic Republic of Congo (DRC) today.

However, positioning them like this is also potentially a trap if it does not allow them to adapt and evolve in historical time, as they have shown great aptitude for doing over many centuries. The legitimacy of the claims of aboriginal or autochthonous ("original") peoples to land or natural resources are often tied to the idea that their cultural identities and livelihoods do not change over time. This is an idea that appropriates one aspect of their assumed identity, their enduring cultural distinctness from their neighboring peoples, and uses it to try to suppress another, namely, their dedication to preserving their autonomy from the attempts of others to impose their ideas and belief systems on them.

The basing of the recognition of peoples as Indigenous on their having occupied a given territory before other populations, on cultural distinctiveness, on self-identification, and often on the claim that they are sidelined (or persecuted) minorities has been contested in several Central African countries. These countries have not distinguished between Indigenous and other local peoples but have rather treated all as communities with specific customs and beliefs and reject the idea that a given group deserves privileged access to natural resources on the grounds of ancestral prior occupation. That is, they have tended to group all such peoples together as "local," recognized as peoples with specific beliefs and customs who claim specific rights over natural resources located around their settlements, setting aside claims about territorial rights based on prior occupation. The Forestry Code of DRC states: "local populations are village populations settled in forest areas, who organize their lives on the basis of custom and tradition and who are united by bonds of solidarity and kinship that underpin their cohesion and ensure their continuity in space and time."

While only Batwa, Bacwa, and Bambuti (Mbuti) pygmy peoples were recognized by the African Commission on Human and Peoples' Rights, and the International Work Group for Indigenous Affairs, as Indigenous peoples in the DRC, until 2021 the DRC government did not accord them specific rights, different to other local African communities. There is, therefore, some dissonance between international initiatives—focused on the legal protection of indigenous peoples, arguably prioritizing their rights (over other locals)—and regional approaches which do not.[5]

Fluid Identities

Anthropologist Stephanie Rupp argues that even the more nuanced accounts of pygmy identity tend to amount to unhelpful simplifications, particularly the insistence that pygmies have hunter-gatherer lifestyles that are incompatible with their being farmers or villagers. Her work in the Lobéké region of southern Cameroon reveals a more fluid and complex reality than outsiders' stereotypes admit. Rupp acknowledges the usefulness of such binary distinctions between pygmy hunter-gatherers and others for deciding whom to study, determining the beneficiaries of conservation and development interventions, of medical care, and who deserves tourist attention. Looking through these outsider lenses, African villagers are seen as late-coming opportunists, less worthy of study and less deserving of access to land and natural resources.[6]

African peoples of the Congo basin rainforests have organized and represented themselves through a variety of initiatives and forums. Two prominent regional entities, the Réseau des Populations Autochtones et Locales pour la Gestion des Ecosystèmes Forestiers d'Afrique Centrale (REPALEAC), and the Commission des Forêts d'Afrique Centrale (COMIFAC) refer to the "autochthonous" and "indigenous peoples" of the region in their mission statements, with no mention of "pygmies." Only in 2022, did DRC president Felix Antonie Tshisekedi sign a law proposed in 2021 which specifically aims to promote and protect the rights of the country's Indigenous pygmy peoples. It remains to be seen what effect this law will have.[7]

Ituri Locals' Perspectives on the Okapi Wildlife Reserve

In his account of his interviews with farmers and Mbuti people around Epulu in the Ituri forest in the mid-1990s published in his *Conversations in the Rainforest* (2000), Richard Peterson describes the complex relationship between outsiders and locals and local villagers and pygmy peoples and the attendant difficulties for conservation managers.

Extending over some 27,000 square miles (70,000 square kilometers) of the Ituri province (established in 2015) in northeast DRC, the Ituri forest has long been a stronghold for okapi. It is where the Mbuti peoples live alongside various African farming communities. It is home to the famous okapi capture station at Epulu, inside the Okapi Wildlife Reserve that extends over a fifth of the forest. In 1985, the Wildlife Conservation Society (WCS) set up the Centre de Formation et Recherche en Conservation Forestière, where Congolese students are trained in conservation science and management, at Epulu.

Peterson grew up the son of American missionaries at a small mission post in the DRC. He worked for WCS at Epulu in the late 1980s, then conducted graduate research in the area on human immigration into the region (by then he was living in the United States) and returned in the mid-1990s to interview locals about their conceptions of the environment and conservation. His book is a relatively small-scale but thickly detailed and self-reflective study.

According to Peterson, many challenges Epulu faced at the time he was there were typical of those faced by tropical regions worldwide where the biodiversity was rich and socioeconomic conditions poor. Limited economic development may have benefited conservation because it meant less pressure was placed on natural environments and resources, but villagers complained to Peterson about restrictions the Okapi Wildlife Reserve put on where they could farm, arguing that the forest was vast enough to allow them more land, and they claimed conservation authorities negotiated access to natural resources only with their leaders, excluding them. (The ways in which conservation NGOs associate local farmers with environmental destruction is not supported by the long history of sustainable subsistence farming in the forest—in fact, the regeneration of important canopy trees has been linked to clearing of the forest by farmers.[8]) These leaders, they asserted, were selling the forest, but the forest was created by God for all. They regarded the idea of a game reserve with a "closed forest" as a foreign one and as an injustice without precedent in their traditions. They denied that their ancestors had imposed hunting restrictions in precolonial times, arguing that God made animals and plants for people to eat.[9]

Locals, Peterson reports, were very aware, nevertheless, of the negative impacts of hunting on the forest. Many were concerned that the use of nylon traps (which long outlast traps made from natural fibers) or firearms would decimate forest animals. Some wanted okapi to be protected from commercial hunters. An Mbo village elder named Heri Bakana argued against the hunting of okapi because "that one is an animal with traditional meaning. . . . They do not attack people nor hurt people, they only go their way," and asserted that his ancestors had forbidden the killing of okapi and other large animals like leopards. Okapi killed by a leopard, however, could be taken, he said (this was also claimed by Mapoli, a Bali elder Peterson interviewed). It was the smaller faster-breeding animals that were killed for food, and the traps set for them were not capable of catching okapi.[10]

Another Mbo elder named Debo recalled the long history of interest in the okapi in the area and maintained that Europeans' efforts to conserve okapi and the establishment of the okapi capture unit by Jean de Medina in 1947 were triggered by a photograph of an okapi that Patrick Putnam had sent to Europe. Debo was adamant that de Medina had not forbidden locals from trapping wildlife, nor had he designated no-go areas in the forest. Rather, it was "the Americans" who had imposed these restrictions.[11]

Several villagers from local communities similarly maintained that the new breed of *wazungu* (white people) regulated the use of the forest in a way that was much less favorable for locals than that of the *wazungu* of the colonial era. They were particularly angered by restrictions on the eating of smaller animals like duiker, since, as Debo explained, villagers living in the forest (unlike those on the savanna) could keep very little livestock, and so they had no other way to get their protein. Villagers along with Masalito, an Mbuti man who worked as a guide in Epulu, argued that no traditional restrictions on killing okapi had existed in the past, but that they modified their practices in the face of regulations such as Medina's designation of the okapi capture season and the restriction of the hunting season to six months of the year. Their traditions had absorbed outside influences and adapted to changing circumstances, but now, they complained, the forests were closed year-round. According to them, even the Mbuti now required permission to hunt, and they hesitated to do so out of fear of the ecoguards.[12]

Peterson worried at the time that preventing the Mbuti from entering the forest (contrary to the official policy of the Okapi Wildlife Reserve) posed an existential threat to Mbuti culture. For the Mbuti, net hunts teach the balance between individual prowess (and property) and community cooperation and sharing. However, villagers too were concerned because hunting was culturally important to them as well, for example in certain communities as part of boys' initiation periods. These traditional hunts were subsistence hunts, not commercial ones, and prohibiting them has led to cultural loss. The rise of commercial hunting (the sale or bartering of hunted meat for more than subsistence needs) has certainly created major problems for conservation in the region. However, allowing the Mbuti but not villagers to hunt seems unfair, and it has caused strife between conservation authorities and villagers and between Mbuti and villagers. At the same time, the Mbuti's idealization by conservationists as hunter-gatherers led those who wanted to take up farming to complain that the Okapi Wildlife Reserve's regulations prevent them from doing so.[13]

Peterson argues that development and environmental education initiatives were not enough and that conservationists working at the Okapi Wildlife Reserve needed to study local cultural traditions and beliefs so that they could design conservation principles and practices that made sense to locals, although he concedes that the significant gulf between Western ideas regarding the relationship between humans and nature and those of African forest-dwelling peoples in the region was a barrier. Locals, he remarks, believed human immersion in the natural world was critical to maintaining the local life balance and therefore also the survival of the animals. Many of them also felt the Okapi Wildlife Reserve cheated them of their forest and its resources, and they resented the export of okapis, which to them amounted to the conservationists denying them a share in the benefits of the conservation effort. They felt their voices were not being heard and they had lost their authority over the forest, which was being administered according to rules and traditions alien to their ways of thinking.[14]

However, there were also villagers who thought the Okapi Wildlife Reserve system was working. They said meat could be obtained from the Mbuti and argued the regulations were important for conserving the wildlife including okapi especially because, they maintained, immigrants from other parts of the country hunted indiscriminately and without restraint. The term bakpala (villagers) was used for immigrant villagers, and some Mbuti complained both about their hunting and their farming practices, which they believed were more intensive and extensive than those of Indigenous farmers in the Ituri.[15]

Both the pygmies and the villagers interviewed by Peterson felt that the conservation regulations restricted their livelihoods in unacceptable ways, leaving a conundrum for wildlife conservationists.

Clashing Worldviews in the Congo Basin

The separation between humans and pristine nature reified in the form of national parks was not shared by local peoples Peterson interviewed; they rather thought of forests and rivers as gardens and of forest animals as livestock to be managed sustainably. Traditional approaches to sustainable use of natural resources are guided by an ethics that demands taking not just past traditions but future generations into account.

The hybrid nature and dynamism of African beliefs and traditions suggests it is unhelpful to try to pin down a static past or moribund indigenous knowledge unsuitable for present conditions. Ethiopian philosopher Workineh Kelbessa argues that African peoples' Indigenous knowledge synthesizes internally generated knowledge and externally borrowed knowledge. It combines old and new ideas and constantly adapts to new events, challenges, and ideas. Peasant farmers maintain biodiversity as a source of resilience to difficult and changeable environmental conditions. They implement diverse land use approaches and techniques like intercropping of compatible plants, agroforestry, and shifting cultivation and fallowing to spread risk out over different locations and time.[16]

As dramatized in the pygmy character Isookanga, protagonist of Congolese novelist In Koli Jean Bofane's *Congo Inc.: Bismarck's Testament*, however, there are intergenerational conflicts over whether and to what extent to adhere to past traditions and heed ancestral voices. The agendas of corrupt politicians, locals, and interfering foreigners bump up against those of young people who are plugged in to social media and drawn to urban lifestyles in this fast-developing country.[17]

Many (but not all) of the university-educated project staff Peterson interviewed were skeptical of the idea of appealing to what they understood to be "dead" traditions to influence people to value nature. They also pointed out that Indigenous knowledge was specific to particular social and ecological contexts and may not apply elsewhere, and was developed in particular historical moments and so was not necessarily relevant to current environmental challenges. Furthermore, certain ancestral ways of thinking were linked to falsehoods, threats and accusations of sorcery and so were undesirable, and others were detrimental to particular species. They might also be irrelevant to immigrants, who had no ties to local places, and were responsible for more intense exploitation than that practiced by locals.[18]

Given these problems regarding traditional practices and the novel, grave, and urgent threats faced by wildlife and the rainforests in their time, some Okapi Wildlife Reserve staff believed that incorporating Indigenous values into conservation efforts was a waste of time. They identified a social shift to a more commercial orientation that had resulted in waning adherence to cultural traditions. The belief that nature is imbued with spiritual significance was fading, an erosion to which Christian evangelism contributed. Park staff attributed the breakdown in the sustainability of human interactions with the environment in the region to population growth, poverty, economic inequality, and state corruption. Peterson called for dialogue and mutual learning, but park staff saw environmental education as a one-way street from educated park staff to uneducated communities. Only a minority recognized that lessons could be learned from traditional practices.[19]

Okapi Conservation Status Reassessed

Since 1964, the International Union for the Conservation of Nature has produced a red list of threatened species, providing a growing inventory of the conservation status of the world's wildlife. The 2008 red list assessment by the organization's antelope specialist group estimated the wild okapi population at 10,000–35,000 animals and categorized the species as near threatened based on data collected up to 1998. The stability of the population in the Okapi Wildlife Reserve was the primary reason okapi were not designated as threatened at this time. However, many conservationists believed this seriously misrepresented the situation for this understudied species, which lacked any coherent conservation strategy.[20]

Researchers compared transect surveys between 1993 and 1995 and between 2005 and 2007 and discovered a 43% decline in dung density in the reserve, which they attributed to the civil war. In response, the Zoological Society of London in partnership with the DRC's Institut Congolais pour la Conservation de la Nature and other wildlife NGOs initiated a range-wide okapi conservation project in 2010.[21]

The attack on Okapi Wildlife Reserve headquarters in 2012, during which all fourteen captive okapi were slaughtered, was followed by occupation of the park by armed groups, illegal miners, and poachers. This was a shocking blow for okapi conservation, as it struck at the heart of its historic operations, in the reserve created specifically for okapi protection. George Rabb, who had a long history with okapi and had chaired the International Union for the Conservation of Nature's (IUCN) Species Survival Commission from 1989 to 1996, initiated an emergency motion on the okapi at the 2012 IUCN congress calling for urgent support for okapi conservation. This provided more impetus to produce a formal conservation plan for the species.[22]

A giraffe and okapi specialist group was founded by the IUCN in March 2013, cohosted by the Zoological Society of London and the Giraffe Conservation Foundation. A new red list reassessment was high on its founding agenda and was conducted at a conservation strategy workshop in Kisangani in May 2013. Despite concerted efforts by researchers like Stuart Nixon to complete a range-wide assessment in the run-up to the 2013 meeting which aimed to produce a status review and conservation plan based upon it, the survey data collected was very patchy due to the dangerous political situation. Maiko National Park, for example, was also occupied by armed rebels (the Simba Mai Mai). Anecdotal evidence suggested okapi were being hunted in other protected areas. Deforestation and degradation were affecting important regions of okapi range, especially in the southern and eastern Ituri forest.[23]

At a May 2013 meeting in Kisangani, experts, conservation managers, representatives of NGOs, and local communities met to share okapi knowledge. Based on this, they drafted the first conservation strategy for okapi, 112 years after its discovery by the West. A ten-year strategy was drafted in a participatory fashion, designed to respond to direct and indirect threats identified at the meeting. Direct threats included habit loss and fragmentation and hunting. Indirect threats included extractive industries, civil conflict and political instability, human population growth, inadequate protected areas and law enforcement, policy and institutional challenges, and a regional lack of coordination (efforts had focused almost entirely on the Okapi Wildlife Reserve). In light of the precautionary principle conservationists adhere to in making decisions, delegates concluded that the okapi population was in decline. On November 26, 2013, the species was reclassified as endangered, based on the IUCN red list criteria of a reduction in population size of more than 50% over the preceding three generations (twenty-four years). The information from the review and conservation plan was gathered into the IUCN's *Okapi (Okapia johnstoni): Conservation Strategy and Status Review*. Published in 2015, it was heralded as "the most detailed and up-to-date account of the status and biology of the okapi produced to date," including historic and current distribution maps and summaries of field surveys.[24]

In 2013, John Hart had estimated the size of the wild okapi population at 35–50,000, but this was informed guesswork, based on extrapolation of dung-based surveys. This guesstimate was complicated by the fact that okapi density varies significantly and unpredictably across its range, for reasons which are not well understood. The 2015 okapi review and conservation strategy concludes that "there is no reliable estimate of current population size."[25]

In the same year, Hart had also suggested the range of okapi totaled around 77,220 square miles (200,000 square kilometers), and the 2015 conservation plan put this at around 94,365 square miles (244,405 square kilometers) most of which were found north and east of the Congo River with a smaller number located southwest of the river. The key protected areas for okapi conservation in the main area of distribution (see figure 7.1) include the Abumonbanzi reserve in the far west of okapi range. In the center is the Rubi-Tele hunting domain. To the east are the Okapi Wildlife Reserve (protecting one-fifth of the Ituri forest), Mount Hoyo reserve, and parts of the Tayna nature reserve and Virunga National Park. In the south is Maiko National Park. The population southwest of the Congo River and west of the Lomami River is protected in the adjoining Sankuru nature reserve and Lomami National Park.[26]

There have long been rumors of okapi occurring east of the DRC's Virunga National Park, in Uganda's Semuliki National Park. Locals told Johnston that okapi occurred there, but no evidence was produced. Similar claims have been made since, including by renowned zoologist Jonathan Kingdon. Local pygmy people are often cited as sources for okapi occurring there into the 1970s.[27]

From October 2017 to February 2018, Stuart Nixon and Chester Zoo colleagues undertook a survey with the Uganda Wildlife Authority in Semuliki National Park. They searched areas adjacent to known past okapi habitat in Virunga but found no okapi. More recently, Zvi Sever undertook a survey based on a literature review, communications with zoos and museums, interviews with local pygmy people, searches on foot for dung, and camera trapping in 2015, 2017, and 2019 and also found no evidence of okapi in Uganda currently nor any definite evidence of their occurrence there since the 1970s.[28]

Political Challenges for Conservation in the Democratic Republic of Congo

The challenges facing wildlife conservation in the Ituri forest and other parts of northeastern DRC are severe. The regional political context is that the peace agreement signed in Pretoria in 2002 has been ignored in the Ituri forest, where some 15,000 troops from seven armed groups continued fighting for resources in the area. Pygmies were brutally mistreated by these soldiers. There was an attack on the Okapi Wildlife Reserve headquarters in 2012, and it took until 2015 for the authorities to begin to recover control over and expel illegal inhabitants from the park.

In neighboring North Kivu province, under the influence of the Forces Démocratiques de Libération du Rwanda, charcoal extraction mainly from

Virunga National Park was netting $2.5 million per year in that period. Gold mining, poaching and timber exploitation were major economic activities. Park authorities did not have the capacity to resist the corruption of military commanders and their appropriation of natural resources or to do battle with unscrupulous overseas mining companies.

The political instability following the 2012 invasion persisted in the region until 2013, when it looked like the Forces Démocratiques de Libération du Rwanda might transition to a conventional political party. Sadly, by late 2014 there was again fierce fighting and massacres of civilians in Beni territory. This was where Johnston had first sought okapi a century before. These considerable challenges for conservation are dramatically recounted in Orlando von Einsiedel's Oscar-nominated film *Virunga* (2014).[29]

Many of the NGOs who have supported okapi conservation on the ground in the DRC have withdrawn because of this violence and political instability. There are no facilities for tourism (it is unsafe) and no research is being done by outside researchers. Okapi have not been kept at Epulu since 2012. The arrangements shoring up okapi conservation are fragile, dependent as they are on the priorities and finances of overseas governments and NGOs.

Okapi conservation requires American support, but getting that support is complicated, and as for other African conservation efforts, funding through USAID for example is worryingly dependent on the whims of incoming political administrations. The United States government has been one of the biggest funders of conservation in the DRC and surrounding countries through USAID's Central Africa Regional Program for the Environment (CARPE), which began in 1995, but as this book goes to press the Trump administration has shut down USAID and frozen foreign funding.[30]

The DRC's nature conservation organization ICCN doesn't have the funds to run the Okapi Wildlife Reserve or okapi conservation efforts elsewhere. Without the continuing involvement of WCS and the Okapi Conservation Project, okapi conservation efforts in the Okapi Wildlife Reserve will collapse. It is worth reflecting on the fragility of such arrangements, dependent as they are on the priorities and finances of overseas governments and NGOs.

Philanthropists and Veterans of the Okapi Cause

The okapi supplies a good example of how the future survival of an iconic species may hang (for a time) on the inclinations of a millionaire philanthropist. Its chances of survival were notably boosted when the millionaire philanthropist Howard Gilman took an interest in it in the 1980s. Gilman inherited a fortune based on the success of his grandfather Isaac's Gilman Paper Company. He was disinherited in 1979, but after his brother died in 1982, he managed to wrest back control. According to a somewhat disapproving article in *Forbes*, Howard spent his inheritance on philanthropy and pet projects rather than on keeping the business competitive.[31]

Gilman's pet project was White Oak Plantation, in Yulee, Florida, which he developed into a dance center for the dancer and choreographer Mikhail Baryshnikov and other famous artists, a conference center, and refuge for sixty species of endangered animals, including okapi. John Lukas (formerly of the Wildlife Conservation Society) joined Gilman's White Oak operation, on the condition it would be scientifically based and conservation oriented. Gilman had okapi at White Oak, which occupies 7,500 acres (thirty square kilometers), and Lukas persuaded him to fund an Epulu okapi project proposed by the Swiss couple Karl and Rosmarie Ruf. Set up in 1987 with funding from Gilman International Conservation, this endeavor became known as the Okapi Conservation Project in the early 2000s.[32]

The Howard Gilman Foundation supported Gilman's philanthropic interests, but following his death in 1998, it withdrew from conservation. Gilman International Conservation was apparently discontinued, and White Oak was put up for sale in 2011. Lukas therefore set up Wildlife Conservation Global in 2012 as an American not-for-profit company designed to manage and fund the Okapi Conservation Project. The organization is registered at Lukas's Jacksonville address in Florida. White Oak was bought by American billionaire businessman Mark Walter in 2013. He and his wife Kimbra are ardent conservationists, and they set up TWF Conservation there, funding wildlife conservation oriented to recovering populations of rare species, including okapi. White Oak has the US's largest single population of okapi, with twenty individuals as of 2024.

The conservation of the okapi does not only depend upon wealthy philanthropists and charitable trusts, however, but also relies on the long-term commitment (and bravery) of individuals like Lukas and Rosmarie Ruf (both veterans of okapi conservation), John Watkin, a French-speaking veteran of conservation in the region who was hired as deputy director of the Okapi Conservation Project in December 2023, and a management team fittingly made up of mostly African Congolese staff. As of 2024, the organization was run by Wildlife Conservation Global out of Jacksonville Zoo.

Lukas's fund-raising efforts, while they are directed at Western audiences and make use of helpful (positive) stereotypes about pygmies and rainforests (possibly less helpful in country), have been nothing short of heroic. A 2017 report on the first thirty years of the Okapi Conservation Project's efforts acknowledged the contributions of 24 foundations, corporations, and related institutions, 48 zoo partners and related organizations, and 354 private donors.[33]

Ongoing Challenges for Okapi Conservation

The vision for the ten-year okapi conservation strategy launched in 2015 is to conserve this "emblematic and endemic species . . . sustainably across its range for the benefit of current and future generations, in collaboration with all stakeholders, and especially with local communities, thanks to the promotion of good governance."[34]

The goals are to reduce threats and protect viable okapi populations in the wild. Conservationists have been tasked with managing the ex situ (captive) populations so as to maximize the help they can provide in conserving wild okapi by coordinating the outreach and education efforts of zoos, conducting research, increasing the number of zoos holding okapis (partly to generate more public awareness), encouraging zoos holding okapi to contribute to in situ conservation work, raising funds from zoos focused on other species in the Ituri forest, and "pushing the okapi's role as a flagship species."[35]

The okapi is still classified as endangered in 2024, with its status uncertain given that it has not been assessed since 2015. However, due to the ongoing and apparently endemic political turmoil and violence, few conservation NGOs and researchers are prepared to work in its remaining habitats. The Wildlife Conservation Society (WCS) and the Okapi Conservation Project continue to work with the Institut Congolais pour la Conservation de la Nature (ICCN) to further conservation goals in the Okapi Wildlife Reserve. These NGOs are working through social media to promote awareness of okapi within DRC, as well as internationally (mainly the United States). WCS puts out regular newsletters and videos, and since its establishment in 2016, it has promoted World Okapi Day, October 18.[36]

Since Peterson's work in the area, the ICCN has worked on improving recognition of Indigenous and local community rights around protected areas. It has set up a framework to comply with international human rights law (as well as the DRC law passed in 2022). This work is very demanding to maintain in the face of armed conflict and professionalized poaching. The Okapi Conservation Project is also working to address the livelihood challenges that Peterson's interviewees identified in the 1990s.

Chris Hamley of WCS is one of the brave souls still working in the Okapi Wildlife Reserve. When I interviewed them in December 2023, they were program manager and chief of party for the reserve project funded by USAID. Chris told me that the reserve hadn't done a census for ten years but that a survey using camera traps and line transects was planned for 2025. Because okapi bones and fat are claimed to have aphrodisiacal properties, the meat is eaten, and skins are reputed to fetch up to US$10,000 abroad, its survival is certainly in the balance. In an interview I conducted with Watkin in June 2024, he warned of an increase in okapi poaching, notably in Maiko National Park, which is troubled by four rebel groups.[37]

The state military defense organization Forces Armées de la République Démocratique du Congo (FARDC) sell arms and ammunition to civilians, so hunting is easy. In a major shift from the 1980s, locals in the Ituri have lost their fear of the deep forest, formerly only visited by the Mbuti. In addition, the loss of tourism income means jobs and money related to conservation have dwindled, making constraints on exploiting wildlife harder to justify to locals. Hamley thinks the greatest challenge for okapi conservation in the reserve remains the presence of minerals, providing a very strong incentive compared with legitimate activities in the reserve. In May 2024, Radio Okapi reported that more than seventy illegal gold-mining sites were operating inside the reserve.[38]

Gabriel Gelin, Central Africa regional communications manager for WCS told me in December 2023 that tourism dried up after the 2012 attacks, despite attempts to reestablish facilities at Epulu in 2015. Given the inaccessibility and political instability in the region, okapi do not enjoy the benefits of tourism that have assisted gorilla conservation. Okapi are much more difficult to see in the wild, and none are kept in captivity at Epulu. While the idea of reestablishing a capture facility at Epulu has support precisely because none are being kept there, the security situation is too unstable to make the idea practical at present.[39]

The militarization of conservation that has been deemed a necessary response to these kinds of challenges to conservation in some regions—characterized by a top-down, intrusive, and often violent approach—has generally been disastrous for community relations and has resulted in human rights abuses. The high stakes game of protecting wildlife in a dangerous region is presented to outsiders and funders of conservation efforts in a simplified way that sets rangers as heroes against locals as villains. The cultural and socioeconomic reasons why locals seek use of protected areas resources, including wildlife, are too often ignored. So are the negative impacts of conservation on local communities. The personal complications for rangers who live in communities but must perform intrusive even aggressive acts in them to fulfill their duties is overlooked. So is the increasing evidence of post-traumatic stress disorder among rangers who are placed in mortal danger and subjected to extreme violence while performing their daily activities.[40]

In 2019, BuzzFeed reported human rights abuses by ecoguards in protected areas comanaged or funded by the World Wide Fund for Nature (WWF) in Cameroon in the Republic of the Congo and in the DRC. World Wide Fund for Nature had been working with partners with a history of abuse of Indigenous peoples and local communities. The histories of natural history collecting I've relayed in this book show that overseas organizations and institutions wanting to work in the region have long faced difficult ethical dilemmas over the compromises and accommodations required to do so.[41]

An independent panel of experts investigated the accusations against the World Wide Fund for Nature. It found that the organization was not directly involved in abuses but ruled it should have done more to ensure partners' compliance with their human rights commitments for working in protected areas. Professionalization of rangers and supervision and monitoring of ecoguards activities was recommended. This case was emblematic of the challenges of wildlife conservation in the region.

Social science commentators have criticized the militarization of conservation, arguing it is antithetical to attempts to achieve just conservation. Dispensing with arms and men trained to use them in combat, however, is difficult in a violent and impoverished region like northeastern DRC. Conservation organizations have had to police peoples, to protect wildlife and forests in protected areas rich not just in wildlife but also in minerals, timber, and other natural resources. Some of these people are desperately poor, some have been displaced, while others are plundering natural resources to fund military aims and political careers. At the very least, these conservation organizations should not be functioning as armies.

Metapopulations and a Transboundary Solution?

John Watkin of the Okapi Conservation Project participated in a rare conservation event earlier in his career—the successful reintroduction of individuals from an ex-situ population of scimitar-horned oryx back into the wild in the Sahel region of the Republic of Chad in Central Africa. Once "extinct in the wild," he told me when I interviewed him, they are now "endangered in the wild." With a rare, site-endemic species like okapi, Watkin suggests, it is not enough to focus either on in situ conservation in northeastern DRC or on breeding "assurance populations" ex situ in the West. Instead, he argues that conservationists need to think about metapopulation management, which would involve factoring all living okapi into plans to ensure the future survival of the species.

Watkin's use of the term "metapopulation" is inspiring, though perhaps in a more metaphorical than technical sense. Essentially, the metapopulation is the aggregation of all local populations of a species. A key challenge for conservation biology is managing the negative effects of humans' fragmentation of wild species' habitat. Fragmenting habitats divides a species' populations into smaller, isolated, and more vulnerable populations. For example, lions once occupied a huge continuous area across Africa, but then human settlement and activities fragmented their historical range into many local populations. These rely on occasional input from outside lions to avoid a loss of genetic variation and local extinction. Lion conservationists thus consider the metapopulation when working to ensure the species survives outside of captivity. They work with other stakeholders to create or maintain corridors between populations surrounded by human settlements or farmlands. They may even consider moving lions between local populations or perhaps eventually reintroducing lions from ex situ populations. These strategies are currently being applied with species like lion tamarins and Tasmanian devils.[42]

This well-established conservation approach has not yet been mobilized in okapi conservation. Okapi are broadly if unevenly distributed within their remaining range, with strongholds in the formally protected areas (see figure 7.1), and it is believed that individuals still move freely between these hotspots. An exception is the genetically distinct and geographically isolated Lomami population, southwest of the Congo River. In the larger region of contiguous okapi range, two areas are at risk of isolation. Settlements along large regional roads are steadily separating the okapi population in the Semuliki forest in northern Virunga National Park from okapi in the Ituri region to the west and the Mt. Hoyo population to the north. The Mt. Hoyo population is also at risk of isolation.

The fragile state of okapi conservation given the region's political instability makes a strong argument for adopting a metapopulation strategy even if a separation among the remaining okapi strongholds can be prevented. It may one day be necessary to reintroduce okapi kept overseas into the wild in the DRC to ensure the viability of remaining wild populations that could at that point be hanging on in a few disconnected patches of suitable habitat.

Watkin emphasized to me how challenging such reintroductions are, however. It is not just the political instability and logistical challenges that make reintroductions hard but also the fact that they are typically extremely expensive. A major challenge for released animals (Watkin prefers "okapi in human care" to "captive okapi") is exposure to unknown diseases. Okapi in care lack antibodies to the diseases they would be exposed to in the wild, where they could no longer receive annual vaccinations.

In celebrating thirty years of the Okapi Conservation Project in 2017, Lukas ruefully admitted that to date the group was at best "holding the line" until things improved in the DRC and its people and government took over conservation efforts. However, as a result of the country's vast mineral wealth too many other countries and companies, African, Chinese, and Western, have an interest in maintaining it in a state of political instability for there to be a major breakthrough in achieving peace and stability any time soon.

Reintroductions are probably worth planning for, and two main ways of doing so have been proposed. First, because evidence was found in 2009 confirming okapi presence east of the Semuliki River, eight kilometers from the DRC/Uganda border, and because good okapi habitat exists in Uganda's Semuliki forest, okapi could be introduced into its Semuliki National Park. Confirmation of habitat suitability would be required first, as there's no recent evidence of okapi presence and the forests are still recovering from sustained exploitation in the twentieth century. Further, these introduced okapi would probably have to be sourced from captive populations, raising the question of whether that would pose genetic problems—given that the genetic profiles of existing captive populations have been shown to differ from those of sampled wild okapi—especially considering released okapi could interbreed with wild okapi from DRC. Further, the DRC's authorities do not seem very enthusiastic about the idea of introducing their iconic national species into Uganda, especially if that involves establishing a captive okapi station like the one that used to exist at Epulu, to attract tourism. Cooperation, however, is possible as a result of the Greater Virunga Transboundary Collaboration agreement between Uganda, Rwanda, and the DRC, though regional relations remain volatile. Finally, whatever the prospects across the border, there is a moral argument that any okapi returned to Africa from the West should go to the DRC.[43]

The second option for releasing captive okapi back into the region would be to restore the stockades at Epulu and restock these with zoo okapi, which would amount to a reintroduction rather than an introduction. Restoring the Epulu capture station and resuming capture of wild okapi to bolster the still small global zoo population has also been floated. This would be true multidirectional metapopulation management at work. However, conservationists I discussed resuming wild capture with were unenthusiastic.

Conclusion

For the foreseeable future, okapi conservation must be undertaken in one of the planet's most volatile and dangerous regions. Park rangers must be heavily armed, and many have died protecting both wildlife and civilians from rebel soldiers, illegal miners, and professional poachers. Conservation NGOs face the challenge of cooperating with regional and local power structures including armed rebel groups, while not compromising their commitments to social as well as environmental justice. Caught up in this mayhem and often held up to the outside world as symbol of an existential struggle for nature conservation are the pygmy peoples, farmers, and villagers who depend on the rainforest to survive.

Perhaps Conrad was right and what lies at the heart of the darkness Western conservationists are combating in the DRC is the destructiveness of the economic system and way of relating to nature that Westerners introduced to the region. The horror they have sought refuge from in idealized rainforests of the mind may really be the horror of self-recognition. Decorating the cover of Edmund Morel's book *Red Rubber* about "the rubber slave trade on the Congo" is a striking adaptation of an illustration from the Egyptian Book of the Dead. On the left-hand pan of the scales that weigh the soul before it enters the afterlife sits King Leopold II of the Belgians depicted as a pharaoh; on the right, dragging down the scales, is a single severed black hand. If we substituted the region's dwindling wildlife and rainforests for the slave trade, the okapi could represent them on the right-hand pan, while who would sit in the other pan is a matter for all of us to consider.[44]

Central African ways of thinking may offer insights into how to live by both using and choosing when not to use nature's gifts and may foster a more holistic approach to living as a community of selves and others, human and nonhuman. Such ways of thinking offer Western conservationists an alternative to the trap of valuing nature either solely for its intrinsic value or for its utility to humans. It offers a conception focused more on relationships and relational value, between humans and their communities, and between humans and other beings and places, in the larger community of the world.[45]

Somewhere in all of this, we can only hope, there will be found a secure space for the okapi to continue to play pooky in the rainforests of the Congo.

Conclusion

It was a thrilling if melancholic experience for me to come face-to-face with the mounts of the first complete okapi seen in the West in a British Natural History Museum storeroom. In the early 1900s, these were exhibited to the upper echelons of British society at a Royal Society conversazione in London and shown to thousands in the Natural History Museum. Now, these taxidermized mounts of a species which graced the cover of a book by Andy Warhol are ghosts, fading from public memory like the photo albums and newspaper cuttings of the forgotten expeditions that triumphally retrieved them from the Congo.

Recovering the story of how the okapi was discovered for Western science has taken me down numerous unexpected trails and byways. My recovering faculties following dengue fever and its aftereffects were tested in sifting through the colorful tales of dauntless Westerners floundering in the rainforests, assisted by many Africans and others usually unacknowledged in such tales of natural history exploration. The caste of a few intrepid natural scientists and collectors who sent back okapi facts and remains swelled to include the cavalcade of others that people the pages of this book.

Untangling the murky colonial and postcolonial histories of this complicated region has been harrowing and a challenge in the face of my own receding brain fog. I reluctantly educated myself on the history of slavery and the rubber trade in the Congo of King Leopold II and his successors. I discovered how okapi had been used as diplomatic currency as Leopold tried to shore up his crumbling reputation in the early 1900s, then by the Belgian government to promote its colony.[1]

I drew maps of routes and rivers and wrote timelines to sort out the complex history and geography. I commiserated with Percy and Hannah Powell-Cotton, shaking with fever in the rainforest, and have come to admire the determination and good humor of Harry Johnston (with all his faults) and of Attilio and Ellen Gatti. I think of Anne Eisner, who was once possibly the only "woman on earth who fertilizes her garden with okapi manure," and of the Mbuti man Sale teaching Patrick Putnam how to feed an okapi. While the memoirs I read were often fascinating and could be charming, they could also be repellent when expressing the prejudices of their times.[2]

Histories (and Futures) of the Sciences

Tracing the process by which current scientific knowledge emerged from a long line of anecdotes, shared traditional knowledge, and the accumulation of studies of okapi, living and dead, wild and captive, uncovers how both historical circumstances and human aims and egos shape science and how the okapi was slowly revealed to Western minds.

"Discovery" tends to be ascribed to one person. Johnston who followed up on stories about a rainforest equid and found the evidence of the existence of the okapi which revealed its existence to Western science, is memorialized in the naming of the species. As I have shown, however, he was part of a network of observers and collaborators. Philip Sclater is remembered in the scientific name, even though he played a small role (though he was institutionally important). Neither Sclater nor Ray Lankester (who gave the okapi its genus *Okapia*) produced a proper monograph, but by happenstance Julien Fraipont did. Fraipont took over the work from Charles Forsyth Major who undertook the pioneering taxonomic work, but is forgotten, along with earlier Western observers like Cuthbert Christy and Attilio Gatti who contributed significant knowledge in their time. They, like the Africans who guided them have been filtered out by the processes of scientific knowledge making.

In tracing the processes through which knowledge becomes filtered, curated and certified as scientific, I have often been reminded of Harriet Ritvo's observation that "it is a wise discipline that can recognize its own forebears." As I assembled this version of the history of okapi science, I tried to provide a coherent narrative while avoiding an oversimplified notion of linear progress steered by universally recognized experts from demarcated disciplines. I have tried to show how institutional politics, disciplinary struggles, and notions of expertise and scientific authority interacted to influence the process and how they in turn collided with world events, political and economic, the evolution of disciplines, passing academic fashions, and the personal ambitions and trajectories of a large caste of disparate characters.[3]

Ambition, media interest, artistic representation, and the availability (or lack) of evidence as much as sober science was behind the initial proliferation of okapi species as named by scientists. The influence of then current interest in "missing links" and "living fossils" colored approaches to categorizing the okapi, which together with new ideas about evolution at the time, shaped the characterization of okapi as primitive and, by extension, of low intelligence.

Okapi taxonomy is still not entirely settled, and I was surprised to discover the extent to which evolutionary relationships and species concepts are still contested. For example, it remains an open question whether there are four giraffe species or just one with nine subspecies. For giraffe conservation, the answer matters, as species trump subspecies when it comes to conservation prioritization. More species means fewer individuals of each, resulting in different levels of endangerment. Despite the conclusion that there was only one species of okapi, based on the physical evidence of skeletons and skins, and more recently of genetic analyses, suggestions that there may at least be subspecies persists. For okapi conservation, saving okapi in general would not be sufficient if we risked losing a distinctive subspecies in one region of their distribution, for example southwest of the Congo River.[4]

The major technologies shaping studies of wildlife in their native habitats were applied relatively late to okapi. The Harts undertook pioneering telemetry studies for a few years from the late 1980s, but subsequent work has been patchy, partly due to violence in the region. Camera trapping yielded the first images only in 2008. Veterinary studies of okapi diseases (in captivity) have a longer history, for obvious reasons, with real progress following the availability (and deaths) of okapi in Western zoos since the 1950s. Much has been learned about okapi reproduction in captivity through, for example, chemical analysis of urine carried out since the 1980s. In 1990, it was discovered that they communicate using infrasound. Genetic studies undertaken since 2010 revealed much about okapi lineages and diversity.

The behavior of captive okapi has been studied, but few significant ethological studies have been published. Much remains to be learned about the behavior of wild okapi. For example, combat between male okapi is suspected, but has never been directly observed. Are they really good swimmers? How do they use infrasound? If their ranges are typically fairly small, when and why do they move between regions across okapi habitat (and do they)?

Only in the 2010s was there a concerted attempt to synthesize scientific knowledge about the okapi's native rainforests in the Congo basin region, knowledge that now informs deeper environmental histories of the region. Understanding of the regional climate is limited, hampered by lack of data. The massive, long running environmental transformations driven by the extraction of natural resources, in particular minerals for manufacturing weapons and electronic devices consumed by the Western world and China, complicate our understanding of these longer environmental trajectories.

Assembling this into a coherent story involved a kind of bricolage, and at times I felt like a child twisting one of those old kaleidoscope tube toys where the patterns come in and out of focus and coherence. Writing this book required a sustained effort to produce as holistic a view of okapi as the sources allow, without either dissolving into incoherence—or overstating the reach and coherence of human knowledge.

Culture, Stereotypes, and Indigenous Knowledge

Okapi, pygmy peoples, and rainforests have long been linked in Western imaginations, and contemporary conservation efforts still mobilize these associations. Older ideas about biological races, primitive peoples and primeval forests are giving way to more nuanced and historically grounded understandings, but a number are deeply engrained and seem to morph into new guises. Claims about identity and priority in the rainforests and misrepresentations of the relations between pygmy peoples and African farmers shape conservation policies and have the potential to be socially disruptive, even if NGOs deploy them strategically and with good intentions.

Even as the knowledge and skills of Indigenous and local peoples along with their rights to their lands and to preserving their traditions are increasingly being acknowledged, and their wisdom in how to relate to the natural world is recognized as a valid alternative to Western framings, African societies, like every other society, are being influenced by globalizing forces of political, economic, and cultural transformation and exchange. There are tensions between acknowledging and defending distinctive identities and preserving the right to social evolution and innovation.

In April 2012, the Intergovernmental Science-Policy Platform on Biodiversity and Ecosystems Services (IPBES) was established as an independent intergovernmental body to strengthen the science-policy interface for biodiversity, as the IPCC did for climate change. IPBES received pushback on its ecosystems services framework from developing world members in its early days, responding by developing a framework with draws on both Western-style ideas of the natural world as a source of ecosystems services and Indigenous traditions like *pachamama.* In these traditions, "services" provided to humanity by Earth are understood as the "gifts" of a generous mother rather than natural resources to be exploited and managed sustainably. Humans are conceived of as members of a community of living beings, not external managers of other, lesser beings.[5]

Reframing a vision for wildlife conservation in the Democratic Republic of Congo requires acknowledging different worldviews and paradigms for natural resource use and biodiversity conservation and embedding diverse values of nature into decision making and policymaking. This includes recognizing the contributions of Indigenous and local knowledge to scientific knowledge. Doing so requires that the peer-reviewed platforms and publications that check, approve, and disseminate Western science should not simply correctly attribute the facts learned or ensure ethical research procedures and informed consent are obtained where locals are the *objects* of study—but should acknowledge their *agency* in the making of knowledge. The contributions of okapi trackers and those with knowledge of their habits and diets could, for example, be represented in publications and presentations by being listed as authors if they substantially contributed to enabling research efforts, just as journals now insist on crediting originators of significant data used in papers as authors, even if they didn't contribute directly to the paper. While local informants may not be interested in such acknowledgments or in abstract academic publications, this will at least provide a more honest account of how the scientific knowledge was acquired.

An emerging call for evidence-based conservation recommends a threefold approach to ensuring decision making is founded on firm foundations: undertaking a systematic review of the scientific evidence, taking experience-based information and local and traditional knowledge into account, and carrying out a study of the local context where a conservation intervention is being considered. These three elements (scientific evidence, experience-based information, contextual knowledge) together inform evidence-based decision making. These elements are separately in place for okapi conservation but need to be integrated. Congolese and foreign conservationists, along with locals with Indigenous knowledge of okapi and their habitats and representatives of all those impacted by and invested in okapi conservation on the ground in the Democratic Republic of Congo should all have a seat at the table. There are many ways of looking at an okapi.[6]

Exporting and Keeping Okapi

This book bears witness to the circumstances under which okapi and other prized wildlife were extracted from the Congo under King Leopold II and Belgian colonial rule. Accommodations were made and human misery was ignored, even denied. The cost of exporting wild animals to captivity overseas was high for the animals, and okapi are emblematic in this respect.

Zoo conservation of okapi after World War II depended to an insufficiently acknowledged extent on Jean de Medina's capture and habituation operation at Epulu. Many captive okapi populations were founded on animals exported at this time. Other lineages were founded on okapi supplied by his eccentric predecessors Franz Hutsebaut and Patrick Putnam, whose stories I have recovered in this connection, and which natural history museums and zoos could share.

The animals zoos exhibit are not exemplars of wild okapi; most are the ancestors of generations of captive-bred okapis traceable to an ancestor captured in a Belgian colony. Gatti claimed that the okapi Congo, whom he visited at the Bronx Zoo in the 1930s, was notable for his "meagre, melancholy appearance," which he ascribed to his being captured as a young calf, being kept "imprisoned in a small cage" and raised on "unnatural food and subjected to terrific changes in climate." In his opinion, this captive "cannot give you even the faintest idea of the majestic appearance, the huge size and fearful might which such animals normally attain in their own natural world." Yet Congo, like Buta and Tele and others, was in his way a remarkable animal who had survived an epic journey to the West and went on to sire offspring.[7]

Perhaps these individual okapi life stories could also be told—the stories of wild okapi who were captured and never left the Congo like Beautiful and Toto and those extraordinary few who survived the trauma of capture, transportation, and adjustment to the routines of zoo life, adapting to new lives in a new environment and interacting with humans and other okapi in enclosed, mostly treeless spaces. Zoos could also reflect on how okapi are kept and shown today, weighing the ongoing struggle of ensuring the animals can be seen by the paying public against efforts to replicate the animals' natural habitats that would enable them to perform the habits and behaviors characteristic of their wild ancestors.

Or would such efforts be a pretense too far? Heini Hediger claimed animals born in captivity have no memory of life in the wild and that in any case many wild animals have restricted territories and follow set routines. This may be true, but reading Hediger, I was reminded of Gatti's observation that in the open landscape of his camp, the okapi calf Toto "showed a deep preference for the spots where the ropes of our tents were most numerous and intricate." His photograph of Toto playing in the guylines (see fig. C.1) haunts me. Providing a more complicated environment, at least a patch of it as I have seen done at London Zoo, would surely enable okapi to explore and enact the kinds of behavior natural to them in the wild.[8]

Zoos have played a vital role in supporting field research in places like the Democratic Republic of Congo. Those that hold okapis should surely support their survival in the wild, and many have, though this is challenging given that their natural habitats are in remote regions troubled by ongoing political instability and violence. Even though key organizations that have historically supported okapi research and conservation have withdrawn from field operations owing to this instability, it is still possible to support outreach and education about the okapi inside and outside the Democratic Republic of Congo, and fund conservation and field research by those intrepid souls who will venture into the region or, even better, by locals. There is certainly much still to learn.

Okapi Conservation

Regional conflicts along with the legacies of Western interferences in Africa complicate okapi conservation, which depends on unlikely collaborations between committed Westerners who care about the survival of this still too-little known species, local African pygmy and farmer peoples, and brave Congolese conservationists like Robert Mwinyihali and Corneille Ewango who work to conserve it on the ground in the Democratic Republic of Congo.

It remains to be seen if the generation of John Lukas and Rosie Ruf, a dedicated American and a Swiss, will be followed by a new generation of Westerners who have the long-term commitment, skills, and outside links to fund raise in the West to support operations in the Democratic Republic of Congo. This seems especially important at the time of writing considering the withdrawal of USAID funds which have previously been very important for conservation efforts in the region. It seems essential that local African champions are supported and emerge as effective global ambassadors and national leaders for okapi conservation.

The efforts of zoos to establish viable ex situ populations of okapi have not succeeded, even though coordinated captive breeding programs have been ongoing in Europe since 1977 and the United States since 1981. The European Association of Zoos and Aquaria's ex situ program for okapi run out of the Antwerp Zoo oversees cooperation on breeding in zoos and supports conservation

Figure C.1. Toto playing in tent guy lines (Attilio Gatti, *Great Mother Forest* [London: Hodder and Stoughton, 1936], opposite 205).

in the Democratic Republic of Congo.[9] Still, the global captive okapi population falls short of the recommended target population of 270 animals believed necessary to guarantee a stable and genetically healthy population over the long term. This is no reason to stop trying, but at the same time, we should not overstate the contributions ex situ okapi breeding can make to ensuring their survival in the wild.

Zoos were initially focused on breeding to maintain captive populations, then as they pivoted toward a conservation mission in the postwar period framed their efforts as breeding for possible release into the wild, and since have mostly quietly gone back to breeding to maintain captive populations who will most likely never be released into the wild. If okapi are to be kept in human care, the only ethical justification is to spread awareness of their uniqueness and the challenges facing okapi conservation and to raise funds for okapi conservation in the Democratic Republic of Congo.

The idea of releasing zoo okapi into the Semuliki National Park in Uganda is an intriguing but remote possibility. It seems unlikely importing zoo okapi to reestablish a captive population at Epulu to attract tourism and raise the species' profile would work given Epulu's remoteness and safety concerns. Still, John Watkin may be right that it is pragmatic to think about the metapopulation of all living okapi when considering their future as a species—to which I would add all the okapi preserved in museums and artworks too. The taxidermized mounts, the skeletons and *küchenreste*, the drawings, paintings, and photographs that have carried okapi into our lives could be far better mobilized to keep them present to us.

The IUCN's Conservation Breeding Specialist Group promotes a One Plan approach, consolidating planning efforts for species living in the wild, in managed reserves, and zoos in an inclusive way.[10] Perhaps an international okapi conference could be convened bringing all the players and stakeholders around a single table: those involved in ex situ and in situ conservation (researchers, zookeepers, veterinarians, conservation managers and policymakers, and those with traditional knowledge), locals (political and traditional leaders and those sharing landscapes with okapi), those studying the social dimensions of conservation in okapi habitat, and those exhibiting and interpreting okapi remains (scholars and representatives of natural history museums knowledgeable about the region's environmental history and its cultures).

In the meantime, zoos and natural history museums outside of the Democratic Republic of Congo can tell the stories of the discovery of the okapi better, drawing on materials presented in this book, including the names, photos, and stories of African hunters and trappers so important to early Western attempts to secure okapi remains and later live okapis. In a few cases, I have been able to link some of these individuals to okapi remains still on display in Western museums. I recommend acknowledging these individuals and their contributions in zoo and museum displays and such materials could be shared (at least digitally) with communities in the regions okapi have been extracted from.[11]

Advocacy for Okapi

My closing plea to you as a reader and to those who have custody of okapi, dead or alive, is to resurrect the public profile of these beautiful beasts. Once famed, they are fading from the public consciousness outside of their native Democratic Republic of Congo. Many of those who are aware of them think of okapi as mystical beasts, forgetting their current predicament as living breathing beings in the rainforests of the Congo basin. The challenges are significant: ongoing violence and instability in their only habitat and Westerners' negative stereotypes about the region are compounded by Westerners' unease about their legacies there, past and present. Despite the Congo Basin's global importance for biodiversity, and as the other green lung of our planet, it is easier to look away.

There are still taxidermized okapis on display in the West, in the American Museum of Natural History in New York, at Tervuren in Belgium, in Basel, Cambridge, Dublin, Frankfurt, Rome and elsewhere, but most okapi remains were collected more than half a century ago. No more will be shot or collected from the wild for exhibition in museums. Whether natural history museums are still prepared to acquire okapi that die in zoos and have them taxidermized for display is unclear. Many existing taxidermized okapi are decaying and gathering dust in storerooms. And yet they are such a charismatic species—visually striking, taxonomically fascinating, and emblematic of the challenges and limitations of current scientific knowledge making.

I would love to see exhibitions and stories about the discovery of this African rainforest unicorn (if that term helps, use it) in the natural history museums and zoos of the world that would celebrate a remarkable animal but also reflect on their own complicated histories of exploration, collection, exhibition, and knowledge making. The Powell-Cotton Museum in Kent, England, now includes photographs and names of many local people who assisted Percy in his explorations. In Berlin, the Museum für Naturkunde's highlights of taxidermy exhibition features the stories of some famed zoo animals, mounted and displayed in the museum after their deaths. This Berlin museum, like a few other museums including the British Museum of Natural History, discusses decolonializing its collections on its website and supports research on the subject. Few permanent exhibitions in physical museums yet reflect this, however. Accounts of Indigenous contributions to sourcing wild animals and knowledge about them are rare, not only at museums but also in zoo displays. The materials exist to remedy this.

Given that the Natural History Museum in London has participated in the history of knowledge making about the okapi for longer than any other institution, it should surely display and discuss this wonderful animal. Okapi were first classified and exhibited within its beautiful terracotta-clad walls. And somewhere in its vaults lie curled up that first pair of okapi bandoliers Harry Johnston bought off a Congolese soldier in 1900 and sent back to Philip Sclater at London Zoo—the scrap of evidence that set this whole story in motion.

Encountering an Okapi

Very few of us will ever see a live, wild okapi. So, I urge you to go and see an okapi in a zoo if you can. And don't just pause a moment to tick off this zoological curiosity before you move on to the giraffe or the zebras or whoever is penned in the next enclosure. Loiter a while to consider the extraordinary tale of how this species got from the rainforests of the Congo to a Western zoo. Admire its strangeness and singularity, its extraordinary coloration, pattern, and shape; it is a much more interesting animal than the unicorn, which, after all, is a horse with a narwhal tusk on its forehead. If the zoo has a calf, you should see some lively antics—look out for the unique pooky.

Consider also what it must be like for a habitually solitary rainforest animal with acute hearing and a keen sense of smell to live in an open pen in a temperate climate, sometimes together with other okapi, next door to exotic animalian neighbors, and be stared at by noisy humans throughout the year. Watching the okapi in their (admittedly secluded) outdoor enclosure at London Zoo, I sometimes fantasize about standing rather on a treetop platform inside a rainforest biome dome, looking down at okapi moving silently through tangles of trees and vines.

The location of Doué-la-Fontaine Bioparc in Anjou in western France has allowed it to create a more naturalistic environment for its okapi in an old quarry that has trees and vines and a large pool the okapi drink from and share with birds and other species. It does not offer a rainforest climate, but online footage suggests the habitat is more naturalistic than most of the paddocks you see in internet videos featuring zoo okapi. I have seen no data on how okapi behave in this simulated rainforest environment, but the zoo's first calf was born in spring 2023. This is one example of what might be possible.[12]

Wouldn't it be good if all those of us with smart phones and electronic devices made with Congolese cobalt and copper together with some of the big brands like Apple and Samsung and Tesla and Daimler who have built fortunes on products incorporating natural resources from the Democratic Republic of Congo could give something back?[13] If we could support the better care and accommodation (in zoos) and the conservation (in the wild) of the Democratic Republic of Congo's national animal, the rare, beautiful, and enigmatic okapi?

ACKNOWLEDGMENTS

I was helped by many people in numerous ways as I strayed into new areas of research. I am grateful for their time and patience and generosity in sharing their expertise. I am responsible for any errors and misinterpretations.

For surviving dengue fever and its long aftermath with me and listening to my ramblings as I stumbled toward the light, I extend my deepest thanks to Susan Pooley and Alexander and Lara. Scott Reynolds had my back. So did Dr. Anna Checkley at University College of London's Hospital for Tropical Diseases. Thanks to Mum (Elsa Pooley) for her father's okapi postcard.

My academic trajectory has been unorthodox, and I have been blessed with mentors who believed in me at critical points along the way. I am indebted to and feel much affection for William Beinart, E. J. Milner-Gulland, and Harriet Ritvo.

Libraries and archives are at the core of this work, and I was fortunate to be welcomed by Ann Sylph, Emma Milnes, and Natasha Wakeley at the Zoological Society of London's Prince Philip Zoological Library and Archives. I am also grateful to the librarians and archivists at the British Natural History Museum, especially Kathryn Rooke, Amy Tjiong in the anthropology division at the American Museum of Natural History, and Wellcome Collection librarians.

Aware of the limitations of the research in English, and my language skills, I consulted colleagues who are fluent and also experts in fields I was straying into: I am grateful to Bernhard Gissibl, Dominik Hünniger, René E. Honegger, Diana Natermann, and Raf de Bont. Thanks to Workineh Kelbessa for sharing his research.

I am grateful to Harriet Ritvo for reading a very early draft, to two anonymous readers who approved my proposal, and to one anonymous reader for in-depth and encouraging feedback. Thanks to Gretchen Walters for reading some anthropological material. I am grateful to Matthew McAdam, Jennifer D'Urso, Hilary Jacqmin, and the entire team at Johns Hopkins University Press.

Curators, collection managers, and researchers at several museums have been helpful, including Jim Middleton, collections manager at Scarborough Museums and Galleries, Catarina Madruga, Katja Kaiser, and Mandy Ullmann at the Museum für Naturkunde in Berlin, Jack Ashby, assistant director at the University Museum of Zoology of the University of Cambridge, Claire Walsh, exhibitions and interpretation manager at the Natural History Museum in Tring, Gie Robeyns, archive and moveable heritage at KU Leuven, Paolo Viscardi, deputy keeper of natural history at the National Museum of Ireland, Rachel Jennings, curator of natural history at the Powell-Cotton Museum, along with Madylene Beardmore and the rest of the team, Roberto Portexla Miguez and (especially!) Natalie Cooper at the British Natural History Museum, who granted me access to okapi remains, David Marques, curator of vertebrates at the Natural History Museum Basel, and Loïc Costeur, curator of the osteological collection at the Natural History Museum Basel.

I am grateful to several key okapi conservationists for speaking with me, including Rosmarie Ruf and John Watkin of the Okapi Conservation Project, Stuart Nixon of Chester Zoo, and Chris Hamley and Gabriel Gelin of the Wildlife Conservation Society. Thanks to Julia Fa of Manchester Metropolitan University for information on bushmeat hunting, Mike Hoffmann of the Zoological Society of London for taxonomic references, Marcus Rowcliffe of the Zoological Society of London for insights on camera trapping, and Jonathan Kingdon of Oxford University for information on okapi at London Zoo and in the Semliki region.

I was delighted to meet London Zoo's okapi and to be taught about the work of zookeepers by Poppy Jewell and Gemma Metcalfe of London Zoo's Hoofstock section.

Preface

1. Attilio Gatti, *Great Mother Forest* (Hodder and Stoughton, 1936), 27.

2. This debate intensified following the killing of George Floyd in May 2020. Institutions in the United Kingdom began rethinking their legacies, and museums came in for criticism, including London's British Museum. See, for example, Dan Hicks, *The Brutish Museums: The Benin Bronzes, Colonial Violence, and Cultural Resistance* (Pluto Press, 2020).

3. John Simons, *Obaysch: A Hippo in Victorian London* (Sydney University Press, 2019).

4. Simon Pooley, "Coexistence for Whom?" *Frontiers in Conservation Science* 2 (2021). https://doi.org/10.3389/fcosc.2021.726991.

5. On some of the challenges and importance of tackling these, see Unai Pascual, William M. Adams, Sandra Díaz, et al., "Biodiversity and the Challenge of Pluralism," *Nature Sustainability* 4, no. 7 (2021): 567–72.

6. On Sigia Gumede, see Tony Pooley, *Mashesha: The Making of a Game Ranger* (Southern Book Publishers, 1992), 195–205, and Elsa Pooley, "Some Notes on the Utilization of Natural Resources by the Tribal People of Maputaland," in *Studies on the Ecology of Maputaland*, ed. Mike N. Bruton and Keith H. Cooper (Natal Branch of the Wildlife Society of Southern Africa, 1980), 467–79.

7. The notion of "Western" including "Western science" or "knowledge" is a useful but vexed convention. As Naoise Mac Sweeny shows in *The West: A New History of an Old Idea* (W. H. Allen, 2023), the term's etymology includes narratives claiming a linear progression from the ancient Greeks through the Romans and their rediscovery during the European Renaissance to flower in "modern" form in the Scientific Revolution usually associated by European scholars with early developments in astronomy, physics, and chemistry in Western Europe. On closer inspection, this story quickly unravels into a more complicated multi-regional and multicultural one. The "West" as a region becomes associated with Western Europe and then the United States, along with linked ideas about superior scientific knowledge, modern commerce, and Christian civilization during a period of ascendant global influence. This is particularly true for the Anglophone world from the mid-1800s to 1914.

In this book, "Westerners" refers to explorers, missionaries, colonists, and imperialists from western Europe and the United States since circa 1860. The term "Western" relates to the metropolitan zoos, museums, and other knowledge institutions (and associated authority structures and means of authorizing and disseminating scientific knowledge) in this period. While acknowledging the global influence of "the West" and the power of scientific inquiry and exploration accompanying it, use of the term does not support narratives asserting its historical coherence or moral/cultural superiority.

8. My thinking on use of the term "pygmy" draws on chapters in *Hunter-Gatherers of the Congo Basin*, ed. by Barry S. Hewlett (Routledge, 2017), and Kairn A. Klieman's *"The Pygmies Were Our Compass": Bantu and Batwa in the History of West Central Africa, Early Times to c. 1900 C. E.* (Heinemann, 2003).

Introduction

1. "Okapi (*Okapia Johnstoni*)," signboard and video at London Zoo, viewed in May 2024.

2. "The Giraffe Family," mammals section, British Natural History Museum, viewed in April 2024.

3. "Rare," display board C41, mammals section, British Natural History Museum. Both the Zoological Society of London and the Natural History Museum insist the okapi was discovered in 1901, but Johnston found evidence of its existence in August 1900, which he sent to London (the society received this evidence in November 1900, as I explain in chapter 2).

Chapter 1. Scientific Authority and Metropolitan Knowledge Institutions

1. Technically, as there are two, they are the syntypes of *Okapia johnstoni* (Philip Sclater, 1901), "syntype" meaning each of a set of type specimens of equal status on which the description and name of a new species is based.

2. This section is based on histories of the zoo, including Adrian Desmond, "The Making of Institutional Zoology in London, 1822–1836," pt. 1, *History of Science* 23, no. 2 (1985): 153–85, and pt. 2, *History of Science* 23, no. 3 (1985): 223–50; John Edwards, *London Zoo from Old Photographs, 1852–1914*, 2nd ed. (Butler Tanner and Dennis, 1996); Takashi Ito, *London Zoo and the Victorians, 1828–1859* (Boydell Press, 2014); Solomon Zuckerman, "The Zoological Society of London: Evolution of a Constitution," in *The Zoological Society of London, 1826–1976, and Beyond*, ed. Solomon Zuckerman (Academic Press, 1976), 1–16; and Isobel Charman, *The Zoo* (Penguin, 2016).

3. "When the Zoo Beasts Perish," *Daily Mail*, August 9, 1898, DF Pub 517/1/1, Natural History Museum and Archives.

4. Nigel Rothfels, *Savages and Beasts: The Birth of the Modern Zoo* (Johns Hopkins UP, 2002), 6.

5. On the orientations of zoology, see Zuckerman, "Zoological Society of London." On ethologists' struggles for recognition, see Richard W. Burckhardt Jr., *Patterns of Behavior: Konrad Lorenz, Niko Tinbergen, and the Founding of Ethology* (U of Chicago Press, 2005).

6. Harry Johnston, *The Story of My Life* (Bobbs-Merrill Co., 1926), 99.

7. On Hagenbeck and Jamrach's sales of animals to Sclater at the Zoological Society of London, see letters to the secretary, BADI-BADJ, Prince Philip Zoological Library and Archives. On Hagenbeck in particular, see Rothfels, *Savages and Beasts*. On avoiding subsuming discussions of the zoo to a framing privileging British imperialism, see Ito, *London Zoo and the Victorians*.

8. On giraffe, see Ito, *London Zoo and the Victorians*, and Charman, *The Zoo*; on hippo, see John Simons *Obaysch: A Hippo in Victorian London* (Sydney UP, 2019).

9. Colin M. Turnbull, *The Mbuti Pygmies: Change and Adaptation* (Holt, Rhinehart and Winston, 1983), 15.

10. Turnbull, *Mbuti Pygmies.*

11. British Natural History Museum background in this section is based on John Thackray and Bob Press, *Nature's Treasure House* (The Natural History Museum, 2013); Nicolaas A. Rupke, "The Road to Albertopolis: Richard Owen (1804–92) and the Founding of the British Museum of Natural History," in *Science, Politics and the Public Good: Essays in Honour of Margaret Gowing,* ed. Nicolaas A. Rupke (Macmillan, 1988), 63–89; *A General Guide to the British Museum (Natural History)* (The British Museum, 1887); "National Collections of Natural History," *The Times,* April 18, 1881, NHM DF Pub 517/1/1, 4.

12. See map in Thackray and Press, *Treasure House,* 43.

13. Thackray and Press, *Treasure House,* 46.

14. Thackray and Press, *Treasure House*; Rupke, "The Road to Albertopolis."

15. Thackray and Press, *Treasure House*; "National Collections," 2–3.

16. Richard Owen, "Report from the Select Committee to the British Museum," *Parliamentary Papers,* vol. 16 (HM Stationery Office, 1860), 262.

17. Thackray and Press, *Treasure House.*

18. *Guide to the British Museum.*

19. *Guide to the British Museum,* 18–19.

20. "National Collections."

21. Richard Milner, "Huxley's Bulldog: The Battles of E. Ray Lankester (1846–1929)," *Anatomical Record* 257, no. 3 (1999): 90–95.

22. "The Natural History Museum," *The Times,* August 1, 1898; James Cossar Ewart, Adam Sedgwick, Sidney H. Hickson, et al., "The Administration of the Natural History Museum," DF Pub 517/1/1, 358, Natural History Museum Library and Archives; letter to the editor, *Morning Post,* April 19, 1909, DF Pub 517/1/2, 244, Natural History Museum Library and Archives.

23. Milner, "Huxley's Bulldog."

24. "Professor Ray Lankester," *Candid Friend,* May 1, 1901, DF Pub 517/1/1, 434, Natural History Museum Library and Archives.

25. Paul Du Chaillu, *Explorations and Adventures in Equatorial Africa* (Harper and Brothers, 1861); Stuart McCook, "'It May Be Truth, But It Is Not Evidence': Paul du Chaillu and the Legitimation of Evidence in the Field Sciences," *Osiris* 11, no. 1 (1996): 177–97.

26. Stuart McCook, "'It May Be Truth,'" 177.

27. C. W. J. Withers and I. M. Keighren, "Travels into Print: Authoring, Editing, and Narratives of Travel and Exploration, c. 1815–c. 1857," *Transactions of the Institute of British Geographers* 36, no. 4 (2011): 560–73; Bruno Latour, *Science in Action: How to Follow Scientists and Engineers through Society* (Harvard UP, 1988).

28. Harriet Ritvo, *The Platypus and the Mermaid and Other Figments of the Classifying Imagination* (Harvard UP, 1997).

29. On giraffe, see Ito, *London Zoo and the Victorians,* and Charman, *The Zoo*; on hippo, see Simons *Obaysch.*

30. James Poskett, *Horizons: A Global History of Science* (Penguin, 2023).

Chapter 2. Discovery of the Okapi

1. The epigraph to this chapter comes from Harry Johnston, *The Uganda Protectorate*, vol. 1 (Hutchinson, 1902), 380.

2. Stanley outlines the alternative routes for reaching Emin Pasha and his reasons for choosing the Congo route, in *In Darkest Africa*, 2 vols. (Sampson Low, Marston, Searle, and Rivington, 1890), 1:44–45.

3. Tim Jeal, *Stanley: The Impossible Life of Africa's Greatest Explorer* (Faber and Faber, 2007).

4. Jeal, *Stanley*.

5. See G. Schweinfurth, F. Ratzel, R. W. Felkin, et al., *Emin Pasha in Central Africa: Being a Collection of his Letters and Journals* (George Philip & Son, 1888); Harry Johnston, "Emin Pasha's Last Collections," *Nature*, November 24, 1921, 398, being a review of E. Pasha, *Die Tagebücher von Dr Emin Pascha: Herausgegeben mit Unterstüzung des Hamburgischen Staates und der Hamburgischen Wissenschaftlichen Stiftung von Dr Franz Stuhlmann*, vol. 6, *Zoologische Aufzeichbungen Emin's und seine Briefe an Dr G. Hartlaub bearbeitet von Prof. Dr H. Schubotz* (Georg Wester-mann, 1921).

6. Alfred E. Pease, "The Okapi," *The Times*, December 5, 1934, 10; Stanley, *Darkest Africa*, 2:228–49.

7. E. Ray Lankester, "On *Okapia*, a New Genus of *Giraffidae*, from Central Africa," *Transactions of the Zoological Society of London* 16, no. 6 (1902): 279–314; Stanley, *Darkest Africa*, 2:83. See also Percy Home, "The Finding of the New African Animal," *Sphere*, August 17, 1901, 184.

8. On asses at Manyema and the Wané-Mpungu people, Maruna warriors, and the King of Ntamo, see Henry M. Stanley, *Through the Dark Continent* (London: Samson Low, Marston, Searle, and Rivington, 1890), 380, 479, 515, 537–38. For more on his first journey down the Congo, see *The Exploration Diaries of H. M. Stanley*, ed. Richard Stanley and Alan Neame (William Kimber, 1961); Jeal, *Stanley*.

9. On wild assess and donkeys, see Evelyn Todd, Laure Tonasso-Calvière, Loreleï Chauvey, et al., "The Genomic History and Global Expansion of Domestic Donkeys," *Science* 377, no. 6611 (2022): 1172–80; International Union for the Conservation of Nature, "African Wild Ass," IUCN Red List of Threatened Species, https://iucnredlist.org/species /7949/45170994 (accessed February 22, 2024); and Roger Blench, "Wild Asses and Donkeys in Africa," revision of a paper presented at School of Oriental and African Studies, May 9, 2012, https://rogerblench.info/Ethnoscience/Animals/Livestock.

10. Stanley, *Darkest Africa*, 2:67–82, 83–87, 89–98.

11. Stanley, *Darkest Africa*, 1:220.

12. While Stanley refers to the Mbuti as the Wambutti, Johnston calls the group the Bambutti. I have used the contemporary Mbuti, following Colin M. Turnbull, *The Mbuti Pygmies: Change and Adaptation* (Holt, Rhinehart and Winston, 1983), and others.

13. Stanley, *Darkest Africa*, 1:198.

14. See appendix 1 in the online resources for further claims to prior discovery. References for this discussion include Wilhelm Junker, *Travels in Africa During the Years 1882–1886*, translated by A. H. Keane (Chapman and Hall, 1892), 268; on Stuhlman, see Herbert Lang, "In Quest of the Rare Okapi," *Zoological Society Bulletin* 21, no. 3 (1911), 1601–13; Auguste Ménégaux, "Sur la presence de l'okapi au Bahr-el-Gazal," *Bulletin du Muséum d'Histoire Naturelle*, 6 (1905), 381–83.

15. Stanley, *Darkest Africa*, 2:490.

16. Harry Johnston, "The Okapi: The Newly Discovered Beast Living in Central Africa," in *Annual Report of the Board of Regents of the Smithsonian Institution* (Government Printing Office, 1902), 661; Harry Johnston, *The Story of My Life* (Bobbs-Merrill Company, 1926), 309. On African unicorns, P. H. Gosse, *The Romance of Natural History*, 6th ed. (James Nisbet & Co., 1863), 285.

17. On discoveries of new species in the period: Charles Darwin, *Voyages of the Adventure and Beagle, Vol. III –Journal and Remarks, 1832–1836* (Henry Colburn, 1839); for Wallace, see "On the Natural History of the Aru Islands," *The Annals and Magazine of Natural History: Zoology, Botany, and Geology* 20, no. 2 (1857), 473–85; Du Chaillu's findings were circulating from 1859, published in 1861 in *Explorations and Adventures in Equatorial Africa* (Harper & Brothers, 1861).

18. Johnston, "The Okapi," 661.

19. Rowland Oliver, *Sir Harry Johnston and the Scramble for Africa* (Chatto and Windus, 1957); James A. Casada, "Sir Harry H. Johnston as a Geographer," *The Geographical Journal* 143, no. 3 (1977), 393–406; and *Sir Harry Johnston: A Bio-Bibliographical Study* (Basler Afrika Bibliographien, 1977). For details of the spat, see the online notes.

20. Thomas Pakenham, *The Scramble for Africa* (Abacus, 1991), 337.

21. Johnston, *My Life*.

22. Johnston, *My Life*, 23.

23. Johnston, *My Life*.

24. For more on Johnston's career and travels, see Harry Johnston, *The River Congo from Its Mouth to Bólóbó* (Sampson Low, Marston, Searle, and Rivington, 1884), and Jeal, *Stanley*, 270–72.

25. Johnston, *My Life*, 92.

26. Jeal, *Stanley*.

27. Adam Hochschild, *King Leopold's Ghost: A Story of Greed, Terror, and Heroism in Colonial Africa* (Mariner Books, 1999); Robert Harms, *Land of Tears: The Exploration and Exploitation of Equatorial Africa* (Basic Books, 2019). The horrors of the rubber trade were yet to come, in the 1890s.

28. Johnston, *River Congo*; Johnston, *My Life*; Jeal, *Stanley*.

29. Johnston, *My Life*.

30. Johnston, *My Life*. My father Tony and I suffered numerous bouts of fever, and I have endured dengue fever and its after-effects.

31. Johnston, *My Life*, 334–35.

32. Johnston, *The Uganda Protectorate*, 379–80.

33. Johnston, *The Story of My Life*, 343–45.

34. Johnston, *My Life*; Sir William Flower, Notes for presentation of ZSL's silver medal to Johnston (untitled), ZSL Prince Philip Zoological Library & Archives, ZSL SEC 7/9/5–7/10/13 BADI-BADJ; memo to Mr. Williams from ZSL, re list of animals sent by Johnston, June 28, 1888; letter to Philip Sclater, from the Collector of Revenues in British Central Africa, re natural history specimens being sent to him, January 2, 1893 (both in ZSL Archive, Sec/7/10/12, Johnston, Sir Harry Hamilton).

35. Christy describes his 1913 visit in *Big Game and Pygmies: Experiences of a Naturalist in Central African Forests in Quest of the Okapi* (Macmillan, 1924), 25.

36. Johnston, *Uganda Protectorate*, 1:380.

37. Harry Johnston, "The Okapi," in *The People's Natural History*, ed. Charles J. Cornish, F. C. Selous, Ernest Ingersoll, et al., vol. 2 (University Society, 1905), 267–70.

38. Johnston, *My Life*, 380, 346–47.

39. Johnston, *My Life*, 346–47; *Uganda Protectorate*, 1:380.

40. Johnston, "The Okapi," 662. Sclater claims the accompanying letter was dated August 21, 1900, and sent from Toru ("On an Apparently New Species of Zebra from the Semliki Forest," *Proceedings of the Zoological Society of London*, no. 1 [1901]: 50). See also Lankester "On *Okapia*," 280. Over the border in Uganda, Johnston discussed the "creature" with locals in the territory of Mboga (southeast of Lake Albert, Western Region), in a British-controlled part of the Semliki Forest. Locals claimed to know it and declared it fairly common, but they did not bring in any specimens. For Johnston's letter and the first use of "okapi" in English, see Harry Johnston, "Sir Harry Johnston on a New Horse," *Proceedings of the Zoological Society of London*, no. 2 (1900): 774–75. Henry Scherren identifies this mention by Sclater as the first time the word "okapi" was used in English (*The Zoological Society of London: A Sketch of Its Foundation and Development* [Cassell, 1905], 220).

41. M.A.C.H., "Obituary: Mr. M. R. Oldfield Thomas," *Nature* 124, no. 3116 (1929), 101–2; C. Hart Merriam, "Revision of the North American Pocket Mice," *North American Fauna* 1 (US Department of Agriculture, 1889).

42. "Sir Harry Johnston's Recent Journeys in the Uganda Protectorate," *The Times*, December 29, 1900, 9. The ' in *o'api* denotes a click sound (Johnston was a keen student of African languages).

43. Sclater, "New Species of Zebra," 50, 52.

44. Johnston, *My Life*, 347. Elsewhere he claims Eriksson got the skin and skull from an African soldier (*Uganda Protectorate*, 380). The provenance of the smaller skull is never documented.

45. Lankester discussed the cloven hoofs of the okapi, including quotes from Johnston, in "On *Okapia*," 280.

46. Lankester confirms receiving a letter written by Johnston in April 1901 suggesting that the okapi be placed in the genus *Helladotherium* ("On *Okapia*," 282). On Johnston's deductions from the bilobed lower canines, see Johnston, *My Life*, 347.

47. Philip Sclater, "On a New African Mammal," *Proceedings of the Zoological Society of London*, no. 2 (1901): 4.

48. Sclater, "On a New African Mammal," 4, 5, plate on 2.

49. "A New Mammal," *The Times*, May 7, 1901, 13.

50. "*Ex Africa semper aliquid novi*," *The Times*, May 10, 1901.

51. As reported in "A New Mammal," *The Times*, May 7, 1901.

52. Auguste Lameere, "L'okapi," *Revue de l'Université de Bruxelles* 2, 1902: 117.

53. Lameere, "L'okapi," 8.

54. Arthur Smith Woodward, "Albert Gaudry," *Nature*, December 10, 1908, 163–64.

55. Woodward, "Albert Gaudry."

56. A. S. Woodward, "Dr. C. I. Forsyth Major, F.R.S.," *Nature* 3(2789): 505; L. Rook and D. M. Alba, "The Pioneering Paleoprimatologist Charles Immanuel Forsyth Major (1843–1923)," *Bollettino della Società Paleontologica Italiana* 51 (1), 2012, 1–2.

57. Woodward, "Dr. C. I. Forsyth Major"; Rook and Alba, "The Pioneering Paleoprimatologist."

58. Woodward, "Dr. C. I. Forsyth Major"; Rook and Alba, "The Pioneering Paleoprimatologist"; and on being approached by the Independent State of the Congo: Julien Fraipont, *Okapia*, in *Contributions a la Faune du Congo*, book 1, *Annales du Musée du Congo* (Independent State of the Congo, 1907).

59. Johnston, "The Okapi," in *The People's Natural History*, 270.

60. Lankester, "On *Okapia*."

61. On the horns of okapi, see John Hart, "*Okapia johnstoni* Okapi," in *The Mammals of Africa. Volume VI: Pigs, Hippopotamuses, Chevrotain, Giraffes, Deer, and Bovids*, ed. Jonathan S. Kingdon and Mike Hoffmann (Bloomsbury Publishing, 2013), 110–15; on the horns of giraffids, see G. Mitchell and J. D. Skinner, "On the Origin, Evolution and Phylogeny of Giraffes *Giraffa camelopardalis*," *Transactions of the Royal Society of South Africa* 58, no. 1 (2003): 51–73.

62. Lameere, "L'okapi," 128, 132.

63. E. Ray Lankester, "Sir Harry Johnston's New Beast," *The Times*, June 18, 1901, 8.

64. New York Zoological Society, "An Important Zoological Discovery," *News Bulletin of the Zoological Society* 5 (1901): 4–5.

65. E. Ray Lankester to Philip Sclater, June 21, 1901 (SEC/7/12/5, 2, Prince Philip Zoological Library and Archives).

66. E. Ray Lankester, "The Specific Name of the Okapi Presented by Sir Harry Johnston to the British Museum," *Annals and Magazine of Natural History* 10, nos. 55–60 (1902): 417–18; Lankester, "On *Okapia*."

67. Jack Ashby, *Platypus Matters: The Extraordinary Story of Australian Mammals* (William Collins, 2022).

68. On taxonomic debates over the classification of the giraffids: Edwin H. Colbert, "The relationships of the Okapi," *Journal of Mammalogy* 19, no. 1 (1938), 47–64.

69. Lankester, "On *Okapia*," 304–5.

70. Lankester, "On *Okapia*," 300.

71. "The Okapi," *Field*, May 1901, 670; "The Naturalist," *Field*, December 1901, 990; Adolf Friedrich von Mecklenburg-Schwerin, *In the Heart of Africa* (Cassell, 1910), 202. The duke's name is presented variously in translation, including spellings and how his title and or given names are incorporated. In naming sources in these notes, I have settled on the German version Adolf Friedrich von Mecklenburg Schwerin, omitting the title.

72. *Proceedings of the Zoological Society of London*, no. 2 (1902): 72–73.

73. New York Zoological Society, "An Important Zoological Discovery," 4.

74. Johnston, *My Life*, 358.

75. Johnston, *My Life*, 358.

76. Johnston, *My Life*, 359.

77. Johnston, *My Life*, 359. On Johnston's reporting of the name "okapi" see also "Sir Harry Johnston's Recent Journeys in the Uganda Protectorate," *Geographical Journal* 17, no. 1 (1901): 39–42, and Fraipont, *Okapia*. According to the latter "many natives of Haut-Ituri designate it under the name of o:api or okapi; the Momus call it dumla, the Mokumus boole, the Kiuvuailia kenghe."

78. "Dr. C. I. Forsyth Major on the Okapi," *Proceedings of the Zoological Society of London*, no. 2 (1902): 73–79; "L'okapi," *Belgique Coloniale*, May 25, 1902, and "Le crâne de l'okapi," *Belgique Coloniale*, June 8, 1902.

79. Charles I. Forsyth Major, "Nouveaux renseignements sur l'okapi," *Belgique Coloniale*, November 9, 1902; "Dr. C. I. Forsyth Major on the Okapi," *Proceedings of the Zoological Society of London*, no. 2 (1902): 73–79.

80. Lankester, "On *Okapia*," 304.

81. E. Ray Lankester, "Dr E. Ray Lankester on the Okapi," *Annals of the Magazine of Natural History*, ser. 7, 10, (1902): 417–18.

82. Lankester, "Dr E. Ray Lankester on the Okapi," 417.

83. Fraipont, *Okapia*, 22–33.

84. Lameere, "L'okapi," 127.

85. E. Ray Lankester, "On Hair Whorls in the Okapi," *Proceedings of the Zoological Society of London*, no. 2 (1903): 337–40, 337.

Chapter 3. Settling Okapi Taxonomy, and the First Monograph

1. The epigraph to this chapter comes from E. Ray Lankester, *From an Easy Chair* (Archibald Constable, 1909), 11.

2. Forsyth Major's publications are discussed in chapter 2. Published literature on okapi was recorded in *The Zoological Record* (see note 6).

3. Lucien Renard, "Julien Fraipont," *Chronique archéologique du Pays de Liége*, March 1910, 5(3): 27–38.

4. Julien Fraipont, *Okapia*, in *Contributions a la Faune du Congo*, book 1, *Annales du Musée du Congo* (Independent State of the Congo, 1907); on Fraipont's dealings with Forsyth Major, and his acknowledgments: 5–6.

5. Fraipont, *Okapia*, 17–18.

6. Summaries of British work on okapi were shared through Deutsche Zoologische Gesellschaft's *Zoologischer Anzeiger* (for example, vol. 24, 1901) and in French in Lameere's summary in *Revue de l'Université de Bruxelles* 2 (1902): 117. Lydekker's surveys of publications on mammals, including okapi, include "Mammalia," in *The Zoological Record*, vol. 38, ed. David Sharp (Zoological Society of London, 1902): 2–40, and "Mammalia," in *The Zoological Record*, vol. 39, ed. David Sharp (Zoological Society of London, 1903), 1–44. See also volumes 39–47.

7. Fraipont, *Okapia*: on the literature, 10–13; on there being only one species, 18.

8. Fraipont, *Okapia*: on differences in color and shape, 18; for a rogues' gallery of taxidermied okapi, 9–23.

9. Fraipont, *Okapia*, 19.

10. Fraipont, *Okapia*, 18, 33.

11. Fraipont, *Okapia*, 85. Fraipont based his map on information from employees of the company, notably Eriksson, the commissioner of the Ubangi (or Oubangui) district, Bertrand, and Commander Sillye (81–83). Sillye drew on African knowledge, reporting that the okapi was "known by all the natives of Rubi-Uele, and as far as Nyangwe in Manyema" and was seen around Adjamu and in Nepoko.

12. Fraipont, *Okapia*, 93–94.

13. A. S. Woodward, "Dr. C. I. Forsyth Major, F.R.S.," *Nature* 2789, no. 3 (April 14, 1923): 505. For a list of Forsyth Major's publications, see David Sharp, ed., *The Zoological Record*, vol. 39 (Zoological Society of London, 1902), 12–13.

14. E. Ray Lankester, "On the Existence of Rudimentary Antlers in the Okapi," *Proceedings of the Zoological Society of London*, no. 1 (1907): 126–35; Lankester's acknowledgment of Forsyth Major is on 127. Two other substantial scientific publications of his on the okapi include

"On Certain Points in the Structure of the Cervical Vertebrae of the Okapi and the Giraffe," *Proceedings of the Zoological Society of London* 78, no. 2 (1908): 320–34, and "Parallel Hair-Fringes and Colour-Striping on the Face of Foetal and Adult Giraffes," *Proceedings of the Zoological Society of London*, no. 1 (1907): 115–25.

15. "Prof. Lankester Asked to Resign: Is a Man of Genius Too Old at 59?," *Daily Express*, August 1, 1906, DF Pub 517/1/2, Natural History Museum and Archives; "Turned Out: Nation's Reward for a Great Scientist," *Evening News*, August 8, 1906, DF Pub 517/1/2, Natural History Museum and Archives.

16. E. Ray Lankester, "The First Photograph of a Living Okapi," *Illustrated London News*, September 7, 1907, DF Pub 517/1/2, 178–79, Natural History Museum and Archives; and "Supposed Horn-Sheaths of an Okapi," *Nature* 95 (1915): 64–65.

17. Lankester, "First Photograph," 178–79; Fraipont, *Okapia*, appendix A, 96–97.

18. Lankester, "Parallel Hair-fringes and Colour-striping on the Face of Foetal and Adult Giraffes," and "On the Existence of Rudimentary Antlers in the Okapi," *Proceedings of the Zoological Society of London*, January–April (1907): 126–35; and "On Certain Points in the Structure of the Cervical Vertebrae of the Okapi and the Giraffe," *Proceedings of the Zoological Society of London* 78, no. 2 (1908): 320–34.

19. Lankester, "First Photograph," 178; E. Ray Lankester with W. G. Ridewood, *Monograph of the Okapi* (British Museum [Natural History], 1910).

20. Harry Johnston, review of *Monograph of the Okapi*, by E. Ray Lankester, *Nature* (December 15, 1910): 209, 210. The Natural History Museum library has never had a copy of Fraipont's monograph.

21. E. Ray Lankester, "Sir Ray Lankester's Book on the Okapi," *Nature* 85, no. 2149 (January 5, 1911): 305–6.

22. Lankester, *Monograph of the Okapi*, v–vi.

23. Harry Johnston, letter to the editor, *Nature* 85, no. 2149 (January 5, 1911): 306.

24. Philip L. Sclater, "Dr P. L. Sclater on the Okapi," *Proceedings of the Zoological Society of London*, no. 2 (1906): 761; Richard Lydekker, *The Game Animals of Africa* (Rowland Ward, 1908); Richard Lydekker, "Hornless Okapis," *Annals of the Magazine of Natural History* 6 (1910): 224–26; Richard Lydekker, *Catalogue of the Ungulate Mammals in the British Museum (Natural History)*, vol. 3 (British Museum, 1914).

25. Henri Schouteden, "Notes sur l'okapi," *Revue Zoologique Africaine* 2 (1912–13): 482–85; Maurice de Rothschild and Henri Neuville, "Sur l'okapi et les girafes de l'est africain," pt. 1, *Annales des Sciences Naturelles Zoologie* 10 (1909): 2, and on the differences in skulls and hides, 86. Maurice, a politician and financier who had traveled in Africa and had an interest in zoology, was from the French branch of the Rothschilds.

26. Fraipont, *Okapia*, 17. Fraipont mentions Sillye's escaped okapi.

Chapter 4. Possession, Exhibition, and Dissemination

1. This chapter's epigraph comes from "Okapia Johnstoni," *Westminster Gazette*, August 14, 1901, DF Pub 517/1/1, Natural History Museum Library and Archives. Quotes in this paragraph come from an untitled June 19, 1901, article published in the *Birmingham Gazette* that is available in DF Pub 517/1/1, Natural History Museum Library and Archives.

2. The title phrase "Race for Specimens" comes from the website of the AfricaMuseum: "The Okapis," (March 10, 2022), https://www.africamuseum.be/en/learn/provenance

/okapis (accessed November 9, 2022). This interesting page contains one error—Johnston sought the okapi in 1900, not 1899.

3. On Johnston's article alerting the Belgian Authorities, Julien Fraipont, *Okapia*, in *Contributions a la Faune du Congo*, book 1, *Annales du Musée du Congo* (Independent State of the Congo, 1907), 10.

4. On these exhibitions and the history of the AfricaMuseum discussed in this section: Matthew G. Stanard, *Selling the Congo* (U of Nebraska Press, 2012); Nell Boeykens, "De Okapi als inzet van en voor de Koloniale Machtsstructuur" (MA thesis, University of Ghent, 2020), 5–13; "Museum History," Royal Museum for Central Africa, Tervuren, n.d., https://www.africamuseum.be/en/discover/history.

5. Stanard, *Selling the Congo*, quote on 38.

6. Fraipont, *Okapia*, 11.

7. Fraipont, *Okapia*, 11.

8. On colonial abuses of Africans in the Congo and reporting of this: Adam Hochschild, *King Leopold's Ghost: A Story of Greed, Terror, and Heroism in Colonial Africa* (Houghton Mifflin Harcourt, 1999). Joseph Conrad, *Heart of Darkness* (Penguin Books, 1995).

9. On diplomacy and rare species: Jack Ashby, *Platypus Matters: The Extraordinary Story of Australian Mammals* (William Collins, 2022). On King Leopold II's propaganda campaigns: Hochschild, *King Leopold's Ghost*.

10. Fraipont lists many of the okapi remains gifted by King Leopold II in *Okapia*, 14–16. The first to reach the US (mounted by Rowland Ward) was gifted to the Museum of Comparative Anatomy at Harvard by its founder, Louis Agassiz in 1905—reported by Samuel Henshaw in "Annual Report of the Curator of the Museum of Comparative Zoology at Harvard College for 1905–1906" (Harvard UP, 1906), 6. For the full list of sources, see online notes to chapter 4.

11. For example, mount 503, the skin of an okapi was gifted to France, while mount 501, the skeleton of the same animal that was donated to Sweden (Fraipont, *Okapia*, 14–16).

12. Fraipont also acknowledges commanders Arnold, Bertrand, and Ensch, chiefs of post Jadoul, Siffer, van Hulde, a Dr. David (presumably the Swiss J. J. David), and M. Mertens (*Okapia*, 6).

13. Casement's damning report was published by the government in February 1904. To head off the subsequent calls to revise the 1885 Berlin Act that had granted him his territories, King Leopold II established his own commission of inquiry as a public relations stunt. But the stunt backfired, and the commission delivered a critical report in October 1905. Emile Vandervelde raised the question of taking control of the Congo away from Leopold in the Belgian house of representatives in February 1906. Spurred on by these discussions and talk of an international Congo conference supported by European governments and Theodore Roosevelt, the Belgian government took over the colony on November 15, 1908. See Robert Harms, *Land of Tears: The Exploration and Exploitation of Equatorial Africa* (Basic Books, 2019), 442–47, 452–59.

14. Harry Johnston, "Introductory," in Edmund D. Morel, *Red Rubber: The Story of the Rubber Slave Trade Flourishing on the Congo in the Year of Grace 1906*, 1st ed. (Nassau Print, 1906), vii–xvii.

15. Johnston, "Introductory"; on thanking the Belgian colonial authorities, see for e.g. Percy Powell-Cotton, "A Journey Through the Eastern Portion of the Congo State," *The Geographical Journal* 30, no. 4 (1907), 382; Cuthbert Christy, *Big Game and Pygmies: Experiences of a Naturalist in Central African Forests in Quest of the Okapi* (MacMillan and Co., 1924), xi.

16. It is beyond the scope of this book, but the situation is complicated in that different parts of the Independent State of the Congo were run by private companies, one of which belonged to the King. "Happy Congo Natives: Report of the British Museum Expedition," *Daily News*, October 16, 1909; this story was reproduced in the *Glasgow Herald*, October 16, 1909, under the less obscene heading "Englishmen in the Congo: Important Exploration Work."

17. On Leopold and the American Museum of Natural History, see Enid Schildkrout and Curtis A. Keim, *African Reflections: Art from Northeastern Zaire* (U of Washington Press, 1990), chapter 3; Lyle Rexer and Rachel Klein, *American Museum of Natural History: 125 Years of Expedition and Discovery* (Harry N. Abrams, 1995), 100–104.

18. Schildkrout and Keim, *African Reflections*, 48, 50, 52.

19. Schildkrout and Keim, *African Reflections*, 58.

20. Philip L. Sclater, "On a New African Mammal," *Proceedings of the Zoological Society of London*, no. 2 (1901), plate on 2.

21. Sclater, "New African Mammal," 316.

22. Johnston's letter is reproduced in Sclater, "New African Mammal," 50, 52; "The Royal Society Conversazione," *The Times*, May 9, 1901, and "*Ex Africa semper aliquid novi*," *The Times*, May 10, 1901.

23. "The Okapi: A Newly-Discovered African Mammal," *Field*, May 11, 1901, 670.

24. E. Ray Lankester, "The New Giraffe-Like Animal," *Tatler*, no. 8, August 21, 1901, 368–69.

25. Rowland Ward, *A Naturalist's Life Study in the Art of Taxidermy* (Rowland Ward, 1913), 44, 145.

26. When exactly the first okapi went on display at the Natural History Museum is not clear. At a time when the okapi always made the news, *The Times* reports in "The International Congress of Zoology in Berlin" on August 9, 1901, only that "the public will shortly be afforded an opportunity of seeing this remarkable ruminant" (6). The *Standard* reports the okapi on show on August 27, 1901, but *The Times* is silent until October 19, 1901, in an article titled "The Okapi" when it reports that "the okapi . . . is now to be seen at the Natural History Museum" (12).

27. Percy Home, "The Finding of the New African Animal," *Sphere*, August 17, 1901, 184; E. Ray Lankester, "The New Giraffe-Like Animal," *Tatler*, August 21, 1901, 369. Lankester remarks on the mutual influence of Johnston and Ward in depicting the okapi.

28. Harry Johnston, "The Okapi," *McClure's Magazine*, September 1901, 497–501; Harry Johnston, "The Okapi: The Newly Discovered Beast Living in Central Africa," in *Annual Report of the Board of Regents of the Smithsonian Institution* (Government Printing Office, 1902), 663; Harry Johnston, "The Okapi: A Newly Discovered Animal," *Scientific American* 85, no. 16 (1901): 250.

29. Harry Johnston, *The Uganda Protectorate*, vol. 1 (Hutchinson, 1902), 381.

30. Johnston, *Uganda Protectorate*, 380–81; "The Okapi," *Field*, October 26, 1901, 2548.

31. "The International Congress of Zoology in Berlin," *The Times*, August 9, 1901, 6; Philip L. Sclater, "A Skull and a Strip of the Newly Discovered African Mammal (*Okapia Johnstoni*)," in *Verhandlungen des V Internationalen Zoologen-Congresses zu Berlin*, ed. Paul Matschie (Gustav Fisher, 1902), 545–47.

32. Sclater, "Skull and a Strip," 545.

33. Charles I. Forsyth Major, "L'okapi," *Belgique Coloniale*, May 25, 1902, 244–45; Charles I. Forsyth Major, "Le crâne de l'okapi," *Belgique Coloniale*, June 8, 1902, 260.

34. E. Ray Lankester, "On *Okapia*, a New Genus of *Giraffidae*, from Central Africa," *Transactions of the Zoological Society of London* 16, no. 6 (1901): 310.

Chapter 5. Okapis Take Shape in the Western Imagination

1. "The Haunts of the Okapi," *Nature*, May 24, 1906, 88. The epigraphs to this chapter that convey popular media representations of okapi come from "*Ex Africa aliquid semper novi*," *The Times*, May 10, 1901, 9, and Frank E. Beddard, "The Okapi: The New Quadruped from Central Africa," *Pall Mall Magazine*, August 1901, 570.

2. The British Museum, like other European museums, has in recent years begun returning cultural possessions snatched from other countries during the colonial period, including, famously, the Benin Bronzes and the Elgin Marbles. For natural history museums, the matter of returning the physical remains to their places of origin is less straightforward, particularly in the case of taxidermized remains because taxidermy is an art and a craft, and most of the recreated "animal" is in fact comprised of scaffolding and stuffing that shape and fill out a preserved skin and that is fitted with outer features like teeth, horns, and hooves. In the case of those in European and other Western natural history museums, these were usually created in the country of exhibition, not the country of collection of the specimen.

3. Karen Jones, "The Rhinoceros and the Chatham Railway: Taxidermy and the Production of Animal Presence in the 'Great Indoors'," *History* 101, no. 348 (2016), 710–35.

4. Julien Fraipont, *Okapia*, in *Contributions a la Faune du Congo*, book 1, *Annales du Musée du Congo* (Independent State of the Congo, 1907). Fraipont uses these numbers (for example, no. 488) to refer to particular specimens, but nowhere explains their provenance. The numbers can be used to identify illustrations of these specimens in his monograph.

5. E. Ray Lankester with W. G. Ridewood, *Monograph of the Okapi* (British Museum [Natural History], 1910). The Edinburgh and Tring okapi mounts with quotes appear on plates 45 and 46.

6. Natural History Museum Standing Committee, minutes, March 27, 1915, 3,385, June 26, 1915, 3,409–10, printed minutes, 1915–18, 3,356–3,625, Natural History Museum Library and Archives.

7. "Okapi Skull," https://powell-cottonmuseum.org/collection-items/okapi-skull/ (accessed on February 28, 2024).

8. "The Capture of Live Okapi: Specimens of the Beast," *Illustrated London News*, December 31, 1910. E. Ray Lankester, "The First Photograph of a Living Okapi," *Illustrated London News*, September 7, 1907, DF Pub 178–79, Natural History Museum Library and Archives.

9. Fraipont, *Okapia*, annex F; "British Association," *The Times*, August 6, 1907, 8. On Ribotti, see Spartaco Gippoliti, "Okapi, Italy, and the Heart of Darkness: The Politics behind Early Specimens of *Okapia Johnstoni* in Italy," *Quaderni del Museo Civico di Storia Naturale di Derrara* 10 (2022): 65–66.

10. Adolf Friedrich von Mecklenburg-Schwerin, *In the Heart of Africa* (Cassell, 1910).

11. Adolf Friedrich von Mecklenburg-Schwerin, *From the Congo to the Niger and Nile*, vol. 2 (John Winston, 1914). Hermann Schubotz, "Vorläufiger Bericht über die Reise und die zoologischen Ergfebnisse der Deutschen Zentralafrika-Expedition 1907–1908," *Sitzungsberichte der Gesellschaft Naturforschender Freunde zu Berlin* (1909): 383–410; and Hermann Schubotz, "Auf den Fahrten des Okapi," *Die Woche* 14, no.1 (1912): 6–8, photos on 12–13.

12. The Austro-Hungarian empire also produced taxidermized okapi. Explorer and zoologist Rudolf Grauer led an expedition to Central Africa from December 1909 to February 1910, during which he collected okapi remains, including two skulls, two hides, and one entire skeleton. Grauer donated all but one skull (which went to Stuttgart) to the natural history museum in Vienna, where two taxidermized okapi from 1910 are still on display. For sources on Grauer see online notes and on the display of okapi at the museum, see "säugetiere" (mammals), https://www.nhm-wien.ac.at/ausstellung/dauerausstellung__schausammlung /erster_stock.

13. Herbert Lang, "In Quest of the Rare Okapi," *Zoological Society Bulletin* 21, no. 3 (1918): 1600–1614, 1600.

14. Lang, "In Quest," 1605, 1611.

15. Lang, "In Quest," photos of the calf on 1605, 1606, 1608.

16. Roland Baetens, *The Chant of Paradise: The Antwerp Zoo* (Royal Zoological Society of Antwerp, 1993), 52, 190. The photo of Willy with okapi is on 59.

17. Cuthbert Christy, *Big Game and Pygmies: Experiences of a Naturalist in Central African Forests in Quest of the Okapi* (MacMillan and Co., 1924). Christy's two okapi are shown on the plate opposite 54, Reid's opposite 60, 66, the head shot opposite 70, the quote is on 70.

18. The first of Ward's series to include okapi is J. G. Dollman and J. B. Burlace (eds.), *Rowland Ward's Records of Big Game*, 9th ed. (Rowland Ward, 1928).

19. Violette Pouillard, "Conservation et captures animales au Congo belge (1908–1960)," *Revue Historique* 679 (2016): 577–604. See Baetens, *Chant of Paradise*, 182, for a reproduction of a photo of Hutsebaut with okapi.

20. "A living okapi in Europe," *Illustrated London News*, March 2, 1929, 349.

21. Cornelius P. Bezuidenhout, "The Okapi 'Snapshotted' at Home by a Photographer in a Hog's Skin," *Illustrated London News*, July 11, 1931, front page and 48–49.

22. See for example *The Times*, August 1, 1935, 18 and *The Times*, August 6, 1935, 14.

23. Attilio Gatti, *Great Mother Forest* (Hodder and Stoughton, 1936), and *South of the Sahara* (Robert M. McBride & Company, 1946).

24. Reginald I. Pocock, "The Okapi," *Zoo Life* 1, no. 1 (1946): 3–7; Agatha Gijzen, *Das Okapi* (A. Ziemsen Verlag, 1959).

25. Editorial, *Zoo Life* 1, no. 1 (1946): 1; Pocock, "The Okapi," 3–7; Editorial, *Zoo Life* 1, no. 2 (1946): 1. The caption to the photo of Buta evolved from "one of the only two Okapis still living in captivity" in the spring issue to "the only Okapi still living in captivity in Europe" in the summer issue.

26. "Okapi: Die Unbekannte Waldgiraffe," Zoo Basel blog entry, February 8, 2021, https://www.zoobasel.ch/de/aktuelles/blog/3/zoo-geschichte/178/. The photo of de Medina is included in this blog entry.

27. Baetens, *Chant of Paradise*, 112.

28. On Besobe, see Agatha Gijzen and Stefan Smet, "Seventy Years Okapi, *Okapia johnstoni*," *Acta Zoologica et Pathologica Antverpiensia* 59 (1974): 22, for photographs of Mafuta, 29–31.

29. On camera trapping and the first okapi captured on a camera trap, see J. Marcus Rowcliffe and Chris Carbone, "Surveys Using Camera Traps: Are We Looking to a Brighter Future?," *Animal Conservation* 11, no. 3 (2008): 185–86; Franceso Rovero and Fridolin Zimmermann, *Camera Trapping for Wildlife Research* (Pelagic, 2016); and James Morgan,

"Rare African Okapi Seen in the Wild," BBC News Channel, September 11, 2008, http://news.bbc.co.uk/1/hi/sci/tech/7609393.stm.

30. Martin Johnson, *Congorilla: Adventures with Pygmies and Gorillas in Africa* (Brewer, Warren and Putnam, 1931), 44, 50, 60.

31. See *The Times*, March 11, 1938, 12, for a review of the film.

32. William Durant Campbell, William D. Campbell African Expedition, 1938, finding aid, Gottesman Research Library, http://libcat1.amnh.org/record=b1140445.

33. Armand Denis, *On Safari: The Story of My Life* (Collins, 1963), 224–32.

34. *Bwana Kitoko*, described as "un film tourné au cours du voyage du Roi des Belges au Congo et dans le Ruanda-Urundi" was first screened in October 1955. It was directed by André Cauvin for Century Pictures, Belgium. It can be viewed on several YouTube pages.

35. *Les Seigneurs de la Forêt* was directed by Henry Barndt and Heinz Sielmann (credited to King Leopold III, 1958), screened as *Masters of the Congo Jungle* in the United States (1960).

36. Alan Root, *Heart of Brightness*, Anglia Television (1991); Alan Root, *Ivory, Apes, and Peacocks: Animals, Adventure, and Discovery in the Wild Places of Africa* (Chatto and Windus, 2012), 266.

37. "A New Mammal," *The Times*, May 7, 1901, 13.

38. "*Ex Africa aliquid semper novi*," 9.

39. "*Ex Africa aliquid semper novi*"; "The Okapi," *The Times*, October 19, 1901, 12.

40. *The Times*, October 29, 1938, 17; "Johnny the Okapi," *Star*, September 10, 1943; "Fighting Sparrows Scare the Okapi," *Evening Standard*, May 11, 1945; Alan Coren, "Pariah—But Don't Rub It In," *The Times*, July 11, 1989, 14.

41. Philip Howard, "Something to Bear with Units of Expression," *The Times*, February 1, 1980, 14; "Edmund Akenhead," *The Times*, December 24, 1990, 10.

42. Edward Lucie-Smith, "The Sociology of Art," *The Times*, November 1, 1966, 5; Matthew Parris, " . . And, Moreover," *The Times*, June 23, 1990, 12; Matthew Parris, "Political Sketch," *The Times*, July 17, 1997, 2; Lode Willems, "No 'Dead-End' for Belgian Okapi," *The Times*, August 2, 1997, 19.

43. Andy Warhol and Kurt Benirschke, *Vanishing Animals* (Springer, 1986), okapi featured on 40–45, quote on 1.

44. Warhol and Benirschke, *Vanishing Animals*, 40, 43.

45. Warhol and Benirschke, *Vanishing Animals*, 43, 44.

Chapter 6. Okapis in African Art, Ancient and Modern

1. The chapter's epigraph comes from Sandra Swart, "O is for Okapi," in *Animalia: An Anti-Imperial Bestiary for Our Times*, ed. Antoinette Burton and Renisa Mawani (Duke UP, 2020), 133.

2. Alfred Wiedemann, "Das Okapi im alten Aegypten," *Die Umschau*, no. 51 (1902), 1002.

3. Wiedemann, "Das Okapi im alten Aegypten," 1002.

4. M.-G. Gaillard, "L'okapi et Set-Typhon," *Bulletin de la Société d'anthropologie de Lyon* 22 (1903), 15, 18.

5. Gaillard, "L'okapi et Set-Typhon," 24; Eugène Lefébure, "L'Animal typhonien," *Sphinx: Revue critique d'égytpologie* 2 (1898), 63–74.

6. Gaillard, "L'okapi et Set-Typhon," 18.

7. Julien Fraipont, *Okapia*, in *Contributions a la Faune du Congo*, book 1, *Annales du Musée du Congo* (Independent State of the Congo, 1907), 13; Paul Hippolyte Boussac, "Set-Typhon et l'okapi," *Le Naturaliste* 29 (1907), 41–43, 54–57, 67–69. See also Paul Hippolyte Boussac and Edouard Trouessart, "L'animal sacré de Set-Typhon et ses divers modes d'interprétation," *Revue de l'histoire des religions* 82 (1920): 189–209.

8. James Henry Brested, *A History of the Ancient Egyptians* (Charles Scribner's Sons, 1908), 30; Ludwig Keimer, "Die fälschlich als Okapi gedeuteten altägyptischen Darstellungen des Gottes Seth," *Acta Tropica* 7 (1950): 112.

9. Agatha Gijzen, *Das Okapi* (A. Ziemsen, 1959), 9; Susan Lindsey, Mary Green, and Cynthia Bennett, *The Okapi: Mysterious Animal of Congo-Zaire* (U of Texas Press, 1999), 44. Giraffes are thought to have evolved long necks twelve to fourteen million years ago, but the oldest rock art in southern African has been dated at 5,743 BP (Brian Switek, "Why Do Giraffes Have Long Necks?," *Wired Science*, June 21, 2017, https://www.wired.com/story /why-do-giraffes-have-long-necks); Adelphine Bonneau, David Pearce, Peter Mitchell, et al., "The Earliest Directly Dated Rock Paintings from Southern Africa," *Antiquity* 91, no. 356 (2017): 322–33.

10. Lloyd Llewellyn-Jones, "Keeping and Displaying Royal Tribute Animals in Ancient Persia and the Near East," in *Interactions Between Animals and Humans in Graeco-Roman Antiquity*, ed. Thorsten Fogen and Edmund V. Thomas (De Gruyter, 2017), 312, 319–20; Raul Valdez and Robert G. Tuck Jr. "Persepolis: Nilgai, Not Okapi," *Cryptozoology* 8 (1989), 146–49.

11. S. M. Perlmann to Walter Rothschild, June 30, 1905, TR/1/1/26/425-431, Natural History Museum Library and Archives; S. M. Perlmann, "Is the Okapi Identical with the 'Thahash' of the Jews?," *Zoologist* 12 (1908): 256–60.

12. Sandra Swart, *The Lion's Historian* (Jacana Media, 2023), 20; Bernard Heuvelmans, "What is Cryptozoology?," *Cryptozoology* 1, no. 1 (1982): 11, 12.

13. Bernard Heuvelmans, "The Birth and Early History of Cryptozoology," *Cryptozoology* 3 (1984), 14–15.

14. Enid Schildkrout and Curtis A. Keim, *African Reflections: Art from Northeastern Zaire* (American Museum of Natural History, 1990), 15–27.

15. On Schweinfurth and Mangbetu representational art and confusing fashions with unchanging traditions, see Schildkrout and Keim, *African Reflections*, 17, 19, 25.

16. Schildkrout and Keim, *African Reflections*, 50, 63; Herbert Lang, Anthropology Notebook 1, 1–2, fieldnotes, Department of Anthropology, American Museum of Natural History Archives.

17. Schildkrout and Keim, *African Reflections*, 155, 157, 179, 188.

18. The ivory horn with okapi skin can be found in the American Museum of Natural History's digital collection, asset ID 111908. On the artist Saza, see Schildkrout and Keim, *African Reflections*, 67.

19. Schildkrout and Keim, *African Reflections*, 153; and on African men wearing okapi skin belts or bandoliers, see James J. Harrison, "The Harrison Diaries," n.d., Scarborough Museums and Galleries (.pdf courtesy of Jim Middleton, collections manager; received November 15, 2022); Boyd Alexander, *From the Niger to the Nile*, 2 vols. (Edward Arnold, 1907); and Percy H. G. Powell-Cotton, "Notes on a Journey through the Great Ituri Forest," *Journal of the African Society* 7, no. 25 (1907): 1–12.

20. Adam Kuper, *The Museum of Other People* (Profile Books, 2023), on Picasso and his peers and African art and primitivism, 93–94, 254–55; on Goldwater, 265.

21. Kuper, *Museum of Other People*, 264–68.

22. Anne Eisner Putnam, with Allan Keller, *Eight Years with Congo Pigmies* (Hutchinson, 1955), 19; Christie McDonald, "Anne Eisner's Art and Ethnology, 1946–58," in *Images of Congo: Anne Eisner's Art and Ethnography, 1946–1958*, ed. Christie McDonald (5 Continents Editions, 2005), 19–20.

23. Enid Schildkrout, "Modernism and Ethnology in the Ituri: Anne Eisner, Colin Turnbull, and the Mbuti," in *Images of Congo*, 54–58; Eisner Putnam, with Keller, *Eight Years*, 104–6.

24. Schildkrout, "Modernism and Ethnology," 55–59.

25. Schildkrout, "Modernism and Ethnology," 57–62.

26. Turnbull's photo of a man making bark cloth appears in "The Mbuti Pygmies: An Ethnographic Survey," *Anthropological Papers of the American Museum of Natural History* 50, no. 3 (1965), plate 32.

27. Suzanne Preston Blier, "Mapping the Ituri: Mbuti Bark-Cloth Paintings and the Canvases of Anne Eisner," in *Images of Congo*, 107–12, 116–17.

28. Colin Turnbull, *The Forest People* (Jonathan Cape, 1961); Schildkrout, "Modernism and Ethnology," 67–68.

29. *Bwana Kitoko* (1955) was directed by André Cauvin for Century Pictures, Belgium. *Les Seigneurs de la Forêt* (1958) was directed by Henry Barndt and Heinz Sielmann (credited to King Leopold III). See https://commons.wikimedia.org/wiki/Category:1955_stamps_of _Belgian_Congo for the stamp.

30. The Institut Congolais pour la Conservation de la Nature logo appears on its official site: https://medd.gouv.cd/iccn/ (accessed on May 9, 2024).

31. Radio Okapi's homepage can be found at https://radiookapi.net (accessed on May 9, 2024).

Chapter 7. Catching Okapi, 1901–15

1. The epigraphs for this chapter come from "Okapi Seen Alive: White Man's Unique Experience," *Daily Mail*, May 22, 1906, DF Pub 517/1/2, 104, Natural History Museum Library and Archives; Philip L. Sclater, "The Okapi," *The Times*, May 25, 1906, 10; and Percy H. G. Powell-Cotton, "Okapi," *The Times*, September 27, 1906, 6.

2. Untitled, *Standard*, August 27, 1901; Harry Johnston, *The Story of My Life* (The Bobbs-Merrill Company, 1926).

3. Council minutes, March 5, 1902, GB 0814 FAA, 401, Prince Philip Zoological Library and Archives. The Okapi Committee's minutes are among items that have been lost or destroyed, and so we only have references to them.

4. Council minutes, April 16, 1902 and May 21, 1902, GB 0814 FAA, 419, 425, 430, Prince Philip Zoological Library and Archives; "Mr J. S. Budgett on His Journey to Uganda," *Proceedings of the Zoological Society of London*, no. 1 (1903): 2–10, 3.

5. "Mr J. S. Budgett on His Journey to Uganda," 3.

6. On the Zoological Society of London, Budgett, and his exchange with Sclater, see Peter. A. Jewell, "The Contribution of the Zoological Society of London to Field Studies and Prospects for the Future," in *The Zoological Society of London 1826–1976 and Beyond*, ed. Solomon Zuckerman (Academic Press, 1976), 269–81, 270. Jewell (1976, 270) was of the opinion that "Budgett became so engrossed in his pursuit of *Polypterus* . . . that he redirected his route . . . and did not enter the Congo."

7. Auguste Lameere, "L'okapi," *Revue de l'Université de Bruxelles* 2 (1902): 132; Julien Fraipont, *Okapia*, in *Contributions a la Faune du Congo*, book 1, *Annales du Musée du Congo* (Independent State of the Congo, 1907), 15–16. The expeditions, and Grauer's, are discussed in chapter 5. On Captain Emilio Piola, see Spartaco Gippoliti, "Okapi, Italy and the Heart of Darkness: The Politics behind Early Specimens of *Okapia johnstoni* in Italy," *Quaderni del Museo Civico di Storia Naturale di Derrara* 10 (2022), 65–66.

8. James J. Harrison, "The Harrison Diaries," n.d., Scarborough Museums and Galleries (pdf courtesy of Jim Middleton, collections manager, received November 15, 2022), 132, 136, 147.

9. Harrison, "Diaries," 153–55, 158.

10. Reported in *The Times*, December 30, 1904, 4; James Harrison, *Life among the Pygmies of the Ituri Forest, Congo Free State* (Hutchinson, 1905).

11. Harrison, "Diaries," 174; and *Life among the Pygmies,* 15–16, 18.

12. Harrison, "Diaries," 197; Gippoliti, "Okapi, Italy and the Heart of Darkness."

13. Harrison, *Life among the Pygmies*, 18, photograph of skins on 21. These skins are missing from the Harrison Collection deposited in Scarborough Museums and Galleries (pers. comm., Jim Middleton, November 16, 2022). On the collection, see Gifty Burrows, "History of the Harrison Collection," December 2021, https://www.fromlocaltoglobal.co.uk/history-of-the-harrison-collection; Jeffrey Green, "Colonel James Harrison," December 2021, https://www.fromlocaltoglobal.co.uk/colonel-james-harrison; and Jim Middleton, "The Harrison Collection: Addressing Colonialism in the Collections of a Victorian Big Game Hunter," *Journal of Natural Science Collections* 9 (2021): 29–34.

14. Harrison, "Diaries," 183.

15. Harrison, "Diaries," 197.

16. Harrison, "Diaries," 228, 263; Green, "Colonel James Harrison."

17. W. R. Ogilvie-Grant, "Boyd Alexander and his Ornithological Work," in *The Ibis: A Quarterly Journal of Ornithology*, P. L. Sclater and A. H. Evans (eds.), IV (R. H. Porter, 1910), 716–29. Boyd Alexander, *From the Niger to the Nile*, vols. 1 and 2 (Edward Arnold, 1907). The journey to Chad is told in volume 1, with okapi material and Gosling's death in vol. 2, chapters 29–31.

18. Alexander, "From the Niger, by Lake Chad, to the Nile," *The Geographical Journal* 30, no. 2 (1907), 142; and *Niger to the Nile*, 2, 258.

19. Alexander, *Niger to the Nile*, 2, 259–60; Alexander, "From the Niger, by Lake Chad," 143. See also "Alone across Africa: English Officer's Exploit," *Daily Telegraph*, February 1, 1907, on the Gosling-Alexander expedition.

20. Reported in Alexander, *Niger to the Nile*, 2, 265–66.

21. Alexander, *Niger to the Nile*, 2, 263–65.

22. Alexander, *Niger to the Nile*, 2, 263.

23. Alexander, *Niger to the Nile*, 2, 268.

24. Alexander, *Niger to the Nile*, 2, 287–89, 292; on the final section of his journey, Alexander, "From the Niger, by Lake Chad," 149. Newspaper stories include: "Alone Across Africa"; "The Alexander-Gosling Expedition," *The Times*, February 6, 1907, 9.

25. "The Capture of a Live Okapi," *Scientific American* 95, no. 1 (1906): 4; "Okapi Seen Alive: White Man's Unique Experience," *Daily Mail*, May 22, 1906.

26. "The Alexander-Gosling Expedition: A Living Okapi Seen," *The Times*, May 21, 1906, 8.

27. Sclater, "The Okapi," 10; pers. comm. David Marques and Loïc Costeur, November 16, 2022. Marques and Costeur provided me with photos of museum catalogues and labeled okapi remains.

28. Harry Johnston, review of *From the Niger to the Nile*, by Boyd Alexander, *Geographical Journal* 31, no. 2 (1908): 205–7, 205; E. Ray Lankester, "Sir Harry Johnston's New Beast," *The Times*, June 18, 1901.

29. E. Ray Lankester, "The First Photograph of a Living Okapi," *Illustrated London News*, September 7, 1907; James Gibbon, "The Okapi," *The Times*, July 25, 1935, 10; "Alone Across Africa."

30. Alexander, *Niger to the Nile*, 6.

31. Alexander, *Niger to the Nile*, 5–6, photo on 24. For examples of articles omitting mention of Lopes see "Okapi Seen Alive," and "The Okapi," *Westminster Gazette*, February 13, 1907. The wooden plaque on the NHM's okapi noted here reads: "Obtained by the Alexander-Gosling Expedition 1903–7 & Presented by Boyd Alexander, Esq., 1907." Observed by the author, NHM storage facility, London.

32. Simon Schaffer, Lissa Roberts, Kapil Raj, et al., eds., *The Brokered World: Go-Betweens and Global Intelligence, 1770–1820* (Watson Publishing International, 2009), ix–xxxviii, and Kapil Raj, "Mapping Knowledge: Go-Betweens in Calcutta, 1770–1820," in *The Brokered World*, 105–50.

33. On Percy's life, see Anon, "Our Portrait Gallery," *United Services Gazette*, July 25, 1901; Powell-Cotton Museum Collection, Congo Trip Book I, News Cuttings, Expedition to Congo Free State 1904–1907.

34. On Powell-Cotton's journey to the Congo: Percy Powell-Cotton, "In Search of the Okapi," unpublished manuscript, Powell-Cotton Museum Collection, in a folder labelled only "Okapi."

35. Percy H. G. Powell-Cotton, "A Journey through the Eastern Portion of the Congo State," *The Geographical Journal* 30, no. 4 (1907), 371–82; "Notes on a Journey through the Great Ituri Forest," *Journal of the African Society* 7, no. 25 (1907), 1–12; and "Africa Society Lecture, Congo," unpublished notes for a slideshow (n.d.), Powell-Cotton Museum Collection.

36. Hannah Powell-Cotton, "Honeymooning at Christmas in Central Africa," Congo trip book 2, news cuttings, expedition to Congo Free State 1904–7, 38–39, Powell-Cotton Museum Collection. While the cutting's source is not documented, a note indicates the piece was published in *Keble's Gazette*.

37. Percy H. G. Powell-Cotton, diary no. 38, book no. 4, January 1–September 18, 1906, Congo, Powell-Cotton Museum Collection. For Hannah's telling of the story see Hannah Powell-Cotton, "Honeymooning." My thanks to Rachel Jennings at the Powell-Cotton Museum for deciphering Powell-Cotton's diary abbreviations.

38. Powell-Cotton, "Journey through the Great Ituri Forest," 7.

39. Powell-Cotton, diaries nos. 18–27, October 1899–May 1902, 14, Powell-Cotton Collection.

40. "The Powell-Cotton African Expedition," *The Times*, February 19, 1907, 5.

41. Percy Powell-Cotton, "Okapi," Letter to *The Times*, September 27, 1906, 6.

42. Powell-Cotton, "Okapi."

43. Powell-Cotton, "Okapi."

44. Berlin's Museum für Naturkunde has excellent resources on the "Colonial Contexts" available at https://msueumfurnaturkunde.berlin/en/museum/today/museum/colonial-contexts (accessed June 10, 2024).

45. Adolf Friedrich von Mecklenburg-Schwerin, *In the Heart of Africa* (Cassell and Company, 1910); "Zoologische Schausammlung," *Führer durch die Zoologische Schausammlung des Museums für Naturkunde in Berlin*, 5th ed. (1919).

46. Adolf Friedrich von Mecklenburg-Schwerin, *From the Congo to the Niger and Nile*, 2 volumes (The John Winston Co, 1914), volume 1, 15.

47. Hermann Schubotz, "On the Shari and Ubangi Rivers," in Adolf Friedrich von Mecklenburg-Schwerin, *From the Congo to the Niger and the Nile*, vol. 2; Hermann Schubotz, "Wissenschaftlichen Innerafrika: Expedition S. H. des Herzogs Adolf Friedrich zu Mecklenburg 1910/1911," pt. 2 of "Zoologische Beobachtungen," *Bericht über die Senckenbergische naturforschende Gesellschaft* (1912), 324–58. My thanks to Bernhard Gißibl and Diana Natermann for information on Schubotz's career.

48. Schubotz, "Shari and Ubangi Rivers," 6.

49. Schubotz, "Shari and Ubangi Rivers," 20–21, 23; W. R. Ogilvie-Grant, "Boyd Alexander and his Ornithological Work," in *The Ibis: A Quarterly Journal of Ornithology*, P. L. Sclater and A. H. Evans (eds), IV (R. H. Porter, 1910), 716–29.

50. Hermann Schubotz, "The Home of the Okapi," in Adolf Friedrich von Mecklenburg-Schwerin, *From the Congo to the Niger and the Nile*, 2, 25.

51. Schubotz, "Home of the Okapi," 25, 26.

52. Schubotz, "Home of the Okapi," 26.

53. Schubotz, "Home of the Okapi," 27.

54. Schubotz, "Home of the Okapi," 32.

55. Schubotz tells the story slightly differently, in two publications, "Auf den Fährten des Okapi," *Die Woche* 14, no. 1 (January 6, 1912), 6–8, photos on 12–13; and "Home of the Okapi," 32.

56. Schubotz, "Auf den Fährten des Okapi," 7; "Home of the Okapi," 32, 35–36.

57. Schubotz, "Auf den Fährten des Okapi," 12, 13.

58. Schubotz, "Home of the Okapi," 36, 38, plates 6 and 7 after page 14, the painting is after 16, and Senckenburg mount after 22.

59. "Obituary: Dr. Christy, Naturalist and Explorer," *The Times*, June 7, 1932, 16; Cuthbert Christy, *Big Game and Pygmies: Experiences of a Naturalist in Central African Forests in Quest of the Okapi* (Macmillan, 1924).

60. Christy, *Big Game and Pygmies*, xi, 54.

61. Christy, *Big Game and Pygmies*, 36.

62. Christy, *Big Game and Pygmies*, 37.

63. Christy, *Big Game and Pygmies*, 39.

64. Christy, *Big Game and Pygmies*, 54, 61, 64, photos after 54, 60, 66.

65. Christy, *Big Game and Pygmies*, 55.

66. Christy, *Big Game and Pygmies*, 62–63.

67. Christy, *Big Game and Pygmies*, 35.

68. Herbert Lang, "In Quest of the Rare Okapi," *Zoological Society Bulletin* 21, no. 3 (1918): 1604; on the expedition, see Henry Fairfield Osborn, introduction to "The Congo Expedition of the American Museum of Natural History," *Bulletin of the American Museum of Natural History* 34 (1919): xv–xxvii.

69. Lang, "In Quest," 1610.

70. Lang, "In Quest," 1605, 1606.

71. Lang, "In Quest," 1613, photographs of the calf on 1606, 1607, 1608.

72. Lang, "In Quest," 1613; Herbert Lang, "A Belgian Woman First to Achieve Success," *Zoological Society Bulletin* 22, no. 4 (1919), 71–73, photo on 73.

73. On the final stages of the expedition: Osborn, introduction to "The Congo Expedition of the American Museum of Natural History," xxi–xxii. I address Lang's ethnographic acquisitions in chapter 13.

74. This phonograph is on display in Quex House, with an information board including details of its use in the Ituri Forest.

Chapter 8. Pursuing Okapi in the Interwar Years

1. The epigraph to this chapter comes from Ellen Gatti, *Exploring We Would Go* (Robert Hale, 1950), 123.

2. Delia J. Akeley, "Among the Pigmies in the Congo Forest," *Brooklyn Museum Quarterly* 13, no. 1 (1926): 2.

3. Delia J. Akeley, *Jungle Portraits* (Macmillan, 1930), 1.

4. Akeley, *Jungle Portraits*, 160.

5. Akeley, "Among the Pigmies," 4.

6. Akeley, *Jungle Portraits*, 219.

7. Akeley, *Jungle Portraits*, 221–22, photo opposite 220.

8. Spartaco Gippoliti, "Okapi, Italy and the Heart of Darkness. The Politics behind Early Specimens of *Okapia johnstoni* in Italy," *Quaderni del Museo Civico di Storia Naturale di Derrara* 10 (2022): 65–66.

9. Information on Atillio Gatti pieced together from his books *Great Mother Forest* (Hodder and Stoughton, 1936) and *South of the Sahara* (Robert M. McBride, 1945), and the film *The Famous 1939 "Attilio Gatti" Expedition of Belgian Congo*, International Harvester Company (1939). On his rendering of the name, *Great Mother Forest*, 159–60.

10. On Hornaday's offer: William Bridges, *A Gathering of Animals: An Unconventional History of the New York Zoological Society* (Harper & Row, 1974), 354.

11. Gatti, *Exploring*, 1, 2.

12. Gatti, *Great Mother Forest*, chapters 14, 16, 18.

13. Gatti, *Great Mother Forest*, 219–27, quote on the dual nature of the okapi on 223. Ellen Gatti remembered the attempted capture of Beautiful differently from her husband: in her version it was Kaluèse with the other two men who tracked down the escaped okapi and her husband who knew it would go to the river to wash off the mud (Attilio doesn't mention Kaluèse in his account). See *Exploring*, 170–72; her note on the gas mark on 164.

14. Gatti, *Great Mother*, quotes on 223, 226–27.

15. For the story of the capture of Toto, and his short life in captivity, see Gatti, *Great Mother Forest*, 265–74 and 275–84.

16. A photo of Gatti bottle-feeding Toto appeared in "In the Belgian Congo: Pygmies and Okapi," *The Times*, February 8, 1936, 16.

17. Gatti, *Great Mother Forest*, 284. For another justification for capturing and keeping wildlife in captivity which takes a welfare perspective, see Gatti's statement on the bongo he sent to Bioparco di Roma, which he claimed would soon lose its shyness and forget its

nightmares of being hunted by Africans or leopards in the rainforest, not to mention terrible storms, falling trees, and venomous snakes (*Great Mother Forest*, 328).

18. Gatti, *Great Mother Forest*, 285. On okapi milk see John Hart, "*Okapia johnstoni* Okapi," in *The Mammals of Africa. Volume VI: Pigs, Hippopotamuses, Chevrotain, Giraffes, Deer, and Bovids*, ed. Jonathan S. Kingdon and Mike Hoffmann (Bloomsbury Publishing, 2013), 110–15.

19. This episode is described in Gatti, *Great Mother Forest*, 287–90, quote on 290.

20. Gatti, *Great Mother Forest*, 289–90.

21. Gatti, *Great Mother Forest*, 290; Anne Eisner Putnam, with Allan Keller, *Eight Years with Congo Pigmies* (Hutchinson, 1955), 84–85.

22. Gatti, *Great Mother Forest*, footnote, 292–93.

23. Gatti, *Great Mother Forest*, 292.

24. Gatti, *Great Mother Forest*, 293.

25. Gatti, *Great Mother Forest*, 294–95. On okapi being strong swimmers: Interview with Stuart Nixon of WCS, on Teams, June 14, 2023.

26. Attilio Gatti, "Okapi and Bongo," *The Times*, March 24, 1936, 12; Guy Dollman, "The Okapi," *The Times*, April 3, 1936, 12. Julien Fraipont, *Okapia*, in *Contributions a la Faune du Congo*, book 1, *Annales du Musée du Congo* (Independent State of the Congo, 1907), plate XVIII, unnumbered page; E. Ray Lankester with W. G. Ridewood, *Monograph of the Okapi* (British Museum [Natural History], 1910), vi.

27. Gatti, *South of the Sahara*, 96 for the photograph.

28. Gatti, *South of the Sahara*, the okapi material is in chapters 10, 13–17. Nell Boeykens, "De Okapi als inzet van en voor de Koloniale Machtsstructuur" (MA thesis, University of Ghent, 2020), 283. Swaluë may have left unpublished notes on okapi, but I have not found them. They would be worth tracking down; if they exist, they are most likely at the AfricaMuseum in Tervuren.

29. Herbert Lang, "In Quest of the Rare Okapi," *Zoological Society Bulletin* 21, no. 3 (1918), 1604.

30. David van Reybrouck, *Congo: The Epic History of a People* (Fourth Estate, 2014), see his introduction for a discussion on oral history and the problem of sources. On Mbuti oral history, see Richard Peterson, *Conversations in the Rainforest: Cultures, Values, and the Environment in Central Africa*, rev. ed. (Westview Press, 2017), Appendix B.

Chapter 9. Capture, Transport, and Survival of Okapi After 1918

1. The epigraph to this chapter comes from "Bank Holiday," *The Times*, August 6, 1935, 13. For other stories on Congo at the zoo, see "An Okapi for London: Royal Gift to the Zoo," *The Times*, July 20, 1935, 13, and "The Okapi Settled at the Zoo," *The Times*, August 3, 1935, 12.

2. On Belgium and the Belgian royal family: Paul Belien, *A Throne in Brussels: Britain, the Saxe-Coburgs and the Belgianisation of Europe* (Imprint Academic, 2006); Herman van Goethem, *Belgium and the Monarchy: From National Independence to National Disintegration* (ASP Editions, 2011).

3. "Zoo Okapi Dead," *The Times*, November 5, 1935, 13.

4. On the early American zoo and circus okapi, see Richard J. Reynolds III, "The Only Circus Okapis," *Bandwagon* 31, no. 2 (1987): 18–21. It is one of the best-researched okapi publications I read.

5. "Historical Studbook Listing," http://www.theokapi.org/Studbook/Hist_studbook. aspx. The birth dates are less reliable, while the estimated arrival dates in zoos are accurate but less useful than the death dates.

6. On the price of okapi, see Roland Baetens, *The Chant of Paradise: The Antwerp Zoo* (Royal Zoological Society of Antwerp, 1993), 181.

7. "Census at the Zoo: How the Animals Are Valued," *The Times*, December 30, 1939, 4; "£1000 Okapi Will Get Bamboo Leaves," *Evening Standard*, October 1, 1943; "Fighting Sparrows Scare the Okapi," *Evening Standard*, May 11, 1945; "Stocktaking at the Zoo," *The Times*, January 24, 1949, 2.

8. "Stocktaking at the Zoo," 2; "The Zoo: Specimens Valued at £81,000," *The Times*, January 21, 1952, 3.

9. On zoos' functions and proclaimed aims: Violette Pouillard, "Le jardin zoologique et le rapport à la faune sauvage: Gestion des 'collections zoologiques' au zoo d'Anvers (1843–vers 2000)," *Revue Belge de Philologie et d'Histoire / Belgisch Tijdschrift voor Filologie en Geschiedenis* 89, nos. 3–4 (2011): 1193–232; Nell Boeykens, "De Okapi als inzet van en voor de Koloniale Machtsstructuur" (MA thesis, University of Ghent, 2020).

10. Harry Johnston, *The Uganda Protectorate*, vol. 1 (Hutchinson, 1902), 383.

11. Auguste Lameere, "L'okapi," *Revue de l'Université de Bruxelles* 2 (1902): 126.

12. On Belgian colonial permit fees, see Boeykens, "De Okapi," 100.

13. Pouillard, "Le jardin zoologique," 1197; Boeykens, "De Okapi," 100. The restrictions on okapi hunting were known in England and were published in *Correspondence Relating to the Preservation of Wild Animals in Africa, Presented to Both Houses of Parliament by Command of His Majesty* (Wyman and Sons, 1906). On Cordier, see René E. Honegger, "Erinnerungen an den bedeutenden Tierfänger Charles Cordier (1897–1994) von Zürich," *Zürcher Taschenbuch* 137 (2017): 225–81.

14. Ellen Gatti, *Exploring We Would Go* (Robert Hale, 1950), 25.

15. Attilio Gatti, "Pygmy and Okapi," *The Times*, February 8, 1936, 13–14.

16. See Gatti, "Pygmy and Okapi," and Lord Onslow, "Protection for the Okapi," *The Times*, April 21, 1936, 10.

17. Violette Pouillard, "Conservation et captures animales au Congo Belge (1908–1960)," *Revue Historique*, no. 679 (2016): 590.

18. The original deal offered to Gatti was £1,000 for two okapi on certification of safe departure and safe arrival: £250 immediately, £250 at shipping at Mombasa, £250 on safe arrival in London, and £250 on survival to one month after arrival. Summarized in Chalmers Mitchell's note "Okapis: For Consideration at the Garden Committee and Council on September 18 & 19" in which he shares Gatti's letter of August 30, 1934, SEC/9/2/27/4, Prince Philip Zoological Library and Archives.

19. Whipsnade is the Zoological Society of London's zoo in the countryside northwest of London that has extensive grounds and space for animals.

20. Boeykens, "De Okapi," 87–89; "Protection of the Okapi," *The Times*, March 9, 1936, 13.

21. Boeykens, "De Okapi," 87–88.

22. R. T. Leiper, "The Okapi: Risks from Internal Parasites," Letter to *The Times*, February 13, 1936, 8; on Leiper's career: "Obituary: Prof. R. T. Leiper," *The Times*, May 23, 1969, 12.

23. Peter J. Stephenson and John E. Newby, "Conservation of the Okapi Wildlife Reserve, Zaire," *Oryx* 31, no. 1 (1997): 49–58.

24. "Telegrams in Brief," *The Times*, February 6, 1936, 13.

25. For more on the okapi Buta and Mrs. van Landeghem, see Baetens, *Chant of Paradise*, 182.

26. Smets's observations are communicated by Herbert Lang, "A Belgian Woman First to Achieve Success," *Zoological Society Bulletin* 22, no. 4 (1919): 71.

27. "A Live Okapi in Europe," *The Times*, August 15, 1919, 9.

28. Baetens, *Chant of Paradise*, 182; P. L. G. Benoit, "Hutsebaut, Joseph," *Biographie Belge d'Outre-Mer*, 6 (1968): 518.

29. Armand Denis, *On Safari: The Story of My Life* (Collins, 1963), 226.

30. "A Living Okapi in Europe: The Second Attempt to Acclimatise this Rarest of African Animals," *The Illustrated London News*, March 2, 1929, 349.

31. For the Hutsebaut quote: Denis, *On Safari*, 226. On the Zoological Society of London medal, see "Report of the Council for the Year Ended 31st December, 1935," 12, Reports of the Zoological Society, 1934–37, Prince Philip Zoological Library and Archives.

32. Pouillard, "Conservation et captures," 585–45; Denis, *On Safari*, 96.

33. Denis, *On Safari*, 225–26.

34. Anne Eisner Putnam, with Allan Keller, *Eight Years with Congo Pigmies* (Hutchinson, 1955), 85.

35. Boeykens, "De Okapi," 109.

36. Boeykens, "De Okapi," 117. On Gatti's claim, and rationale for the portable cage, see *South of the Sahara* (Robert M. McBride, 1946), 164.

37. Boeykens, "De Okapi," 94.

Chapter 10. Zoo Conservation and the Deadly Journey to the West

1. The epigraph to this chapter comes from "Keeping Animals in Captivity," *The Times*, March 14, 1959, 7.

2. Violette Pouillard, "Le jardin zoologique et le rapport à la faune sauvage: Gestion des 'collections zoologiques' au zoo d'Anvers (1843–vers 2000)," *Revue Belge de Philologie et d'Histoire/Belgisch Tijdschrift voor Filologie en Geschiedenis* 89, nos. 3–4 (2011): 1221.

3. On okapi exports, see "Historical Studbook Listing," http://www.theokapi.org /Studbook/Hist_studbook.aspx.

4. Bernhard Grzimek, *No Room for Wild Animals* (Thames and Hudson, 1956), 204.

5. Grzimek, *No Room*, 205.

6. Grzimek, *No Room*, 4, 21–22.

7. This section is based on Gijzen and Smets, "Seventy Years Okapi, *Okapia Johnstoni*," *Acta Zoologica et Pathologica Antverpiensia* 59 (1974): 7–111 (see in particular 8–14). As it is based on the zoo's bulletins and society documents, this account reflects their points of view. Okapi survival rates and ages from "Historical Studbook Listing."

8. On "systematic obstruction," see Gijzen and Smets, "Seventy Years Okapi," 9. This attempt by the Antwerp Zoo to in effect establish a monopoly on exports from northeastern Belgian Congo, with a zoo/capture station in Stanleyville, is also explored by Pouillard, "Le jardin zoologique," 1207–8.

9. For the correspondence beginning with this letter from Vevers to de Quidt on April 5, 1946, see superintendent's correspondence, Belgian Ministry for Colonies, SUP/5/1/2/25, Prince Philip Library and Archives.

10. Gijzen and Smets, "Seventy Years Okapi," 11, 13. For Basel Zoo's position on how they acquired their okapi, see "Okapi: Die Unbekannte Waldgiraffe," Zoo Basel blog, available at: https://www.zoobasel.ch/de/aktuelles/blog/3/zoo-geschichte/178/okapi-die-unbekannte-waldgiraffe/ (accessed on January 5, 2023). The director of the Basel zoo, Ernst Lang, traveled to the Congo in 1955 to replace their first okapi, Bambe. In sum, the zoo's first male died, and was replaced in 1955. The first female died shortly after arrival in 1956, and was replaced in 1957. The surviving pair successfully bred three times.

11. Gijzen and Smets, "Seventy Years Okapi," 14.

12. Roland Baetens, *The Chant of Paradise: The Antwerp Zoo* (Royal Zoological Society of Antwerp, 1993), 184.

13. Pouillard, "Le jardin zoologique," 1210.

14. Grzimek thought de Medina had been officially catching okapi since 1946, but I am following Pouillard on this, in terms of the official start date.

15. Pouillard, "Le jardin zoologique," 1205. Aspects of this story are told in Sophy Roberts, *A Training School for Elephants* (Doubleday, 2025), though Offerman is surprisingly absent from her account.

16. Armand Denis's account of his visit is in *On Safari: The Story of My Life* (Collins, 1963), 227–32, quotes on 228.

17. Grzimek, *No Room*, 215; Raf de Bont, "Moving Across the Zoo–Field Border: Heini Hediger in Congo," *Isis* 113, no. 3 (2022): 506.

18. I am following Agatha Gijzen, *Das Okapi* (A. Ziemsen, 1959) 66–70 regarding de Medina's genera approach, including the structure of his capture teams. Nell Boeykens, by contrast, claims there were twenty-five natives, each of whom was responsible for ten traps ("De Okapi als inzet van en voor de Koloniale Machtsstructuur" [MA thesis, University of Ghent, 2020]), 110. On Swakuë's innovation, see Boeykens, "De Okapi," 89.

19. Gijzen, *Das Okapi*, 69. On the treatment of wounds, see Bernhard Grzimek, "Giraffes," in *Grzimek's Animal Life Encyclopedia*, vol. 13, Mammals IV (Van Nostrand Reinhold Company), 247–54 on okapi specifically.

20. *Bwana Kitoko*, directed by André Cauvin (1955). On films of the Belgian Congo, and *Bwana Kitoko*, see Matthew G. Standard, *Selling the Congo* (U of Nebraska Press, 2012).

21. Gijzen, *Das Okapi*, 69–73, 80, 103.

22. Gijzen, *Das Okapi*, 66.

23. This account of Wendnagel's journey comes from Gijzen, *Das Okapi*, 84–85.

24. On de Medina's distress: Violette Pouillard, "Conservation et captures animales au Congo belge (1908–1960)," *Revue Historique* 679 (2016): 577–604, 598.

25. The "Historical Studbook Listing" gives Zendy's birth as January 1947, but *The Times* article reporting her death states that "she was known to be at least fourteen years old" ("Tidying up at the Zoo," *The Times*, December 29, 1952, 9).

26. Grzimek, *No Room*, 236–37; Grzimek, "Giraffes," 249; on the New York flight, see René E. Honegger, "Erinnerungen an den bedeutenden Tierfänger Charles Cordier (1897–1994) von Zürich," *Zürcher Taschenbuch* 137 (2017): 238.

27. Gijzen, *Das Okapi*, 86–87.

28. Gijzen, *Das Okapi*, 84, and culled from 88–101; Gijzen and Smet, "Seventy Years Okapi," 55–56; Attilio Gatti, *Great Mother Forest* (Hodder and Stoughton, 1936), 290.

29. Statistics derived from "Historical Studbook Listing."

30. Pouillard, "Conservation et captures," 599.

31. Boeykens, "De Okapi," 114.

32. This section draws on: Gijzen and Smet, "Seventy Years Okapi," 23; Walter Van den Bergh, "Birth of an Okapi in the Antwerp Zoo (Stripfilm)," International Union of Directors of Zoological Gardens, minutes, Basel, September 5–6, 1955, unnumbered page, NZSL/IDZ/3/5, Prince Philip Zoological Library and Archives; Susan Lindsey, Mary Green, and Cynthia Bennett, *The Okapi: Mysterious Animal of Congo-Zaire* (U of Texas Press, 1999), 57–58.

33. On feeding calves: Steve Taylor, "Hand-raising Hoofed Animals: Part 2," *The Keeper: National AAZK Bulletin* 7, nos. 1&2 (1974): 7. On the 1974 zoo okapi census: Anon, "Okapis at Brookfield," *Animal Keepers' Forum* 3, no. 1 (1976): 92. "An Eight-Week-Old Okapi Photographed for the First Time with Its Mother at Bristol Zoo," *The Times*, September 23, 1966, 22.

34. On the political history, Kris Berwouts, *Congo's Violent Peace* (Zed Books, 2017). Grzimek, "Giraffes," 251.

35. This section is based on my interview with Rosmarie Ruf, September 14, 2023, via Zoom.

36. Interview with Rosmarie Ruf.

37. "Keeping Animals in Captivity," *The Times*, March 4, 1959, 7.

38. On the changing image of the hunter, and rise of new role models for European interaction with wildlife in Africa: William Beinart and Dominique Schafer, "Hollywood in Africa 1947–62: Imaginative Construction and Landscape Realism," in *Wild Things: Nature and the Social Imagination*, ed. William Beinart, Karen Middleton, and Simon Pooley (White Horse Press, 2013), 44–66.

39. On animal rights in conservation: A. D. Wallach, C. Batavia, M. Bekoff, et al., "Recognizing Animal Personhood in Compassionate Conservation," *Conservation Biology* 34 (2020): 1097–106.

40. IUCN, *African Wildlife Laws*, IUCN Environmental Law Centre Occasional Paper 3 (1986), 141–47.

41. Lindsey, Green, and Bennett, *The Okapi*, 5; "Okapi's Half-Century," *Zooquaria* 85 (2014): 7.

42. On zoo conservation of okapi, see Lindsey, Green, and Bennett, *The Okapi*, 104–5. On okapi births in US zoos, see Connie Carson, "San Diego Zoo / Wild Animal Park Lists Births and Hatchings," *Animal Keepers' Forum* 9, no. 1 (1982): 98, and "San Diego Wild Animal Park," *Animal Keepers' Forum* 15, no. 1 (1988): 175. On the SSPs: Robert J. Wiese and Michael Hutchins, "Evolution of the AZA Species Survival Plan," *Animal Keepers' Forum* 23, no. 2 (1996): 58–62.

43. Critiques of captive breeding summarized in John F. Oates, *Myth and Reality in the Rain Forest* (U of California Press, 1999), 216–19.

44. On genetic variations: David Stanton, Philippe Helsen, Jill Shephard, et al., "Genetic Structure of Captive and Free-ranging Okapi (*Okapia johnstoni*) with Implications for Management," *Conservation Genetics* 16, no. 5 (2015): 1115–26.

45. Oates, *Myth and Reality*, 203, 222, 225; on the European Association of Zoos and Aquaria's ex situ program, see https://www.eaza.net/eep-pages.

46. Lindsey et al., *The Okapi*; "Okapi's Half-Century," *Zooquaria* 85 (Spring 2014): 7; Marlowe Starling, "We Know How Many Okapi Live in Zoos. In the Wild? It's Complicated," *Mongabay*, September 18, 2024, https://news.mongabay.com/2024/09/we-know-how-many-okapi-live-in-zoos-in-the-wild-its-complicated/.

Chapter 11. Nature of the Beast

1. Attilio Gatti, *Great Mother Forest* (Hodder and Stoughton Limited, 1936), 223.

2. James J. Harrison, "The Harrison Diaries," n.d., Scarborough Museums and Galleries (pdf courtesy of Jim Middleton, collections manager, received November 15, 2022), 183; "The Alexander-Gosling Expedition: A Living Okapi Seen," *The Times*, May 21, 1906, 8.

3. Percy Powell-Cotton, "Okapi," Letter to *The Times*, September 27, 1906, 6.

4. James J. Harrison, "The Okapi," *The Times*, October 5, 1906, 13.

5. Julien Fraipont, *Okapia*, in *Contributions a la Faune du Congo*, book 1, *Annales du Musée du Congo* (Independent State of the Congo, 1907), 81, 84.

6. Fraipont, *Okapia*, 81–84. For orientation see figures 3.1 and 7.1.

7. Hermann Schubotz, "On the Shari and Ubangi Rivers," in Adolf Friedrich von Mecklenburg-Schwerin, *From the Congo to the Niger and the Nile*, vol. 2 (The John Winton Co., 1914), 3–12; Herbert Lang, "In Quest of the Rare Okapi," *Zoological Society Bulletin* 21, no. 3 (1918): 1606, regarding the gait of okapi.

8. Lang, "In Quest," 1605, 1610.

9. Lang, "In Quest," 1601–14, quote on 1610, map on 1602.

10. Lang, "In Quest," 1607, 1608, 1610, 1613. On Mrs. Landeghem's okapi, see Herbert Lang, "A Belgian Woman First to Achieve Success," *Zoological Society Bulletin* 22, no. 4 (1919): 71–73, photo on 73.

11. Henri Schouteden, "Un Okapi Vivant en Belgique," *Revue Zoologique Africaine* 7 (1919): 199–201, 200.

12. Edwin H. Colbert, "The Relationships of the Okapi," *Journal of Mammology* 19, no. 1 (1938): 47–64, 62, 63.

13. On the sending of okapi viscera to Europe, see Henri Schouteden, "Notes sur l'Okapi," *Revue Zoologique Africaine* 2 (September 1912–May 1913): 482; on the British viscera, see R. H. Burne, "Notes on Some of the Viscera of an Okapi (*Okapia johnstoni* Sclater)," *Proceedings of the Zoological Society of London* 87, no. 1 (1917): 187–208. On Burne, see Arthur Keith, "Richard Higgins Burne, 1868–1953," Royal Society Obituary Notices (1953), 26–33.

14. Papers relating to the dissection are collected in the folder "Okapi dissection 1935" (British Natural History Museum Archive, NHM DF232/7/4/58). The quote is from Martin Hinton, Letter to G. Seccombe-Hett, November 9, 1935, 1 page. Unfortunately in their documentation both the NHM and ZSL just refer to "Gerrard," and it is unclear whether they're referring to one of the Edward Gerrards who led the London-based taxidermy firm Gerrard & Sons, or to the company.

15. On handling the corpse: Hinton, Letter to Brinton, January 17, 1936, 1 page; on the liver: R. T. Leiper, Letter to Martin Hinton, December 4, 1935, 1 page; reply from Hinton, December 7, 1935, 1 page. From: NHM Archive, Okapi dissection 1935, DF232/7/4/58.

16. R. I. Pocock, "Some Additional External Characters of the Okapi (*Okapia johnstoni*) that Died in the Society's Gardens," *Proceedings of the Zoological Society of London* B113, 1–2 (1943): 31–35, 31.

17. The full list of papers is included in the online notes for this book.

18. Cuthbert Christy, *Big Game and Pygmies: Experiences of a Naturalist in Central African Forests in Quest of the Okapi* (MacMillan and Co., 1924), 55.

19. Christy, *Big Game and Pygmies*, 70.

20. Gatti, *Great Mother Forest*, 210.

21. Gatti, *Great Mother Forest*, 215.

22. Gatti, *Great Mother Forest*, 212.

23. On diet, Gatti, *Great Mother Forest*, 211; on eyes and coat, 214; on vocalisations, 215; Lang, "In Quest."

24. Gatti, *Great Mother Forest*, 268.

25. Gatti, *Great Mother Forest*, 283.

26. On the dual nature of the okapi: Gatti, *Great Mother Forest*, 223, and on Toto, 276; see also Attilio Gatti, *South of the Sahara* (Robert M. McBride & Company), 135.

27. Gatti, *Great Mother Forest*, 289–90.

28. Gatti, *Great Mother Forest*. The "new race" is discussed in chapter 25, and see photographs opposite 228 (skulls) and 229 (heads). On variability in skulls, E. Ray Lankester with W. G. Ridewood, *Monograph of the Okapi* (British Museum [Natural History], 1910), vi.

29. Gatti, *Great Mother Forest*, 298–99.

30. Attilio Gatti, "Okapi and Bongo," *The Times*, March 24, 1936, 12; Guy Dollman, "The Okapi," *The Times*, April 3, 1936, 12.

Chapter 12. Okapi Science, 1946–2015

1. The epigraph to this chapter comes from Agatha Gijzen, *Das Okapi* (A. Ziemsen, 1959).

2. Heini Hediger, *Wild Animals in Captivity* (Butterworths Scientific Publications, 1950).

3. On the limits of scientific knowledge about wild okapi behavior, see Agatha Gijzen, *Das Okapi* (A. Ziemsen Verlag, 1959), 104.

4. Reginald I. Pocock, "The Okapi," *Zoo Life* 1, no. 1 (1946): 3. On Hediger's outlook on keeping wild animals and reproducing their habitats, see Matthew Chrulew, "My Place, My Duty: Zoo Biology as Field Philosophy in the Work of Heini Hediger," *Parallax* 24, no. 4 (2018): 480–500. On Hediger's Congo expedition, see Raf de Bont, "Moving Across the Zoo-Field Border: Heini Hediger in Congo," *Isis* 113, no. 3 (2022): 491–512. On von Uexküll, see Richard W. Burkhardt Jr., *Patterns of Behavior: Konrad Lorenz, Niko Tinbergen, and the Founding of Ethology* (Chicago University Press, 2005).

5. Pocock, "The Okapi," 3–7.

6. Pocock, "The Okapi," 6; Geoffrey Vevers, Letter to Pocock, January 4, 1946, and reply from Pocock on January 9, ZSL Prince Philip Zoology Library & Archives Library, QBAB, SUP/5/1/2/36.

7. "The Gate Was Open So Buta Walked Out," *Evening Standard*, August 2, 1943; "Johnny the Okapi," *The Star*, September 10, 1943; "£1000 Okapi Will Get Bamboo Leaves," *Evening Standard*, October 1, 1943; "Okapi Should Be in Observer Corps," *Evening Standard*, November 25, 1943; "Buta, the Okapi, Has Two 'Danger Months,'" *Evening Standard*, January 3, 1945 (this article also discusses the decision to not evacuate Buta); "Fighting Sparrows Scare the Okapi," *Evening Standard*, May 11, 1945.

8. Gijzen, *Das Okapi*, 104. On her life: A. C. van Bruggen, "In memoriam Dr Agatha Gijzen (1904–1995), eminent museum historian and zoo biologist," *Zoölogische Mededelingen*, 70 (16) 1996, 235–47.

9. Gijzen, *Das Okapi*, 104.

10. Gijzen, *Das Okapi*, 106.

11. On Bambe the okapi: "Okapi: Die Unbekannte Waldgiraffe," Zoo Basel blog, https://www.zoobasel.ch/de/aktuelles/blog/3/zoo-geschichte/178/okapi-die-unbekannte-waldgiraffe/ (accessed on January 5, 2023). According to the Okapi Studbook, Bambe was actually a little older: he was born in January 1945 and died in August 1949. For scientific papers on Bambe, see note 15.

12. J. Nouvel, J. Rinjard, and M. A. Pasquier, "Evolution de la famille d'okapis au Parc Zoologique de Paris," *Mammalia* 34 (1970): 320–23. Papers on okapi from the 1950s include Anonymous, "La reproduction des okapis en captivité," *Zoo Anvers* 19 (1953), 20 and 22 (1956), 57–58; Anonymous, "Naissance et trépas d'un petit okapi," *Zoo Anvers* 22 (1954), 43–44; J. G. Baer, "Etude critique des helminthes parasites de l'okapi," *Acta Tropica* 7, 164–86; R. Brückner, "Das Auge des Okapi," *Acta Tropica* 7 (1950), 123–32; L. M. G. Geurden, "Les facteurs physio-pathologiques influencant l'acclimation de l'okapi," *Bulletins de la Société Royale de Zoologie d'Anvers* 3 (1953), 2–18; "Notice sur la reproduction de l'okapi (*Okapia johnstoni* Sclater) au Jardin Zoologique d'Anvers," *Bulletins de la Société Royale de Zoologie d'Anvers* 8 (1958), 3–62; J. Nouvel, "Remarques sur la function genitale et la naissance d'un okapi," *Mammalia* 22 (1958), 107–11; Heini Hediger, "Das Okapi al sein Problem der Tiergartenbiologie," *Acta Tropica* 7 (1950), 97–109; J. de Landsheere, "Waarnemingen betreffende de vangst, het fokken en de verzorging van de okapi," *Zoo Antwerpen* 23 (1957), 12–25; E. M. Lang, "Klinische Beobachtungen am Okapi 'Bambe,'" *Acta Tropica* 7 (1950), 118–20; A. Portmann and K. Wirz, "Die cerebralen Indices beim Okapi," *Acta Tropica*, 7 (1950), 120–22; S. Scheidegger, "Pathologisch-anatomische untersuchung des Okapi 'Bambe,'" *Acta Tropica* 7 (1950), 133–50.

13. Bernhard Grzimek, "Giraffes," in *Grzimek's Animal Life Encyclopedia*, vol. 13 (Van Nostrand Reinhold, 1956), 247–54 on okapi specifically.

14. Grzimek, "Giraffes," 251, 252.

15. Grzimek, "Giraffes," 254.

16. Terese Hart, "Ituri Story: Before This War," pt. 1, https://www.bonoboincongo.com/2014/06/14/ituri-story-before-this-war-part-1. The impacts of these changes are the subject of Colin M. Turnbull, *The Mbuti Pygmies: Change and Adaptation* (Rinehart and Winston, 1983).

17. Terese Hart, "Ituri Story," pt. 7, https://www.bonoboincongo.com/2014/08/19/ituri-story-dead-and-preserved-part-7.

18. Hart, "Ituri Story," pt. 7.

19. On their findings, see John A. Hart and Terese B. Hart as follows: "A Summary Report on the Behaviour, Ecology and Conservation of the Okapi (*Okapia johnstoni*) in Zaïre," *Acta Zoologica et Pathologica Antverpiensia* 80 (1988), 19–28, republished in *Nature et Faune* 6, no. 3 (1990), 21–28; "Ranging and Feeding Behaviour of Okapi (*Okapia johnstoni*) in the Ituri Forest of Zaïre: Food Limitation in a Rainforest Herbivore?," *Symposia of the Zoological Society of London* 61 (1989), 31–50.

20. Terese Hart, "Lesser Known Facts about Okapis and the Leaves They Eat," https://www.bonoboincongo.com/2009/01/06/lesser-known-facts-about-okapis-and-the-leaves-they-eat; Andy Warhol and Kurt Benirschke, *Vanishing Animals* (Springer, 1986), 44.

21. On the instability in DRC, see Kris Berwouts, *Congo's Violent Peace: Conflict and Struggle Since the Great African War* (Zed Books, 2017); Terese Hart and Robert Mwinyihali, "DR Congo's Ituri Forest Has Survived Seven Years of Civil War," *Wildlife Conservation* 108, no. 4 (2005): 42–45; and Corneille Ewango, "A Hero of the Congo Basin Forest," Ted Talk, July 8, 2008, https://www.youtube.com/watch?v=DRmWo3Rd8xM.

22. G. B. Rabb, "Birth, Early Behavior, and Clinical Data on the Okapi," *Acta Zoologica et Pathologica Antverpiensia* 71 (1978), 233–51; Richard E. Bodmer and George B. Rabb, "*Okapia johnstoni*," *Mammalian Species* 422, no. 10 (1992), 1–8. Biographical information from Joseph C. Mitchell, Joseph R. Mendelson III, and Margaret M. Stewart, "George Bernard Rabb," *Copeia* no. 4 (2015), 1086–92.

23. Rabb quoted on the zoo installation "Habitat Africa" by William Mullen, "African Rain Forest Grows In, of All Places, Brookfield," *Chicago Tribune*, April 23, 2000, https://www.chicagotribune.com/news/.

24. Bodmer and Rabb, "*Okapia Johnstoni*," 5; D. Geraads, "Remarques sur la systématique et la phylogénie des Giraffidae (Artiodactyla, Mammalia)," *Geobios* 19 (1986), 465–77.

25. Bodmer and Rabb, "*Okapia Johnstoni*," 2, 5.

26. Grzimek, "Giraffes," 249; Anne Eisner Putnam, with Allan Keller, *Eight Years with Congo Pigmies* (Hutchinson, 1955), 105.

27. On US zoo research, see Lynette Shirley, "Hillkeeper Peepers or Urine-for-Life (A Fine Art)," *Animal Keepers' Forum* 10, no. 1 (1983), 5354; N. M. Loskutoff, J. E. Ott, and B. L. Lasley, "Urinary Steroid Evaluations to Monitor Ovarian Function in Exotic Ungulates: I. Pregnanediol-3-Glucuronide Immunoreactivity in the Okapi (*Okapi johnstoni*)," *Zoo Biology* 1 (1982), 45–53.

28. "Problem species" quote from an advertisement for minutes for a 1994 workshop: European Endangered Species Program, "Research and Captive Propagation Workshop Held in Germany," *Animal Keepers' Forum* 21, no. 7 (1994), 349.

29. Susan Lyndaker Lindsey, Mary Neel Green, and Cynthia L. Bennett, *The Okapi: Mysterious Animal of Congo-Zaire* (U of Texas Press, 1999), 38.

30. John Hart, "*Okapia johnstoni* Okapi," in *The Mammals of Africa. Volume VI: Pigs, Hippopotamuses, Chevrotain, Giraffes, Deer, and Bovids*, ed. Jonathan S. Kingdon and Mike Hoffmann (Bloomsbury Publishing, 2013), 110–15.

31. Hart, "*Okapia johnstoni* Okapi," 111.

32. On genetic studies of okapi: David Stanton, John Hart, Peter Galbusera, et al., "Distinct and Diverse: Range-Wide Phylogeography Reveals Ancient Lineages and High Genetic Variation in the Endangered Okapi (*Okapia johnstoni*)," *PLoS ONE* (July 2014), https://doi.org/10.1371/journal.pone.0101081; David Stanton, Philippe Helsen, Jill Shephard, et al., "Genetic Structure of Captive and Free-Ranging Okapi (*Okapia johnstoni*) with Implications for Management," *Conservation Genetics* 16, no. 5 (2015), 1115–26. On okapi being strong swimmers: Interview with Stuart Nixon of WCS, on Teams, on June 14, 2023.

33. Maryvonne Leclerc-Cassan, *Vivre Avec Eux* (France Loisirs, 1978). She was, it should be said, a highly qualified academic as well as an experienced veterinarian. Quote on 123.

34. Quote from "Husbandry," The Okapi Management Website, theokapi.org/Husbandry/enclosure.aspx.

35. Jonathan Kingdon, by email, November 25, 2023; Hediger, *Wild Animals in Captivity*; interview with Gemma Metcalf and Poppy Jewell, London Zoo, November 10, 2021.

36. Harriet Ritvo, "Beasts in the Jungle (or Wherever)," in *Noble Cows and Hybrid Zebras*, by Harriet Ritvo (U of Virginia Press, 2010), 203–12.

37. Irus Braverman, *Wild Life: The Institution of Nature* (Stanford UP, 2015).

38. Emma Marris, *Wild Souls: Freedom and Flourishing in the Non-Human World* (Bloomsbury, 2021).

39. G. Mitchell and J. D. Skinner, "On the Origin, Evolution, and Phylogeny of Giraffes *Giraffa camelopardalis*," *Transactions of the Royal Society of South Africa* 58, no. 1 (2003), 51–73.

40. See, for example, "Okapi nutrition," and the references page at https://okapinutrition.weebly.com/background-information.html; and Mads Bertelsen's comment that "Available information on okapi anatomy and physiology is limited" in "Giraffidae," *Fowler's Zoo and Wild Animal Medicine* 8 (2014), doi: 10.1016/B978-1-4557-7397-8.00061-X.

Chapter 13. Indigenous Africans as "Primitive Experts" on Okapi

1. The epigraph to this chapter comes from Attilio Gatti, *South of the Sahara* (Robert M. McBride, 1946), 58.

2. My thinking on use of the term "pygmy" draws on chapters in *Hunter-Gatherers of the Congo Basin*, ed. Barry S. Hewlett (Routledge, 2017); and Kairn A. Klieman, *"The Pygmies Were our Compass": Bantu and Batwa in the History of West Central Africa, Early Times to c. 1900 C.E.* (Heinemann, 2003). On how some peoples so described view the term, and negative connotations, see Mark Dowie, *Conservation Refugees: The Hundred Years Conflict between Global Conservation and Native Peoples* (MIT Press, 2009), 65–66.

3. The claim that the Mbuti told Stanley about the okapi is discussed in chapter two.

4. Philip L. Sclater, "On a New African Mammal," *Proceedings of the Zoological Society of London* no. 2 (1901): 3.

5. Harry Johnston, "The Okapi: The Newly Discovered Beast Living in Central Africa," in *Annual Report of the Board of Regents of the Smithsonian Institution* (Government Printing Office, 1902), 663.

6. Quoted in Sclater, "On a New African Mammal," 3.

7. Harry Johnston, quoted in Julien Fraipont, *Okapia*, in *Contributions a la Faune du Congo*, book 1, *Annales du Musée du Congo* (Independent State of the Congo, 1907) 9.

8. "A New Mammal," *The Times*, May 7, 1901, 13; "*Ex Africa semper aliquid novi*," *The Times*, May 10, 1901, 9.

9. "*Ex Africa semper aliquid novi*," 9.

10. "*Ex Africa semper aliquid novi*," 9.

11. James J. Harrison, "The Harrison Diaries," n.d., Scarborough Museums and Galleries (pdf courtesy of Jim Middleton, collections manager, received November 15, 2022), discussed in chapter two.

12. Boyd Alexander, *From the Niger to the Nile*, 2 vols. (Edward Arnold, 1907), 2, discussed in chapter two.

13. Percy H. G. Powell-Cotton, "Okapi," *The Times*, September 27, 1906, 6; Percy H. G. Powell-Cotton, "A Journey through the Eastern Portion of the Congo State," *Geographical Journal* 30, no. 4 (1907): 7.

14. Philip L. Sclater, "Dr P. L. Sclater on the Okapi," *Proceedings of the Zoological Society of London*, no. 1 (1906): 760–61, 761.

15. Sclater, "Dr P. L. Sclater," 761.

16. Fraipont, *Okapia*, 81.

17. Fraipont, *Okapia*, 86.

18. Richard Lydekker, "Hornless Okapis," *Annals of the Magazine of Natural History* 6 (1910): 226.

19. Fraipont, *Okapia*, 85–87.

20. Hermann Schubotz, "The Home of the Okapi," in Adolf Friedrich von Mecklenburg-Schwerin, *From the Congo to the Niger and the Nile*, vol. 2 (John Winton, 1914), discussed in detail in chapter two.

21. Herbert Lang, "In Quest of the Rare Okapi," *Zoological Society Bulletin* 21, no. 3 (1918): 1608, 1609.

22. Lang, "In Quest," 1609, 1610.

23. Lang, "In Quest," 1610; Harrison, "The Harrison Diaries," discussed in chapter two.

24. Lang, "In Quest," 1609.

25. Lang, "In Quest," 1610, 1604, photos on 1604, 1611, respectively.

26. Lang, "In Quest," 1611, 1613.

27. Lang, "In Quest," 1613.

28. Lang, "In Quest," 1613.

29. Lang, "In Quest," 1613.

30. Lang, "In Quest," 1613, photographs of the okapi calf on 1606, 1607, and 1608.

31. Cuthbert Christy, *Big Game and Pygmies: Experiences of a Naturalist in Central African Forests in Quest of the Okapi* (Macmillan, 1924), 37–39; see also 42–43, 46–51.

32. Attilio Gatti, *Great Mother Forest* (Hodder and Stoughton, 1936), 21.

33. Gatti, *Great Mother Forest*, 22, 23.

34. Gatti, *Great Mother Forest*, 23–24.

35. Gatti, *Great Mother Forest*, 26, 27, 28.

36. Gatti, *Great Mother Forest*, chapters 14 and 18, and on the battering ram, 121.

37. Gatti, *Great Mother Forest*, 212.

38. Gatti, *Great Mother Forest*, 217–18.

39. Gatti, *Great Mother Forest*, 222.

40. Gatti, *Great Mother Forest*, 270–71.

41. Gatti, *Great Mother Forest*, 272–73.

42. Gatti, *Great Mother Forest*, 275, 282.

43. Gatti, *South of the Sahara*, 58.

44. Gatti, *South of the Sahara*, 59.

45. Gatti, *South of the Sahara*, 60. On Gatti's interpretation of the different zones, and degrees of true pygmy identity, see 96.

46. Gatti, *South of the Sahara*, this episode told on 126 28.

47. Gatti, *South of the Sahara*, 130–31.

48. Gatti, *South of the Sahara*, 132–37.

49. Gatti, *South of the Sahara*, this story told in chapters 16 and 17.

50. Gatti, *South of the Sahara*, 152–53, 155.

51. Anne Eisner Putnam, with Allan Keller, *Eight Years with Congo Pigmies* (Hutchinson, 1955), 104.

52. Eisner Putnam, with Keller, *Eight Years*, 105.

53. Swaluë and de Medina are discussed in detail in chapter nine.

54. Colin Turnbull, *The Forest People* (Pimlico, 1993), 160.

55. Colin M. Turnbull, "The Mbuti Pygmies: An Ethnographic Survey," *Anthropological Papers of the American Museum of Natural History* 50, no. 3 (1965): 164.

56. Turnbull, "Mbuti Pygmies," 202. On trapping okapi in pits, see *In Darkest Africa*, 2 vols. (Sampson Low, Marston, Searle and Rivington, 1890), 2: 490; Lang, "In Quest," 1609, and Gatti, *Great Mother Forest*, 177.

57. Turnbull, "Mbuti Pygmies," 202–4.

58. Turnbull, "Mbuti Pygmies," 230.

59. The standard work on bushmeat and hunting and its impacts on biodiversity in the tropics is Julia Fa, Stephan Funk, and Robert Nasi, *Hunting Wildlife in the Tropics and Subtropics* (Cambridge UP, 2022). On differential extraction rates and kinds of prey, Julia E. Fa, Jesús Olivero, Miguel Angel Farfán, et al., "Differences between Pygmy and Non-Pygmy Hunting in Congo Basin Forests," *PLoS ONE* 11(9) (2016): e0161703.

60. Giuseppe M. Carpaneto and Francesco P. Germi, "The Mammals in the Zoological Culture of the Mbuti Pygmies in North-Eastern Zaire," *Hystrix* 1 (1989): 40.

61. Interview with Rosmarie Ruf, September 14, 2023, via Zoom; Alan Root, *Heart of Brightness*, Anglia Television (1991); Terese Hart, "Lesser Known History of Okapi," October 22, 2008, https://www.bonoboincongo.com/2008/10/22/lesser-known-history-of-okapi.

Chapter 14. *Western Framings of the Peoples and Forests of the Congo*

1. The epigraph to this chapter comes from Cuthbert Christy *Big Game and Pygmies: Experiences of a Naturalist in Central African Forests in Quest of the Okapi* (Macmillan, 1924), 50–51.

2. Harry Johnston, *The Uganda Protectorate*, vol. 2 (Hutchison, 1902), 470.

3. S. Das and M. Lowe, "Nature Read in Black and White: Decolonial Approaches to Interpreting Natural History Museums," *Journal of Natural Science Collections* 6 (2018): 5.

4. Sven Lindqvist, *"Exterminate all the Brutes"* (Granta, 1992), 131.

5. For the human skeletons in the NHM, see *Skeleton of a Man, a Woman from the Akka Tribe and a Gorilla (Provided by the Hunter-Naturalist Paul Du Chaillu), Displayed in a Row.* Photograph by Gambier Bolton, 1890. Wellcome Collection, 577014i. See also *Guide to the Specimens Illustrating the Races of Mankind (Anthropology), Exhibited in The Department of Zoology, British Museum (Natural History), Cromwell Road, London, S. W.* (The Trustees of the British Museum, 1908), 27.

6. Harry Johnston, *A History of the Colonization of Africa by Alien Races* (Cambridge UP, 1899), 101.

7. Johnston, *Alien Races*, 100–102.

8. Johnston, *Uganda Protectorate*, 471.

9. Johnston, *Uganda Protectorate*, 471, 544–45.

10. Johnston, *Uganda Protectorate*, 539, 544.

11. Paul Schebesta, *Among Congo Pigmies*, trans. Gerald Griffin (Hutchinson, 1933), 12, 21.

12. James J. Harrison, *Life among the Pygmies of the Ituri Forest, Congo Free State* (Hutchinson, 1905). See also "African Pygmies at the Hippodrome," *The Times*, June 6, 1905, 12; Jeffrey Green, "Exhibiting the Six Africans," From Local to Global, Scarborough Museums and Galleries, https://www.fromlocaltoglobal.co.uk/exhibiting-the-six-africans.

13. Percy H. G. Powell-Cotton, "A Journey through the Eastern Portion of the Congo State," *Geographical Journal* 30, no. 4 (1907): 376; Herbert Lang, "Report from the Congo Expedition," *American Museum Journal* 11 (1911): 48.

14. Adolphus Frederick von Mecklenburg-Schwerin, *In the Heart of Africa* (Cassell, 1910), 201–3.

15. Maano Ramutsindela, Frank Matose, and Tafadzwa Mushonga, "Conservation and Violence in Africa," in *The Violence of Conservation in Africa: State, Militarization, and*

Alternatives, ed. Maano Ramutsindela, Frank Matose, and Tafadzwa Mushonga (Edward Elgar, 2022), 6.

16. Turnbull admitted he found "the very romantic appeal of the Mbuti provocative when turning from them to our own society" (*The Mbuti Pygmies: Change and Adaptation* [Holt, Rinehart and Winston, 1983], 153). He thought his book *The Forest People* (Jonathan Cape, 1961) in the West had had the impact it did "because the near-Utopia described rang true, and showed that certain voids in the lives of many of us could indeed be filled" (153). Notable in this respect were the Mbuti's stubborn focus on the quality of community life over economic success, on values rather than goods, on "Spirit rather than Matter" (153). On the US transformation of the world in this period, see, for example, Richard P. Tucker, *Insatiable Appetite: The United States and the Ecological Degradation of the Tropical World* (U of California Press, 2000).

17. Turnbull, *The Forest People*, on its influence on Rupp, Stephanie Rupp, *Forests of Belonging: Identities, Ethnicities, and Stereotypes in the Congo River Basin* (U of Washington Press, 2011), 6.

18. On the family of humankind and the hunter hypothesis in the context of Western framings of pygmy peoples, see Kairn A. Klieman, *"The Pygmies Were Our Compass": Bantu and Batwa in the History of West Central Africa, Early Times to c. 1900 C.E.* (Heinemann, 2003), 13–20.

19. Klieman, *"The Pygmies Were Our Compass,"* 17; Adam Kuper, *The Reinvention of Primitive Society: Transformations of a Myth* (Routledge, 2017), 6–7.

20. Kuper, *Reinvention*, 169–70.

21. Kuper, *Reinvention*; see chapter 10.

22. Kent Redford, "The Ecologically Noble Savage," *Cultural Survival Quarterly* 15, no. 1 (1991): unnumbered page, available at https://www.culturalsurvival.org/publications/. See also Rupp, *Forests of Belonging*.

23. Redford, "Ecologically Noble Savage"; Jared Diamond, *Collapse: How Societies Choose to Fail or Survive* (Penguin, 2011), especially chapter 5 on the Maya collapse. This debate continues in Raymond Hames, "The Ecologically Noble Savage Debate," *Annual Reviews in Anthropology* 36 (2007): 177–90.

24. "Indigenous and Tribal Peoples Convention, 1989 (No. 169)," General Conference of the International Labour Organisation (1989), available at: https://ohchr.org/en/instruments-mechanisms/instruments/indigenous-and-tribal-peoples-convention-1989-no-169. On pygmy peoples and definitions of "indigenous peoples" see Mitsuo Ichikawa, "Forest Conservation and Indigenous Peoples in the Congo Basin: Net Trends toward Reconciliation between Global Issues and Local Interest," in *Hunter-Gatherers of the Congo Basin*, ed. Barry S. Hewlett (Routledge, 2017): 326–30; Rupp, *Forests of Belonging*, chapter one "Paradigms: The Forest and its Peoples."

25. Rupp, *Forests of Belonging*.

26. Serge Bahuchet, "Cultural Diversity of African Pygmies," in *Hunter-Gatherers*, 1–30.

27. Kuper, *Reinvention*, 18.

28. Henry M. Stanley, *In Darkest Africa*, 2 vols. (Charles Scribner's Sons, 1890); Tim Jeal, *Stanley: The Impossible Life of Africa's Greatest Explorer* (Faber and Faber, 2007); Herbert Lang, "In Quest of the Rare Okapi," *Zoological Society Bulletin* 21, no. 3 (1918): 1604.

29. Harriet Ritvo, *The Platypus and the Mermaid and Other Figments of the Classifying Imagination* (Harvard UP, 1997), 209.

30. Attilio Gatti, *South of the Sahara* (Robert M. McBride, 1946), 95–96. On Ota Benga, see Pamela Newkirk, *Spectacle: The Astonishing Life of Ota Benga* (Amistad, 2015).

31. This section is based on Serge Bahuchet, "Cultural Diversity," 1–30; Paul Verdu, "Population Genetics of Central African Pygmies and Non-Pygmies," in *Hunter-Gatherers*, 31–58.

32. Verdu, "Population Genetics."

33. This section on the deep history of pygmy peoples is based on Robert E. Moïse, "'Do Pygmies Have a History?' Revisited: The Autochthonous Tradition in the History of Equatorial Africa," in *Hunter-Gatherers*, 85–116.

34. This section is based on Moïse, "'Do Pygmies Have a History?'"

35. This section draws mainly on David Van Reybrouck, *Congo: The Epic History of a People* (Fourth Estate, 2014). Van Reybrouck's acclaimed history is based on archival research, personal experience, and extensive interviews with Africans. On the territorialization of ethnicity, see Kris Berwouts, *Congo's Violent Peace* (Zed Books, 2017), 34–38.

36. Van Reybrouck, *Congo*, 106–10.

37. Van Reybrouck, *Congo*, 110–16.

38. Wilkie and Curran, "Historical Trends in Forager and Farmer Exchange in the Ituri Rain Forest of Northeastern Zaire," *Human Ecology* 21, no. 4 (1993): 389–417.

39. Turnbull, *Mbuti Pygmies*.

40. Turnbull, *Mbuti Pygmies*, 136–40.

41. Joseph Conrad, *Heart of Darkness* (Penguin, 1995), 59; Johnston, *My Life*, 380; Stanley, *In Darkest Africa*, 2:75; Boyd Alexander, *From the Niger to the Nile*, vol. 2 (Edward Arnold, 1907), 260; Lang, "In Quest," 1606; Delia J. Akeley, *Jungle Portraits* (Macmillan, 1930), 185.

42. This section is based on a special issue summarized in Yadhvinder Malhi, Stephen Adu-Bredu, Rebecca A. Asare, et al., "African Rainforests: Past, Present, and Future," *Philosophical Transactions of the Royal Society* 368B, no. 1625 (2013): 20120312. The papers I consulted include Katherine A. Abernethy, Lauren Coad, Gemma Taylor, et al., "Extent and Ecological Consequences of Hunting in Central African Rainforests in the Twenty-First Century"; Simon L. Lewis, Bonaventure Sonké, Terry Sunderland, et al., "Above-Ground Biomass and Structure of 260 African Tropical Forests"; Yadvinder Malhi, Stephen Adu-Bredu, Rebecca A. Asare, et al., "African Rainforests: Past, Present, and Future"; Philippe Mayaux, Jean-François Pekell, Baudouin Desclée, et al., "State and Evolution of the African Rainforests Between 1990 and 2010"; Richard Oslisly, Lee White, Ilham Bentaleb, et al., "Climatic and Cultural Changes in the West Congo Basin Forests over the Past 5000 Years"; and Thomas K. Rudel, "The National Determinants of Deforestation in Sub-Saharan Africa."

43. James Fairhead and Melissa Leach, *Misreading the African Landscape: Society and Ecology in a Forest Savanna Mosaic* (Cambridge UP, 1996).

44. Paul W. Richards, *The Tropical Rainforest: An Ecological Study*, 2nd ed. (Cambridge UP, 1996); Gretchen Walters, James Angus Fraser, Nicolas Picard, et al., "Deciphering African Tropical Forest Dynamics in the Anthropocene: How Social and Historical Sciences Can Elucidate Forest Cover Change and Inform Forest Management," *Anthropocene* 27 (2019): 100214.

45. Oslisly, White, Bentaleb, et al., "Climatic and Cultural Changes." Slightly different periodizations for human impacts exist by rainforest region; for a discussion, see Romaric Ndonda Makemba, Christian Moupela, Félicien Tosso, et al., "New Evidence on the Role

of Past Human Activities and Edaphic Factors on the Fine-Scale Distribution of an Important Timber Species: *Cylicodiscus gabunensis* Harms," *Forest Ecology and Management* 521 (2022): 120440.

46. Julie Morin-Rivat, Adeline Fayolle, Charly Favier, et al., "Present-Day Central African Forest Is a Legacy of the 19th Century Human History," *eLife* 6 (2017): e20343. See also Makemba, Moupela, Tosso, et al., "New Evidence."

47. Mayaux, Pekell, Desclée, et al., "State and Evolution of the African Rainforests."

48. Julia Fa, Stephan Funk, and Robert Nasi, *Hunting Wildlife in the Tropics and Subtropics* (Cambridge UP, 2022).

49. Malhi, Adu-Bredu, Asare, et al., "African Rainforests."

50. Regarding the relative visibility of the Amazon and the Congo rainforests, a simple Google Scholar search returns three times more results for the Amazon. Conrad, *Heart of Darkness*; Francis Coppola, *Apocalypse Now* (Omni Zoetrope, 1979). On minerals, mining, and munitions from the Democratic Republic of Congo, see Van Reybrouck, *Congo*, Orlando von Einsiedel, *Virunga* (Netflix, 2014), and Siddharth Kara, *Cobalt Red: How the Blood of the Congo Powers Our Lives* (St. Martin's Press, 2023).

51. Discussed in chapter 15.

52. Rupp, and discussed in the context of okapi habitat in chapter 15.

53. On biocultural diversity, see J. Peter Brosius and Sarah L. Hitchner, "Cultural Diversity and Conservation," *International Social Science Journal* 61, no. 199 (2010): 141–68; Peter Bridgewater and Ian D. Rotherham, "A Critical Perspective on the Concept of Biocultural Diversity and Its Emerging Role in Nature and Heritage Conservation," *People and Nature* 1, no. 3 (2019): 291–304.

Chapter 15. Clashing Worldviews in a Crucible for Wildlife Conservation

1. The epigraph to this chapter comes from Cosma Wilungula, foreword to Noëlle F. Kümpel, Alex Quinn, Elise Queslin, et al., eds, *Okapi* (Okapia johnstoni): *Conservation Strategy and Status Review* (International Union for the Conservation of Nature and Institut Congolais pour la Conservation de la Nature, 2015), 1.

2. On the creation of a reserve for okapi: Peter J. Stephenson and John E. Newby, "Conservation of the Okapi Wildlife Reserve, Zaire," *Oryx* 31, no. 1 (1997): 49–58; Kümpel, Quinn, Queslin, et al., *Okapi*.

3. On the history of the Wildlife Conservation Society's and the Okapi Conservation Project's involvement in the region, see "Wildlife Protection" and other links at https://okapiconservation.org.

4. Evan Hale and Lucas Meere, "Mbuti Pygmies Assist with Monitoring Okapis," January 2020, Wildlife Conservation Network, https://wildnet.org/wildlife-programs/okapi. The UNESCO World Heritage Convention site for the reserve is at https://whc.unesco.org/en/list/718.

5. On issues around indigeneity, local communities, and attempts to recognize their rights in the forests of the Congo basin region: Richard Eba'a Atyi, François Hiol, Guillaume Lescuyer, et al., *The Forests of the Congo Basin: State of the Forests 2021* (CIFOR, 2022), especially 172, 342–44. The DRC Forestry code is quoted on 344.

6. Stephanie Rupp, *Forests of Belonging: Identities, Ethnicities, and Stereotypes in the Congo River Basin* (U of Washington Press, 2011), 7–10.

7. Réseau des Populations Autochtones et Locales pour la Gestion des Ecosystèmes Forestiers d'Afrique Centrale, for example, has contributed Indigenous and local peoples' perspectives to the Commission des Forêts d'Afrique Centrale's 2015–25 convergence plan designed to coordinate interventions related to the conservation and sustainable management of Central Africa's forest ecosystems in *Stratégie 2018–2025 pour le développement durable des peuples autochtones et des communautés locales en Afrique centrale* (2020). Patrick Saidi Hemedi, "After 14 Years of Advocacy, the DRC President Finally Signs New Indigenous Peoples Law," *Mongabay*, November 16, 2022.

8. Romaric Ndonda Makemba, Christian Moupela, Félicien Tosso, et al., "New Evidence on the Role of Past Human Activities and Edaphic Factors on the Fine-scale Distribution of an Important Timber Species: *Cylicodiscus gabunensis* Harms," *Forest Ecology and Management* 521 (2022): 120440.

9. Richard Peterson, *Conversations in the Rainforest*, rev. ed. (Westview Press, 2017), introduction and chapter five.

10. Peterson, *Conversations*, 165, 178.

11. Peterson, *Conversations*, 166.

12. Peterson, *Conversations*, 166–67, 173, 176.

13. Peterson, *Conversations*, 177–78.

14. Peterson, *Conversations*, 170–71.

15. Peterson, *Conversations*, 178–80, 195–96.

16. Workineh Kelbessa, "African Worldviews, Biodiversity Conservation, and Sustainable Development," *Environmental Values* 31, no. 5 (2022): https://doi.org/10.3197/0963 27121X16328186623922.

17. Koli Jean Bofane's novel is available in English as *Congo Inc.: Bismarck's Testament*, translated by Marjolijn de Jager (Indiana UP, 2018).

Ecocritic Kenneth Toah Nsah draws out concerns over fortress conservation, green colonialism, neoliberal environmental destructiveness, development-driven interethnic conflict, cultural appropriation, and epistemicide (violence against Indigenous ways of knowing the natural world, with colonialist overtones) in "Conserving Africa's Eden? Green Colonialism, Neoliberal Capitalism, and Sustainable Development in Congo Basin Literature," *Humanities* 12, no. 3 (2023), https://doi.org/10.3390/h12030038.

18. Peterson, *Conversations*, 227–28.

19. Peterson, *Conversations*, 229–39.

20. Kümpel, Quinn, Queslin, et al., *Okapi*, 20.

21. Kümpel, Quinn, Queslin, et al., *Okapi*, 15.

22. Kümpel, Quinn, Queslin, et al., *Okapi*, 11, 2.

23. Kümpel, Quinn, Queslin, et al., *Okapi*, 2; interview with Stuart Nixon of WCS, on Teams, on June 14, 2023.

24. Kümpel, Quinn, Queslin, et al., *Okapi*, 20, 2. As of the last assessment of July 25, 2015, the okapi are still listed as endangered; see https://iucnredlist.org/species/15188 /51140517.

25. Kümpel, Quinn, Queslin, et al., *Okapi*, 20.

26. Kümpel, Quinn, Queslin, et al., *Okapi*, 14.

27. Jonathan Kingdon, *Origin Africa: Safaris in Deep Time* (William Collins, 2023), 167; Zvi Sever, "Searching for the Okapi (*Okapia johnstoni*) in Semuliki National Park, Uganda," *African Journal of Ecology* 59, no.1 (2021): 288.

28. Stuart Nixon, Magloire Kambale Vyalengerera, and Naomi Matthews, *Explorations for Okapi and a Pilot Camera Trap Survey of Large Mammals in the Semuliki National Park, Uganda* (Chester Zoo, 2018), 17; Sever, "Searching," 290.

29. Kümpel, Quinn, Queslin, et al., *Okapi*; Kris Berwouts, *Congo's Violent Peace* (Zed Books, 2017); Orlando von Einsiedel, *Virunga* (Netflix documentary, 2014).

30. Ashoka Mukpo, "Across the World, Conservation Projects Reel after Abrupt US Funding Cuts," *Mongabay*, February 14, 2025, available at: https://news.mongabay.com /2025/02/across-the-world-conservation-projects-reel-after-abrupt-us-funding-cuts/

31. "The Fall of the House of Gilman," *Forbes*, August 11, 2003, https://www.forbes.com /forbes/2003/0811/068.html.

32. On White Oak and the setting up of the Okapi Conservation Project: interview with Rosmarie Ruf, September 14, 2023, via Zoom; interview with John Watkin, June 17, 2024, via Google Meet; John Lukas, "Thirty Years of Okapi Conservation," WCS, October 31, 2018, available at https://wildnet.org/thirty-years-of-okapi-conservation/ (accessed September 14, 2023); "Fall of the House of Gilman"; "About White Oak," White Oak Conservation, https://www.whiteoakwildlife.org/conservation.

33. *Okapi Conservation Project 2017 Annual Report* (Okapi Conservation Project, 2017), https://www.okapiconservation.org/annual-reports.

34. Kümpel, Quinn, Queslin, et al., *Okapi*, 37.

35. Kümpel, Quinn, Queslin, et al., *Okapi*, 24.

36. IUCN Red List information on okapi: https://iucnredlist.org/species/15188 /51140517 (accessed on September 22, 2024); Interview with Gabriel Gelin, on Teams, December 2023.

37. Interview with Chris Hamley, on Teams, December 2023; Interview with John Watkin, June 17, 2024, via Google Meet.

38. Interview with Chris Hamley; "Ituri: plus de 70 sites d'exploitation d'or fonctionnent illégalement dans la Réserve de faune à Okapi," May 28, 2024, https://www.radiookapi.net.

39. Interview with Gabriel Gelin, on Teams, December 2023.

40. On arguments against militarization in conservation: Rosaleen Duffy, Francis Massé, Emile Smidt, et al., "Why We Must Question the Militarisation of Conservation," *Biological Conservation* 232 (2019): 66–73. On wildlife conservation, indigenous peoples, and evictions from protected areas including those in the Congo Basin Region: Mark Dowie, *Conservation Refugees: The Hundred Years Conflict between Global Conservation and Native Peoples* (MIT Press, 2009), notably chapter five.

41. Eba'a Atyi, Hiol, Lescuyer, et al., *The Forests of the Congo Basin*, 363.

42. Stephanie Dolrenry, Jennifer Stenglein, Leela Hazzah, et al., "A Metapopulation Approach to African Lion (*Panthera leo*) Conservation," *PLoS ONE* 9, no. 2 (2014): e88081. For a good discussion of metapopulation management with examples including those noted here, Irus Braverman, *Wild Life: The Institution of Nature* (Stanford UP, 2015).

43. Interview with John Watkin, June 17, 2024; Stuart Nixon, pers. comm., June 19, 2024; *Okapi Conservation Project 2017 Annual Report*, 1; Nixon, Vyalengerera, and Matthews, *Explorations*, 17–19; Sever, "Searching," 290.

44. Joseph Conrad, *Heart of Darkness* (Penguin Books, 1995); Edmund D. Morel, *Red Rubber: The Story of the Rubber Slave Trade Flourishing on the Congo on the Year of Grace 1906* (Nassau Print, 1906). On the massive impacts of Western overseas empires on the natural world see Corey Ross, *Ecology and Power in the Age of Empire: Europe and the*

Transformation of the Tropical World (Oxford UP, 2017). On the drastic environmental transformations of the twentieth century: John McNeill, *Something New Under the Sun: An Environmental History of the Twentieth-Century World* (W. W. Norton & Co., 2001). On the development and impacts of global capitalism more generally, Mark Stoll, *Profit: An Environmental History* (Polity Press, 2023).

45. On relational value, see Kai M. A. Chan, Patricia Balvanera, Karina Benessaiah, et al., "Why Protect Nature? Rethinking Values and the Environment," *PNAS* 113, no. 6 (2012): 1462–65; Robin Wall Kimmerer, *Braiding Sweetgrass: Indigenous Wisdom, Scientific Knowledge, and the Teachings of Plants* (Milkweed Editions, 2013); and the volumes in the series *Kinship: Belonging in A World of Relations*, Gavin van Horn, Robin Wall Kimmerer, and John Hausdoerffer, eds. (Center for Humans and Nature Press, 2021).

Conclusion

1. Apparently, the Democratic Republic of Congo's government has been advised to take the same stance with respect to okapi as China has done with pandas that is, to insist okapi are the sovereign property of the Democratic Republic of Congo and extract huge fees for loans to foreign institutions. Interview with John Watkins, June 17, 2024, via Google Meet.

2. Anne Eisner Putnam, with Allan Keller, *Eight Years with Congo Pigmies* (Hutchinson, 1955), 106. There are many difficult passages exhibiting racist ideas, but perhaps the most troubling to me was coming to terms with the fact that Harry Johnston could write both the interesting and sometimes charming volume 1 (physical geography, botany, and zoology) and the odious volume 2 (anthropology, languages, and history) of his *The Uganda Protectorate* (Hutchinson & Co., 1902).

3. Harriet Ritvo, "Animal Consciousness: Some Historical Perspective," in *Noble Cows and Hybrid Zebras: Essays on Animals and History*, by Harriet Ritvo (University of Virginia Press, 2010), 64.

4. On giraffe evolution and taxonomy, see G. Mitchell and J. D. Skinner, "On the Origin, Evolution, and Phylogeny of Giraffes *Giraffa Camelopardalis*," *Transactions of the Royal Society of South Africa* 58, no. 1 (2003): 51–73, and Fred B. Bercovitch, Philip S. M. Berry, Anne Dagg, et al. "How Many Species of Giraffe Are There?", *Current Biology* 27 (2017): R123–38.

5. Maud Borie and Mike Hulme, "Framing Global Biodiversity: IPBES between Mother Earth and Ecosystem Services," *Environmental Science and Policy* 54 (2015): 487–96; Unai Pascual, Patricia Balvanera, Michael Christie, et al., eds., *Summary for Policymakers of the Methodological Assessment Report on the Diverse Values and Valuation of Nature of the Intergovernmental Science-Policy Platform on Biodiversity and Ecosystem Services* (IPBES Secretariat, 2022).

6. On evidence-based conservation, see William J. Sutherland, ed., *Transforming Conservation: A Practical Guide to Evidence and Decision Making* (OpenBook Publishers, 2022).

7. Attilio Gatti, *South of the Sahara* (Hodder and Stoughton, 1946), 97–98.

8. Heini Hediger, *Wild Animals in Captivity*, trans. Geoffrey Sircom (Butterworth Scientific Publications, 1950), 4–5, 27–28. As Hediger notes (27–28), "by capturing [a wild animal] we utterly destroy the animals' previous world, and put it into a different environment. The animal must construct an entirely fresh subjective world." Those born in

captivity are spared this, at least; Attilio Gatti, *Great Mother Forest* (Hodder and Stoughton, 1936), caption to plate opposite 205.

9. European Association of Zoos and Aquaria, okapi ex situ program, https://strapi.eaza .net/uploads/Okapi_EEP_EAZA_10_29_2024_3_03_12_PM_ec887d7d00.html; *EAZU Best Practice Guidelines: Okapi* (*Okapia johnstoni*) (European Association of Zoos and Aquaria, 2024), https://strapi.eaza.net/uploads/2024_EAZA_Okapi_Best_Practice_Guidelines _APPROVED_c901c668ed.pdf. The target for captive okapi given here is 220, with 90 in European zoos.

10. Summarized in Onnie Byers, Caroline Lees, Jonathan Wilcken, et al., *WAZA Magzine* 14 (2013): 2–5. See also "One Plan Approach" at https://www.cpsg.org/our-work /our-approach/one-plan-approach.

11. I have discussed the challenges of competing for interpretation space in natural history museums with Isabel Davis of the Natural History Museum in London. Museum scientists are already frustrated by the limitations of interpretative display space, so adding this kind of material in physical space will be difficult. However, with the use of QR codes and other digital innovations, it is possible to provide virtual portals and guided walks through the physical spaces of museums which have the added advantage of making collections more widely accessible online.

12. https://www.bioparc-zoo.fr/espace/le-sanctuaire-des-okapis.

13. Siddharth Kara, *Cobalt Red: How the Blood of the Congo Powers Our Lives* (St. Martin's Press, 2023).

Page numbers in italics refer to figures.

9 781421 452487